Two week

Please return on or before the last
date stamped below.

TANGLED TREES

TANGLED TREES

Phylogeny, Cospeciation, and Coevolution

EDITED BY Roderic D. M. Page

THE UNIVERSITY OF CHICAGO PRESS

Chicago and London

RODERIC D. M. PAGE is a Reader in the Division of Environmental and Evolutionary Biology, Institute of Biomedical and Life Sciences, at the University of Glasgow. He is the coauthor of *Molecular Evolution: A Phylogenetic Approach.*

THE UNIVERSITY OF CHICAGO PRESS, CHICAGO 60637
THE UNIVERSITY OF CHICAGO PRESS, LTD., LONDON
© 2003 by The University of Chicago
All rights reserved. Published 2003
Printed in the United States of America

12 11 10 09 08 07 06 05 04 03   1 2 3 4 5

ISBN: 0-226-64466-9 (cloth)
ISBN: 0-226-64467-7 (paper)

Library of Congress Cataloging-in-Publication Data

Tangled trees : phylogeny, cospeciation, and coevolution / edited by Roderic D. M. Page.
    p.   cm.
Includes bibliographical references.
ISBN 0-226-64466-9 (hardcover : alk. paper) – ISBN 0-226-64467-7 (pbk. : alk. paper)
1. Phylogeny.   2. Host-parasite relationships.   3. Coevolution.   I. Page, Roderic D. M.
QH367.5 .T36   2003
576.8′—dc21   2002002821

CONTENTS

CONTRIBUTORS

Sarah Al-Tamimi
Department of Biology
University of Utah
Salt Lake City, UT 84112-0840
United States of America

Michael A. Charleston
Department of Zoology
University of Oxford
South Parks Road, Oxford OX1 3PS
United Kingdom

Dale H. Clayton
Department of Biology
University of Utah
Salt Lake City, UT 84112-0840
United States of America

James W. Demastes
Department of Biology
University of Northern Iowa
Cedar Falls, IA 50614
United States of America

Russell D. Gray
Department of Psychology
University of Auckland
Private Bag 92017, Auckland
New Zealand

Mark S. Hafner
Museum of Natural Science
and Department of Biological
Sciences
Louisiana State University

Baton Rouge, LA 70803
United States of America

John P. Huelsenbeck
Department of Biology
University of Rochester
Rochester, NY 14627
United States of America

J.-P. Hugot
Muséum National d'Histoire
Naturelle
Institut de Systématique, FR 1541 du
CNRS, Laboratoire des Mammifères
et Oiseaux
(Biogéographie et Coévolution)
55, rue Buffon, 75231 Paris cedex 05
France

Kevin P. Johnson
Illinois Natural History Survey
607 East Peabody Drive
Champaign, IL 61820
United States of America

Peter Kabat
Institute of Virology
Slovak Academy of Sciences
Dubravska cesta 9
84246 Bratislava
Slovak Republic

Bret Larget
Department of Mathematics and
Computer Science

Duquesne University
Pittsburgh, PA 15282
United States of America

Joanne Martin
Department of Biology
Imperial College, Silwood Park
Ascot, Berkshire SL5 7PY
United Kingdom

Yannis Michalakis
Centre d'Etude sur le Polymorphisme
des Micro-Organismes
CEPM/UMR CNRS-ORSTROM
9926
Montpellier
France

Roderic D. M. Page
Division of Environmental
and Evolutionary Biology
Institute of Biomedical
and Life Sciences
University of Glasgow
Glasgow G12 8QQ
United Kingdom

Ricardo L. Palma
Museum of New Zealand Te Papa
Tongarewa
P.O. Box 467, Wellington
New Zealand

Adrian M. Paterson
Ecology and Entomology Group
Soil, Plant and Ecological
Sciences Division
Lincoln University
Ellesmere Junction Rd./Springs Rd.
Lincoln, Canterbury 8150
New Zealand

Susan L. Perkins
Department of Environmental,
Population and Organismic Biology
Ramaley C373

University of Colorado
Boulder, CO 80309
United States of America

Andy Purvis
Department of Biology
Imperial College, Silwood Park
Ascot, Berkshire SL5 7PY
United Kingdom

Bruce Rannala
Department of Medical Genetics
University of Alberta
8-39 Medical Sciences Building
Edmonton, Alberta T6G 2H7
Canada

David L. Reed
Department of Biology
University of Utah
Salt Lake City, UT 84112-0840
United States of America

Fredrik Ronquist
Department of Systematic Zoology
Evolutionary Biology Centre
Uppsala University
Norbyvägen 18D, SE-752 36 Uppsala
Sweden

Theresa A. Spradling
Department of Biology
University of Northern Iowa
Cedar Falls, IA 50614
United States of America

Jason Taylor
Department of Biology
Imperial College, Silwood Park
Ascot, Berkshire SL5 7PY
United Kingdom

Michael Tristem
Department of Biology
Imperial College, Silwood Park
Ascot, Berkshire SL5 7PY
United Kingdom

PREFACE

This book owes its origins (in more ways than one) to Mark Hafner. In 1997 he suggested that he and I edit a book on cospeciation. Mark had already assembled a list of contributors and approached some publishers, so I readily agreed to undertake what I though would be a straightforward task as junior editor. Soon afterwards, it became clear that Mark's commitments as director of the Museum of Natural Science at Louisiana State University meant he could not devote as much time to the project as he had originally planned. As a result, I assumed full editorial duties. Mark provided the invaluable initial momentum, contributed to two chapters, and provided a constant source of suggestions and advice on aspects of the book. Without his involvement, the book would never have been put together.

Early on, Mark and I hoped to base the book on a symposium where the authors could get together and exchange ideas. The second biennial meeting of the Systematics Association, held at the University of Glasgow in 1999, provided an ideal venue, and I thank the council of the association for sponsoring the symposium that gives the book its name. Gordon Curry in particular provided much-needed organizational help and support.

In choosing contributors I hoped to achieve a mix of theory and empirical chapters, with a wide range of methodologies and organisms represented. Regrettably, not all the authors initially involved in the project were able to contribute to the final volume. The most noticeable consequence of this in the book is the absence of organisms such as plants and bacteria. I expect this will not go unnoticed by readers and reviewers alike; to both I apologize in advance. At the same time, the dominance of lice in the empirical section of the book is a not unreasonable reflection of the role these organisms have played in the development of the field.

The project benefited from enthusiastic and diligent authors, who also displayed considerable patience as the book slowly came together. I thank them for their excellent contributions, and for keeping me endlessly amused trying to process the myriad graphic and text formats in which their manuscripts were created. In addition to the chapter authors themselves, external reviewers included Stephane Aris-Brosou, François Catzeflis, Rob Cruickshank, Brent Emerson, Martyn Kennedy, Steve Nadler, Diana Percy, Andrew Read, Allen Rodrigo, Mark Siddall, Vince Smith, Jonathan Stoye, Simon Whelan, and Xuhua Xia. For their very helpful comments I am grateful. I also thank Dale Clayton for advice, and for warning me that editing a book is harder than writing one (based on my albeit limited experience, Dale is just plain wrong). Lastly, I thank Antje and Alec for tolerating the lost evenings and weekends spent on "the Chicago book."

Glasgow
March 2001

I

INTRODUCTION

Roderic D. M. Page

These are exciting times in the study of cospeciation. The increasing ease with which phylogenetic data can be obtained, combined with ever more sophisticated tree building methods (Swofford et al., 1996), has resulted in a profusion of evolutionary trees (Pagel, 1999). At the same time, new developments in the theory of comparing host-parasite phylogenies have lead to further insights into the complexity of inferring the history of an association between a parasite and its host. In this chapter I review some basic concepts of cospeciation, and then guide the reader through the diverse contributions that make up the rest of this book.

What Is Cospeciation?

Cospeciation is the joint speciation of two or more lineages that are ecologically associated, the paradigm example being a host and its parasite. The association need not be parasitic; mutualistic, symbiotic, and other relationships may also show cospeciation. However, for convenience I will refer solely to hosts and parasites. If cospeciation were the only process occurring, the host and parasite phylogenies would be exact mirror images of each other. That this is rarely the case establishes that other process also occur (fig. 1.1). Rather than track their host with perfect fidelity, parasites may switch lineages (fig. 1.1b), speciate independently of their host (fig. 1.1c), go extinct (fig. 1.1d), fail to colonize all descendants of a speciating host lineage (fig. 1.1e), or fail to speciate when its host does (fig. 1.1f).

It is important to distinguish between cospeciation and coevolution. If we define coevolution as the evolution of reciprocal adaptations in hosts and parasites, then it is clear that lineages can coevolve without cospeciating. Indeed, the studies that form the core of the coevolution literature do not involve cospeciating taxa (Thompson, 1994). However, it is difficult to imagine that cospeciation can occur without at least some degree

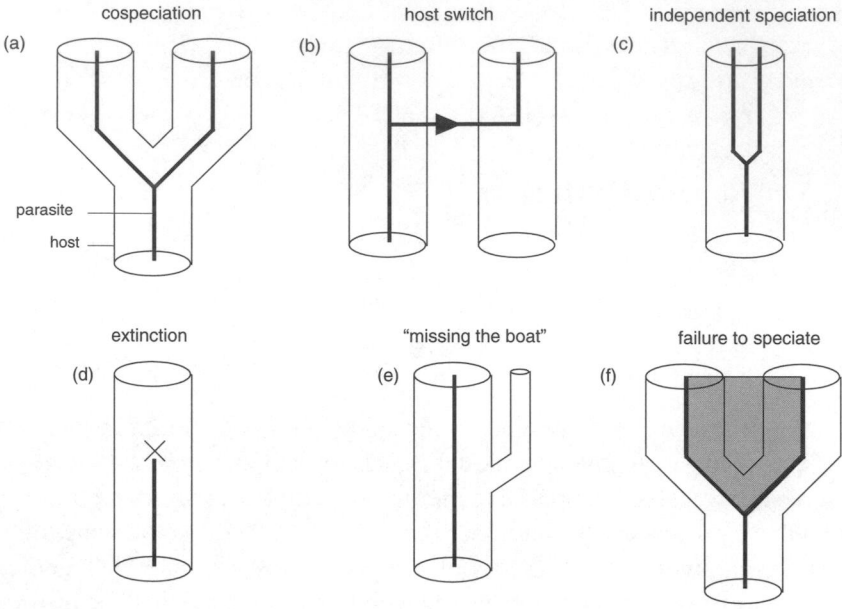

FIGURE 1.1. Processes in a host-parasite association. Host and parasite cospeciate *(a)*, or the parasite may speciate independently of its host *(b, c)*. One or more of the descendant parasites may colonize a new host *(b)*, or the parasite may remain on the original host *(c)*. Absence of a parasite from a host where it would be expected to occur may be due to extinction of that parasite *(d)*, or the ancestors of the host lineage may have not inherited the ancestral parasite *(e)*. Hosts may speciate independently of their parasites, so that the two hosts share the same parasite *(f)*.

of coevolution, although demonstrating such coevolution is another matter. Correlated character change in hosts and parasites may reflect other processes; for example, if both host and parasite genes are evolving in a clocklike manner, then overall amounts of molecular evolution in the two lineages will be correlated without their having coevolved. Better prospects for demonstrating coevolution in cospeciating lineages can be sought in characters directly involved in the host-parasite interaction (Clayton et al., 1999).

Workers have differed on how to employ the terms *cospeciation* and *coevolution* (e.g., Clayton and Moore, 1997). Some prefer to restrict coevolution to mean just reciprocal adaptation between host and parasite, and treat cospeciation as a distinct process. For others (e.g., Brooks and McLennan, 1991, 1993) cospeciation is coevolution that occurs in macroevolutionary time; reciprocal adaptation is coevolution in microevolutionary time ("coadaptation"). Some balk at the process-oriented nature of the term *cospeciation,* especially as the pattern shown in figure 1.1a may occur between lineages at levels other than that of the species (e.g., Funk et al., 2000, and see chap. 5), and opt for terms such as *cophylogeny* or *parallel cladogenesis*. The chapters in this volume reflect this terminological pluralism, but in this chapter I shall use the term *coevolution* only in the sense of reciprocal selection.

Why Is Cospeciation Interesting?

A basic question to ask of any host and its parasite assemblage is, how old is their association—is the parasite an ancient associate or a recent acquisition? This question can be addressed by comparing host and parasite phylogenies to determine whether host and parasite have cospeciated. But beyond this fundamental concern with the tempo and mode of host-parasite evolution, cospeciation is of broader interest for a number of reasons, some of which I consider below.

A Model System for Other Historical Associations

Comparing host and parasite phylogenies is a special case of a more general problem that may be termed the study of "historical associations" (Page and Charleston, 1998). The parallels between biogeography, where we compare organismal phylogenies with cladograms depicting geological history (Rosen, 1978), and parasitology have long been appreciated (Hennig, 1966; Brooks, 1981a). However, the realization of the parallels between these problems and the relationship between gene trees and species trees has been a more recent development (e.g., Page, 1988, 1993; Doyle, 1992). This analogy (fig. 1.2) implies that a single method for reconstructing historical associations could be developed that would be applicable to all three kinds of association. The quest for such a method is the "holy grail" for theoreticians interested in historical associations. Dan Brooks's BPA method (see below) was developed to study host-parasite cospeciation (Brooks, 1981a), but has been applied to biogeography (Wiley, 1988) and molecular systematics (Doyle, 1992). In my own work I have explored the applicability of ideas derived from Goodman et al.'s (1979) method for reconciling gene trees and species trees to problems in biogeography and host-parasite associations.

It is now widely appreciated that phylogenies for individual genes (or organelle genomes) need not be congruent with the organisms those genes were sampled from. Gene duplication, allelic polymorphism, and lineage sorting can result in incongruent gene and species trees. The use of phylogenies in studies of cospeciation rests on the notion that congruent phylogenies implies cospeciation, whereas incongruence implies host switching

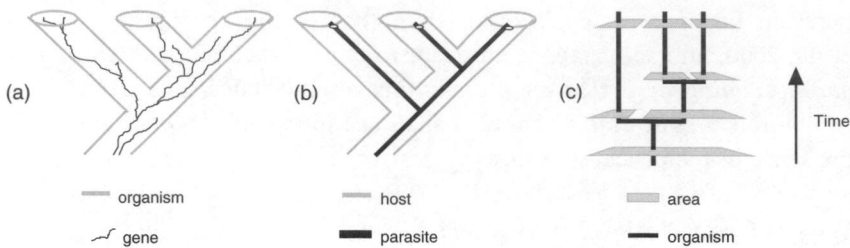

FIGURE I.2. The parallels between comparing trees for genes with organismal phylogeny *(a)*, phylogenies for parasites with their host phylogeny *(b)*, and organismal phylogeny with geographic history *(c)*. In each case one entity can be viewed as tracking another, with a lesser or greater degree of fidelity (after Page and Charleston, 1998, fig. 1).

(Brooks and McLennan, 1991). The analogy with gene and species trees shows how misleading the latter assumption may be. Parasites are likely to be prone to similar processes; lineage sorting may occur due to parasite extinction or failure to colonize both descendants of a speciation event ("sorting events"); gene duplication is equivalent to in situ speciation of parasites on the same host (Page, 1994a). This analogy suggests that much of the theoretical and modeling work on the relationship between gene trees and species could be fruitfully applied to the study of host-parasite cospeciation. Indeed, chapter 5 in this book is a detailed exploration of the implications of population genetic models for the study of cospeciation.

Comparative Biology

If understanding the evolution of an adaptation requires knowledge of the historical context in which that character evolved, then cospeciation in hosts and parasites provides a unique opportunity to study adaptation. The ancestral environment of a parasite is its host, hence reconstructing the environment in which a feature of the parasite evolved corresponds to reconstructing its ancestral host. Although reconstructing ancestors is not necessarily straightforward (Cunningham et al., 1998), in many cases it is likely to be easier than attempting to reconstruct paleoenvironments of free living organisms. Standard comparative methods (Harvey and Pagel, 1991) can be used to compare changes in host and parasite traits. For example, Morand et al. (2000) showed that gopher louse body size is dependent on the size of their gopher hosts. This appears to be a consequence of a simple "lock-and-key" relationship between the thickness of gopher hair and the width of the groove in the louse's head with which it grips the host's hair shaft.

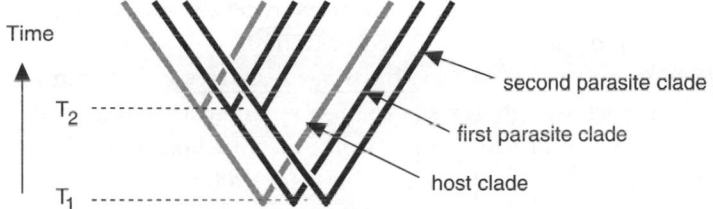

FIGURE 1.3. Two parasite clades cospeciating with the same host clade. The two parasite clades provide replicate lineages of the same age as the corresponding host lineages.

Another reason that parasites are particularly attractive for comparative studies is that hosts frequently harbor several parasite lineages (fig. 1.3). The presence of replicate lineages opens up possibilities of comparative analyses of cospeciation and coevolution between lineages of parasites on the same hosts. Given two or more clades of parasites tracking the same hosts, we can ask whether they cospeciate to the same extent, and whether those cospeciation events occur at the same points on the host tree. Chapter 11 provides an example of just this approach. The presence of multiple parasites also offers the possibility of teasing apart the different contributions of host and parasite biology to the prevalence of cospeciation—for an example of this, see chapter 7.

Rates of Evolution

A long-standing question in parasitology concerns the relative rate of evolution in hosts and parasites. As Brooks and McLennan (1993) note, literature on this question has often failed to clearly distinguish between change within a lineage (anagenesis) and rates of origination and extinction of lineages (cladogenesis). Hence when authors state, for example, that parasites are evolving more slowly than their hosts, it is not always clear whether they mean that the parasites show less character change than their hosts, speciate less often, or both.

Pairs of cospeciating hosts and parasites are of the same antiquity, and hence by comparing amounts of evolutionary divergence in the two lineages we can estimate their relative rate of evolution. While comparisons of amounts of morphological character change in hosts and parasites have led some to conclude that parasites are evolving less rapidly than their hosts (codified as "Manter's first rule": Brooks and McLennan, 1993, p. 15), such comparisons are fraught with difficulties. In contrast with molecular data, there is no single unit of morphological change that can be applied across disparate taxonomic groups. This absence of a common yardstick for morphological characters means that most recent work on rates of evolution in hosts and parasites has concentrated on molecular data. Examples include Moran et al.'s (1993) calibration of the rate of substitution in endosymbiotic bacteria with respect to their aphid hosts, and demonstrations of an elevated rate of substitution in louse mitochondrial DNA (Hafner et al., 1994; Page et al., 1998; Paterson et al., 2000). Cospeciating assemblages provide a unique opportunity to investigate the role factors such as generation time, population size, and metabolic rate may play in determining differences in the rate of molecular evolution among clades.

Inferring Host Phylogeny from Parasites

The use of parasites to infer host phylogeny is a long-standing tradition in parasitology (Klassen, 1992). If parasites are cospeciating with their hosts, then parasite phylogeny will, to a greater or lesser extent, reflect host phylogeny. One rationale for using parasites is the notion, mentioned above, that parasites evolve more slowly than their hosts. If this is the case, then parasites may retain characters indicative of relationships which have been lost in their hosts. However, at the molecular level, the evidence is that parasites are evolving more rapidly than their hosts (e.g., Hafner et al., 1994; Moran et al., 1995).

Although early workers on taxonomy (e.g., Hopkins, 1942) enthused about the potential of lice as phylogenetic markers of their hosts, this tradition fell into disrepute. Only one author in this book attempts to infer host phylogeny using parasites (chap. 6). Paterson et al. (1995) recount the tense exchange between Ernst Mayr and the louse taxonomist Gunther Timmermann that occurred in 1957 during the *Premier symposium sur la spécificité parasitaire des parasites de Vertébrés*, held at the University of Neuchâtel. Mayr accused Timmermann of placing a "child-like faith" in the evidence lice offered concerning avian relationships. At one point Mayr exclaimed:

> Two birds can exchange their parasites, nothing prevents this, but I have not yet seen two birds exchanging their heads, their wings or their legs. These have come down from its ancestors and not from another bird that nested in a hole right next to it! (Timmermann, 1957, p. 170)

Mayr has a point (see chap. 13), but it is ironic that the flood of genomic data in which we are currently awash provides evidence for rampant lateral gene transfer among distantly related lineages (Martin, 1999). This reinforces the parallels between the different kinds of historical association (Page and Charleston, 1998). In one sense, molecular systematists making inferences of organismal phylogeny from gene phylogenies are in much the same boat as parasitologists. We cannot assume that all genes are faithfully tracking species trees as not all genes of a species have "come down from its ancestors." Given this parallel, perhaps it is time to rethink the possible utility of parasites as markers of host phylogeny.

Methodological Issues

In Klassen's (1992) interesting, if somewhat narrow, view of the history of cospeciation research, Dan Brooks is credited with creating a

> synthesis of the North American biogeographic paradigm with its European phylogenetic counterpart to provide a vibrant and ever-changing methodological framework on which practically all coevolution research of the past decade has been based. (p. 584)

Hyperbole aside, Brooks is largely responsible for reinvigorating the field of cospeciation in the late twentieth century. His pioneering work (Brooks, 1979, 1981, 1985, 1988) was the first serious attempt to marry quantitative phylogenetics and cospeciation. My own interest in the field as an undergraduate student was sparked by his paper on freshwater stingrays (Brooks et al., 1981), and by the appeal of a quantitative methodology (Brooks, 1981). Despite Klassen's assertion (p. 585) that "the historical questions of methodology [are] now solved," it soon became apparent that Brooks's method (christened "Brooks Parsimony Analysis," or BPA by Wiley, 1988) suffered from some serious problems (Page, 1990a; Ronquist and Nylin, 1990). Debate over the relative merits of BPA and alternative methods has at times been heated, and has been depicted as a clash between two different research traditions (e.g., Brooks, 1996; Hoberg et al., 1997). In reality, I think the dispute is more accurately (and modestly) represented as a reflection of the technical challenges posed by trying to incorporate horizontal transfer into any method for comparing host and parasite trees. As it turns out, neither BPA nor the solution I proposed (Page, 1994b) and implemented in TREEMAP handled horizontal transfer entirely successfully—neither Brooks nor I were aware of how complicated the problem really is.

The Problem of Horizontal Transfer

If parasites either cospeciated (fig. 1.1a), speciated independently (fig. 1.1c), went extinct (fig. 1.1d), or missed the boat (fig. 1.1c), then the task of reconstructing the evolution of a host-parasite system would be straightforward, and was essentially solved by Goodman et al.'s reconciled tree concept. The simple biological fact that parasites can switch hosts poses

postulated host switches

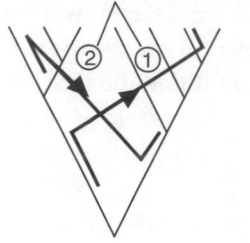 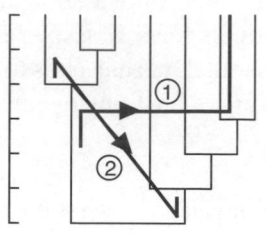 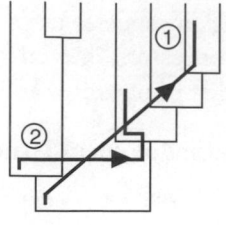

impossible transfer
(2 goes back in time)

impossible transfer
(1 goes forwards in time)

FIGURE 1.4. The problem of horizontal transfers. *(a)* Given this host phylogeny, the two postulated host switches (1 and 2) are incompatible. Ensuring that the source and destination hosts for switch 1 are contemporaneous *(b)* would require switch 2 to have gone backward in time. Likewise, for switch 2 to occur on this tree, switch 1 would have to go forward in time *(c)* (after Page and Charleston, 1997, figs. 11 and 12).

not so simple methodological problems. Ronquist (1995) was one of the first to point out why the problem is difficult and why, for example, the solution that I had implemented in TREEMAP could fail in some cases. Because postulating a horizontal transfer requires that the source and destination hosts are contemporaneous, we have to consider the relative ages of different host lineages. Failure to take this constraint into account can result in postulating transfers that are mutually incompatible (fig. 1.4). The complexity of the problem is reflected in the complexity of some of the proposed solutions, notably the "jungles" method developed by Charleston (1998). This topic is pursued further in chapters 2 and 3.

When Is It Cospeciation?

The most basic test of cospeciation is whether the topology of the host and parasite phylogenies is significantly more similar than would be expected due to chance alone. Given a measure of similarity between the two trees, the distribution of this measure can be obtained by generating a large number of pairs of random host and parasite phylogenies, and for each pair computing the similarity measure between the two trees. If the observed similarity is no greater than that obtained in, say, 95% of the random pairs, then the hypothesis of cospeciation can be rejected. This approach was pioneered in biogeography by Rosen (1978), Platnick and Nelson (1978), and Simberloff (1981), and applied to cospeciation analysis by Simberloff (1987), Hafner and Nadler (1988), and myself (Page, 1990b). The program TREEMAP (Page, 1994b) implements this method.

Agreement between host and parasite trees need not always be due to cospeciation. If host and parasite have responded in a similar way to shared biogeographic history, then their phylogenies will be largely concordant even if little cospeciation has occurred. For example, the southern beech tree *Nothofagus* is host to the fungus *Cyttaria*. Comparisons of phylogenies of these two organisms reveal some concordance, but this appears to reflect their similar response to the breakup of Gondwana, rather than strict cospeciation (Humphries et al., 1986). Similarly, it is possible that if a host and a parasite independently colonized a set of areas in a similar sequence (for example, hopping from one island to another), then their phylogenies will closely resemble each other without cospeciation having occurred. Matches between host and parasite phylogenies that are not due to cospeciation have been termed *pseudocospeciation* (Hafner and Nadler, 1988). Pseudocospeciation poses problems for tests of cospeciation that rely solely on comparing the topology of evolutionary trees. Tests that compare times of divergence will be more discriminating because in most cases of pseudocospeciation the host and parasite lineages will be of different ages.

If host and parasite phylogenies are identical, then we can pose more detailed questions concerning the timing of speciation and the relative rate of evolution in hosts and parasites (fig. 1.5). Even if host and parasite phylogenies have the same topology, host and parasite could have speciated at different times with respect to each other. If we have a molecular clock in both host and parasite (which may "tick" at different rates), the null hypothesis of identical speciation times ("temporal cospeciation": Page, 1996b) can be tested (fig. 1.5b).

The final null hypothesis is that host and parasite are cospeciating at the same time and furthermore are evolving at an identical rate (fig. 1.5c). This is equivalent to the previous hypothesis but with the additional constraint of equality of rates in host and parasite. Comparisons of evolutionary rate require comparable units of measurement, such as nucleotide substitutions per site, and hence must employ molecular data (Hafner and Page, 1995). Furthermore, because different genes may have quite different evolutionary dynamics, homologous genes in host and parasite should be used.

The Expectation of Congruence

The use of phylogenies in studies of cospeciation rests on the notion that congruent phylogenies implies cospeciation, whereas incongruence implies host switching (Brooks and McLennan, 1991). A key development in cospeciation research in the last two decades has been the realization that incongruence between host and parasite phylogenies can be due to a

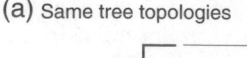

(a) Same tree topologies

(b) Same topologies and times of speciation

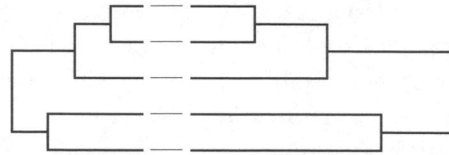

(c) Same topologies, speciation times, and rates of evolution

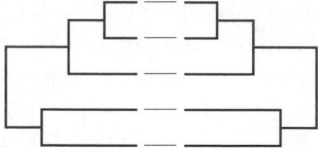

FIGURE 1.5. Three null hypotheses concerning host-parasite cospeciation. The first *(top)* is that the topologies of the host and parasite trees are identical, although the rates of evolution and speciation times may differ. The second *(middle)* is that, given that the topologies are the same, the relative speciation times are identical, even if the rates are different in the two clades. The third *(bottom)* is that topologies, speciation times, and rates are identical in host and parasite (after Huelsenbeck et al., 1997, fig. 3).

range of processes, not just host switching. Independent speciation by parasites (Page, 1990b, 1996b), extinction, and lineage sorting (Paterson and Gray, 1997; Paterson et al., 1999) can yield incongruent host and parasite trees. These processes are analogous to gene duplication, loss, allelic polymorphism, and lineage sorting in gene trees (Page and Charleston, 1998). Application of population genetic tools such as coalescent theory to modeling host-parasite cospeciation shows that incongruent host and parasite phylogenies can readily arise even if host and parasite are cospeciating (see chap. 5).

Because incongruence can have multiple causes, and need not by itself signal a lack of cospeciation, it requires careful analysis. In the past, incongruence has been taken at face value as evidence against cospeciation. For example, consider Kethley and Johnston's (1975) classic paper in which they introduced the term *resource tracking*. They argued that the discordance between a phenetic analysis of feather quill mite morphology and the relationships of the avian hosts refuted the idea that quill mites were cospeciating with their hosts. Instead, they found that mites clustered mainly on the basis of size, which is a function of the width of the quill in which they live. Hence, mite phylogenies were tracking the resource of feather quill size, rather than avian phylogeny. Apart from the obvious objection that phenetic techniques are often poor estimators of phylogeny, it would be worth investigating to what extent this discordance between mite and avian phylogeny is due to processes other than host switching.

Advantages and Limitations of Tree Comparison Methods

Tree-based methods for analyzing host-parasite cospeciation have the advantage of being applicable to many kinds of data. The host and parasite trees can be based on molecular, morphological, behavioral, or other kinds of data in any combination. Indeed, the trees being compared need not even be representations of phylogeny. In an elegant study, Becara (1997) compared the phylogeny of *Blepharida* beetles with a phylogeny of their host plants *(Bursera),* and a dendrogram of the hosts based on chemical similarity. The insects' evolutionary history more closely reflects their host plant chemistry rather than host plant phylogeny.

Like any comparative method, the study of cospeciation stands or falls on the reliability of our estimates of phylogeny. Different trees may imply quite different histories of a host-parasite association. One class of methods for comparing host and parasite trees assumes that the trees for host and parasite are known without error. TREEMAP, for example, requires fully resolved trees for both members of the association. In reality, we rarely have robust, unequivocal trees for the taxa of interest. One approach to this problem is to delimit a "confidence set" (Sanderson, 1989; Rodrigo et al., 1993; Page, 1996a) of trees for the host and the parasite. These sets comprise trees that we cannot reliably reject given the data at hand. Given m host trees and n parasite trees, we would then make all $m \times n$ comparisons between host and parasite phylogenies. However, due to the rapidly increasing number of possible trees with increasing number of taxa (Felsenstein, 1978), this approach may be rapidly swamped by excessive numbers of trees.

As an alternative to comparing trees, John Huelsenbeck and colleagues have been pursuing the use of maximum likelihood and Bayesian methods (Huelsenbeck et al., 1997, 2000; see chap. 4) that operate on the underlying character data supporting the trees. Initially these workers employed likelihood-ratio tests to determine whether host and parasite phylogenies were identical. This approach relies on the availability of mathematical models of character evolution, which in practice means molecular data. The null hypothesis is that the hosts and parasites have the same underlying tree topology (fig. 1.5a). This approach has been applied by Peek et al. (1998) to test cospeciation in chemoautotrophic bacteria and deep-sea clams. Thus, while contemplation of the number of possible evolutionary trees and the uncertainty that often surrounds estimates of phylogenies may lead one to despair, another approach is to explicitly incorporate this uncertainty in our methods of analysis. Bayesian methods explore the history of the host and parasite phylogeny over a range of phylogenies, which are sampled in proportion to their probability of being correct under a given model of evolution.

While model-based methods have considerable promise, they are not without problems. Inferences made from these techniques may be strongly influenced by the adequacy of our models of sequence evolution. Clark et al. (2000) used both tree- and data-based methods to investigate cospeciation between aphids and *Buchnera* bacterial endosymbionts. Although aphid and *Buchnera* trees are nearly identical, likelihood ratio tests rejected the null hypothesis that the two clades have cospeciated. Clark et al. (2000) suggested that inaccurate evolutionary models have resulted in the likelihood-ratio tests incorrectly rejecting the hypothesis of cospeciation. Given the increasing role maximum likelihood and Bayesian methods are likely to play in the study of cospeciation, the sensitivity of these methods to violation of their assumptions needs to be explored further.

This Book

Having set the scene above, I now turn to the book itself. It is divided into chapters that are primarily concerned with methodology, and those that have a particular group of organisms as their central focus. The methodological chapters describe parsimony and statistical approaches to the study of host-parasite assemblages; the empirical chapters provide examples of cospeciation involving vertebrate hosts and nematodes, viruses, and lice.

Methodology

In a series of papers, Fredrik Ronquist (1990, 1995, 1998a, 1998b) has sought to develop a rigorous parsimony-based approach to comparing host and parasite phylogenies. In chapter 2 he defines a set of properties that he argues any event-based method of analyzing host-parasite associations should possess. He then evaluates how well different parsimony methods (such as that implemented in TREEMAP, and the jungle method of Charleston, 1998) satisfy these properties.

In chapter 3, Mike Charleston and Susan Perkins provide a gentle introduction to "jungles," a representation of all feasible reconstructions of the history of a parasite and its host (Charleston, 1998). The complexity of the jungle is a direct result of the need to incorporate host switching into reconstructions of the history of host-parasite assemblages. Charleston and Perkins explain why host switching is so complicated, and then apply the jungles method to investigate the evolution of lizard malaria in the Caribbean.

Whereas most methods for examining host-parasite cospeciation have focused on comparing tree topologies, John Huelsenbeck, Bruce Rannala, and Bret Larget present an alternative, explicitly statistical approach (chap. 4) that uses sequence data directly. Given that any tree topology has associated with it some degree of uncertainty, they combine the analyses of sequence data with cospeciation in a Bayesian framework. Although the authors concede that their model of host switching is simplistic, and that the method currently ignores other events (such as independent speciation or extinction of the parasites), Bayesian methods show great promise. The method leads to some striking visualizations of the distribution of probabilities of parasites switching among different pairs of host taxa, which immediately raises the question why some host lineages are more likely to be involved in host switches than others.

Bruce Rannala and Yannis Michalakis (chap. 5) confront the underlying expectation of much of the cospeciation literature, namely that cospeciating hosts and parasites will have congruent phylogenies. Departures from congruence are usually interpreted as evidence for other processes, such as host switching. Mobilizing some machinery from population genetics, Rannala and Michalakis relate mechanisms of speciation (and population demographic structure) to expected patterns of cospeciation at a range of hierarchical levels above and below the species. They show that even in the "best-case" scenario in which there is only cospeciation, the probability that host and parasite gene trees have identical topologies may be small in some situations. They argue that population demographic factors play a major role in determining whether cospeciation generates congruent host and parasite phylogenies (empirical evidence for this idea is provided by Jim Demastes and coauthors in chapter 9).

Empirical Examples

The empirical examples included in this book are heavily biased toward lice (chaps. 8–13), although other systems are represented, including lizards and malaria (3), rodents and nematodes (6), and vertebrates and viruses (7). This partly reflects the prominent role lice have historically played in the development of the field, from the formulation of rules of parasitism by Fahrenholz and Eichler, through to the seminal gopher louse studies of Mark Hafner and his colleagues. Where possible, the DNA sequence and morphological data sets for each study presented here have been deposited in the phylogenetic database TreeBASE (Sanderson et al., 1994, <http://www.treebase.org>). Accession numbers are given in the relevant chapter.

Jean-Pierre Hugot's study of primate pinworms (Hugot, 1999) is one of the largest examples of host-parasite cospeciation to date, and in chapter 6 he investigates the pinworms of another group of mammals, the hystrico-morph rodents (porcupines, guinea pigs, and their relatives). The phylogeny of these mammals is controversial, prompting Hugot to ask what the pin-worms can tell us about the relationships of their hosts. In contrast with recent molecular studies, Hugot finds that the pinworms support the mono-phyly of New and Old World porcupines, a similarity usually dismissed by mammalian systematists as convergent. It will be intriguing to see whether future work will support the host relationships implied by the pinworms.

Based on our recent experience of human and animal diseases, we tend to think of viruses as readily moving between distantly related hosts. How-ever, some viral phylogenies show a remarkable match to their hosts' phy-logeny (McGeoch et al., 1995). In chapter 7, Jo Martin, Peter Kabat, and Mike Tristem investigate the relationship between phylogenies of retro-viruses and their vertebrate hosts. The broad correspondence between viral and vertebrate hosts is remarkable, but there are some notable exceptions to this pattern. Martin, Kabat, and Tristem explore ways of teasing apart the relative contributions of host and viral biology to the degree of concor-dance of evolutionary histories. The increasing use in virology of tools for analyzing cospeciation should lead to further insights into the origins and epidemiology of viruses (Siddall, 1997; Sharp et al., 2000).

The remaining chapters in this book concentrate on lice, which have played a preeminent role in the development of the study of cospeciation. Appropriately, we begin with Mark Hafner's review of the gopher-louse system (chap. 8). Hafner describes how an encounter with the louse taxonomist Roger Price led to more than a decade of work unraveling the relationships between gophers and their tiny ectoparasites. The publi-cations by Hafner, his postdocs, and his students have provided the data sets theoreticians have cut their teeth on (see the first half of this book). In his chapter, Hafner provides historical and ecological background to the gopher-louse system, and describes the several different lines of research his group is now pursuing. Notable among these is the addition of an addi-tional layer to the gopher-louse association by exploring the phylogeny of endosymbiotic bacteria hosted by the lice.

Chapters 9 and 10 explore cospeciation in mammalian trichodectid lice at two very different scales. Jim Demastes, Theresa Spradling, and Mark Hafner investigate gopher-louse cospeciation at more local geographical scales than their previous work, and find that the pattern of cospeciation so prominent in *Orthogeomys* and its lice (Hafner and Nadler, 1988; Hafner et al., 1994) is more complicated in the gopher genus *Thomomys*. The combination of incomplete lineage sorting, chance extinction events, and variable host population results in a lack of correlation between genetic divergence in hosts and their lice. This chapter lends empirical support to the theoretical analyses presented in chapter 5.

Gopher lice belong to the family Trichodectidae, and Jason Taylor and Andy Purvis place the gopher-lice system within a broader context by investigating trichodectid phylogeny. Chris Lyal (1985, 1986, 1987) was one of the first to apply cladistic methods to the study of louse phylogeny and host-parasite cospeciation. His data matrix was so large that it defeated phylogenetic software available in the early 1980s, and hence Lyal had to assemble the tree by hand. Using a recently developed tree search algorithm, Taylor and Purvis reanalyze Lyal's matrix and discuss the implications of the resulting trees for the hypothesis that trichodectid lice and their mammalian hosts have cospeciated. They find only moderate evidence for cospeciation, the same conclusion reached by Lyal. Surprisingly, they find that the celebrated gopher-louse system does not stand out from other trichodectids as showing noticeably more cospeciation. Complicated host-parasite associations, host switching, and uncertainty concerning basal phylogenetic relationships make it difficult to decipher the deep history of mammals and their lice. Taken together with chapter 9, this chapter suggests that the patterns of cospeciation between hosts and parasites are most likely to be found nearer (but not too near) the tips of the tree.

Kevin Johnson and Dale Clayton (chap. 11) exploit the presence of multiple lineages of lice on the same avian hosts to compare patterns of cospeciation in the two clades. They find that both louse clades show significant cospeciation, but these cospeciation events are uncorrelated in the two lice. Explanations are sought in differences in the dispersal abilities of the two louse clades (this link between parasite biology and cospeciation is further explored by the authors in chapter 13). Johnson and Clayton also suggest that methods for mapping parasite trees onto host trees have overlooked the extent to which failure of parasites to speciate can lead to incongruence between host and parasite phylogenies. They argue that methods such as that implemented in TREEMAP can postulate excessive numbers of sorting events and duplications, when in reality parasite lineages have simply not speciated.

The reality of the events postulated by current methods of cospeciation analysis is the topic of chapter 12. Adrian Paterson, Ricardo Palma, and Russell Gray investigate whether the high numbers of sorting events that are typically postulated in TREEMAP reconstructions are realistic or not. Sorting events may be due to parasite extinction or absence from the ancestral host range ("missing the boat"), or are simply an artifact of our failure to find them when they are present in low numbers (for which the authors introduce the term x-*event*). Although the correlation between collecting effort and numbers of parasite species recovered is a well known phenomenon in parasitology (e.g., Gregory, 1997), Paterson and colleagues argue from studies of museum collections that (above a threshold number of host individuals) sampling effort is an unlikely explanation for many sorting events. They present evidence that missing the boat is a more important cause of sorting events than extinction.

A theme that emerges consistently in the empirical chapters is the role of both host and parasite biology in determining the degree of cospeciation. In the final chapter in this volume, Dale Clayton, Sarah Al-Tamimi, and Kevin Johnson seek to integrate ecological processes with phylogenetic patterns. Surveying cospeciation studies between different bird and louse clades, they explain the differing degrees of matching as a function of parameters of host and parasite ecology, notably the distribution of hosts and their parasites both geographically and ecologically. It is still early days, but chapter 13 points the way to synthesizing ecological and phylogenetic approaches to the study of cospeciation.

REFERENCES

Becerra, J. X. 1997. Insects on plants: Macroevolutionary chemical trends in host use. *Science* 276:253–56.

Brooks, D. R. 1979. Testing the context and extent of host-parasite coevolution. *Systematic Zoology* 28:299–307.

———. 1981. Hennig's parasitological method: a proposed solution. *Systematic Zoology* 30:229–49.

———. 1985. Historical ecology: A new approach to studying the evolution of ecological variation. *Annals of the Missouri Botanical Garden* 72:660–80.

———. 1988. Macroevolutionary comparisons of host and parasite phylogenies. *Annual Review of Ecology and Systematics* 19:235–59.

———. 1996. Explanations of homoplasy at different levels of biological organization. In *Homoplasy: The recurrence of similarity in evolution,* edited by M. J. Sanderson and L. Hufford, 3–36. San Diego: Academic Press.

Brooks, D. R., and D. A. McLennan. 1991. *Phylogeny, ecology, and behavior.* Chicago: University of Chicago Press.

——. 1993. *Parascript: Parasites and the language of evolution.* Washington, D.C.: Smithsonian Institution Press.

Brooks, D. R., T. B. Thorson, and M. A. Mayes. 1981. Freshwater stingrays (Potamotrygonidae) and their helminth parasites: Testing hypotheses of evolution and coevolution. In *Advances in cladistics: Proceedings of the first meeting of the Willi Hennig Society,* edited by V. A. Funk and D. R. Brooks, 147–75. New York: New York Botanical Garden.

Charleston, M. A. 1998. Jungles: A new solution to the host/parasite phylogeny reconciliation problem. *Mathematical Biosciences* 149:191–223.

Clark, M. A., N. A. Moran, P. Baumann, and J. J. Wernegreen. 2000. Cospeciation between bacterial endosymbionts *(Buchnera)* and a recent radiation of aphids *(Uroleucon)* and pitfalls of testing for phylogenetic congruence. *Evolution* 54:517–25.

Clayton, D. H., P. L. M. Lee, D. M. Tompkins, and E. D. Brodie III. 1999. Reciprocal natural selection on host-parasite phenotypes. *American Naturalist* 154:261–70.

Clayton, D. H., and J. Moore. 1997. Introduction to *Host-parasite evolution: General principles and avian models,* edited by D. H. Clayton and J. Moore, 1–6. Oxford: Oxford University Press.

Cunningham, C. W., K. E. Omland, and T. H. Oakley. 1998. Reconstructing ancestral states: A critical reappraisal. *Trends in Ecology and Evolution* 13:361–66.

Doyle, J. J. 1992. Gene trees and species trees: Molecular systematics as one-character taxonomy. *Systematic Botany* 17:144–63.

Felsenstein, J. 1978. The number of evolutionary trees. *Systematic Zoology* 27:27–33.

Funk, D. J., J. Helbling, J. J. Wernegreen, and N. A. Moran. 2000. Intraspecific phylogenetic congruence among multiple symbiont genomes. *Proceedings of the Royal Society of London,* ser. B, 267:2517–21.

Goodman, M., J. Czelusniak, G. W. Moore, A. E. Romero-Herrera, and G. Matsuda. 1979. Fitting the gene lineage into its species lineage: A parsimony strategy illustrated by cladograms constructed from globin sequences. *Systematic Zoology* 28:132–68.

Gregory, R. D. 1997. Comparative studies of host-parasite communities. In *Host-parasite evolution: General principles and avian models,* edited by D. H. Clayton and J. Moore, 198–211. Oxford: Oxford University Press.

Hafner, M. S., and S. A. Nadler. 1988. Phylogenetic trees support the coevolution of parasites and their hosts. *Nature* 332:258–59.

Hafner, M. S., and R. D. M. Page. 1995. Molecular phylogenics and host-parasite cospeciation: Gophers and lice as a model system. *Philosophical Transactions of the Royal Society of London,* ser. B, 349:77–83.

Hafner, M. S., P. D. Sudman, F. X. Villablanca, T. A. Spradling, J. W. Demastes, and S. A. Nadler. 1994. Disparate rates of molecular evolution in cospeciating hosts and parasites. *Science* 265:1087–90.

Harvey, P. H., and M. D. Pagel. 1991. *The comparative method in evolutionary biology.* Oxford: Oxford University Press.

Hennig, W. 1966. *Phylogenetic systematics.* Urbana: University of Illinois Press.

Hoberg, E. P., D. R. Brooks, and D. Seigel-Causey. 1997. Host-parasite co-speciation: History, principles, and prospects. In *Host-parasite evolution:*

General principles and avian models, edited by D. H. Clayton and J. Moore, 212–35. Oxford: Oxford University Press.

Hopkins, G. H. E. 1942. The Mallophaga as an aid to the classification of birds. *Ibis* 1942:94–106.

Huelsenbeck, J. P., B. Rannala, and B. Larget. 2000. A Bayesian framework for the analysis of cospeciation. *Evolution* 54:352–64.

Huelsenbeck, J. P., B. Rannala, and Z. Yang. 1997. Statistical tests of host-parasite cospeciation. *Evolution* 51:410–19.

Hugot, J.-P. 1999. Primates and their pinworms parasites: The Cameron hypothesis revisited. *Systematic Biology* 48:523–46.

Humphries, C. J., J. M. Cox, and E. S. Nielsen. 1986. *Nothofagus* and its parasites: A cladistic approach to coevolution. In *Coevolution and systematics,* edited by A. R. Stone and D. L. Hawksworth, 55–76. Oxford: Clarendon Press.

Kethley, J. B., and D. E. Johnston. 1975. Resource tracking patterns in bird and mammal ectoparasites. *Miscellaneous Publications of the Entomological Society of America* 9:231–36.

Klassen, G. J. 1992. Coevolution: A history of the macroevolutionary approach to studying host-parasite associations. *Journal of Parasitology* 78:573–87.

Lyal, C. H. C. 1985. A cladistic analysis and classification of trichodectid mammal lice (Phthiraptera: Ischnocera). *Bulletin of the British Museum (Natural History), Entomology* 51:187–346.

———. 1986. Coevolutionary relationships of lice and their hosts: A test of Farenholz's Rule. In *Coevolution and systematics,* edited by A. R. Stone and D. L. Hawksworth, 77–91. Oxford: Clarendon Press.

———. 1987. Coevolution of trichodectid lice (Insecta: Phthiraptera) and their mammalian hosts. *Journal of Natural History* 21:1–28.

Martin, W. 1999. Mosaic bacteria chromosomes: A challenge en route to a tree of genomes. *BioEssays* 21:99–104.

McGeoch, D. J., S. Cook, A. Dolan, F. E. Jamieson, and E. A. R. Telford. 1995. Molecular phylogeny and evolutionary timescale for the family of mammalian herpesviruses. *Journal of Molecular Biology* 247:443–58.

Moran, N. A., M. A. Munson, P. Baumann, and H. Ishikawa. 1993. A molecular clock in endosymbiotic bacteria is calibrated using the insect hosts. *Proceedings of the Royal Society of London,* ser. B, 253:167–71.

Moran, N. A., C. D. van Dohlen, and P. Baumann. 1995. Faster evolutionary rates in endosymbiotic bacteria than in cospeciating insect hosts. *Journal of Molecular Evolution* 41:727–31.

Morand, S., M. S. Hafner, R. D. M. Page, and D. L. Reed. 2000. Comparative body size relationships in pocket gophers and their chewing lice. *Biological Journal of the Linnean Society* 70:239–49.

Page, R. D. M. 1988. Quantitative cladistic biogeography: Constructing and comparing area cladograms. *Systematic Zoology* 37:254–70.

———. 1990a. Component analysis: A valiant failure? *Cladistics* 6:119–36.

———. 1990b. Temporal congruence and cladistic analysis of biogeography and cospeciation. *Systematic Zoology* 39:205–26.

———. 1993. Genes, organisms, and areas: The problem of multiple lineages. *Systematic Biology* 42:77–84.

————. 1994a. Maps between trees and cladistic analysis of historical associations among genes, organisms, and areas. *Systematic Biology* 43:58–77.

————. 1994b. Parallel phylogenies: Reconstructing the history of host-parasite assemblages. *Cladistics* 10:155–73.

————. 1996a. On consensus, confidence, and "total" evidence. *Cladistics* 12:83–92.

————. 1996b. Temporal congruence revisited: Comparison of mitochondrial DNA sequence divergence in cospeciating pocket gophers and their chewing lice. *Systematic Biology* 45:151–67.

Page, R. D. M., and M. A. Charleston. 1997. Reconciled trees and incongruent gene and species trees. In *Mathematical Hierarchies in Biology*, edited by B. Mirkin, F. R. McMorris, F. S. Roberts, and A. Rzhetsky, 57–70. Providence, R. I.: American Mathematical Society.

————. 1998. Trees within trees: Phylogeny and historical associations. *Trends in Ecology and Evolution* 13:356–59.

Page, R. D. M., P. L. M. Lee, S. A. Becher, R. Griffiths, and D. H. Clayton. 1998. A different tempo of mitochondrial DNA evolution in birds and their parasitic lice. *Molecular Phylogenetics and Evolution* 9:276–93.

Pagel, M. 1999. Inferring the historical patterns of biological evolution. *Nature* 401:877–84.

Paterson, A. M., and R. D. Gray. 1997. Host-parasite cospeciation, host switching, and missing the boat. In *Host-Parasite Evolution: General Principles and Avian Models*, edited by D. H. Clayton and J. Moore, 236–50. Oxford: Oxford University Press.

Paterson, A. M., R. D. Gray, and G. P. Wallis. 1995. Of lice and men: The return of the "Comparative parasitology" debate. *Parasitology Today* 11:158–60.

Paterson, A. M., R. L. Palma, and R. D. Gray. 1999. How frequently do avian lice miss the boat? *Systematic Biology* 48:214–23.

Paterson, A. M., G. P. Wallis, and R. D. Gray. 2000. Seabird and louse coevolution: Complex histories revealed by sequence data and reconciliation analyses. *Systematic Biology* 49:383–99.

Peek, A. S., R. A. Feldman, R. A. Lutz, and R. C. Vrijenhoek. 1998. Cospeciation of chemotrophic bacteria and deep sea clams. *Proceedings of the National Academy of Science of the USA* 95:9962–66.

Platnick, N. I., and G. Nelson. 1978. A method of analysis for historical biogeography. *Systematic Zoology* 27:1–16.

Rodrigo, A. G., M. Kelly-Borges, P. R. Berquist, and P. L. Berquist. 1993. A randomisation test of the null hypothesis that two cladograms are sample estimates of a parametric phylogenetic tree. *New Zealand Journal of Botany* 31:257–68.

Ronquist, F. 1995. Reconstructing the history of host-parasite associations using generalised parsimony. *Cladistics* 11:73–89.

————. 1998a. Phylogenetic approaches in coevolution and biogeography. *Zoologica Scripta* 26:313–22.

————. 1998b. Three-dimensional cost matrix optimisation and maximum cospeciation. *Cladistics* 14:167–72.

Ronquist, F., and S. Nylin. 1990. Process and pattern in the evolution of species associations. *Systematic Zoology* 39:323–44.

Rosen, D. E. 1978. Vicariant patterns and historical explanation in biogeography. *Systematic Zoology* 27:159–88.

Sanderson, M. J. 1989. Confidence limits on phylogenies: The bootstrap revisisted. *Cladistics* 5:113–29.

Sanderson, M. J., M. J. Donoghue, W. Piel, and T. Eriksson. 1994. TreeBASE: A prototype database of phylogenetic analyses and an interactive tool for browsing the phylogeny of life. *American Journal of Botany* 81:183.

Sharp, P. M., E. Bailes, F. Gao, B. E. Beer, V. M. Hirsch, and B. H. Hahn. 2000. Origins and evolution of AIDS viruses: Estimating the time-scale. *Biochemical Society Transactions* 28:275–82.

Siddall, M. E. 1997. The AIDS pandemic is new, but is HIV *not* new? *Cladistics* 13:267–73.

Simberloff, D. 1987. Calculating the probabilities that cladograms match: A method of biogeographic inference. *Systematic Zoology* 36:175–95.

Simberloff, D., K. L. Heck, E. D. McCoy, and E. F. Connor. 1981. There have been no statistical tests of cladistic biogeographical hypotheses. In *Vicariance biogeography: A critique,* edited by G. Nelson and D. E. Rosen, 40–63. New York: Columbia University Press.

Swofford, D. L., G. J. Olsen, P. J. Waddell, and D. M. Hillis. 1996. Phylogenetic inference. In *Molecular Systematics,* edited by D. M. Hillis, C. Moritz, and B. K. Mable, 407–514. Sunderland, Mass.: Sinauer.

Thompson, J. N. 1994. *The coevolutionary process.* Chicago: University of Chicago Press.

Timmermann, G. 1957. Stellung und Gliederung der Regenpfeifervögel (Ordnung Charadriiformes) nach massgabe des Mallophagologischen befundes. In *Premier symposium sur la spécificité parasitaire des parasites de Vertébrés,* edited by J. B. Baer, 159–72. Neuchâtel: Institut de Zoologie, Université de Neuchâtel.

Wiley, E. O. 1988. Parsimony analysis and vicariance biogeography. *Systematic Zoology* 37:271–90.

Appendix 1.1

Glossary of some terms encountered in the cospeciation literature. For a primer of more general terms in parasitology see Clayton and Moore (1997).

association by colonization	Host acquired its parasite from another host that was not itself ancestral to the host (cf. **association by descent**)
association by descent	Host inherited its parasite from ancestral host (cf. **association by colonization**)
BPA	Brooks Parsimony Analysis, a technique for comparing host and parasite cladograms that uses binary codes (0s and 1s) to represent the parasite phylogeny (see chap. 2)
coadaptation	Reciprocal adaptation in two or more associated organisms, such as a host and its parasite

coevolution	Joint evolution of two or more associated organisms. Often used in the strict sense as a synonym of **coadaptation**
congruence	Agreement between topologies of two or more trees (cf. **incongruence**)
cospeciation	Joint speciation of both host and parasite. Either host or parasite may speciate slightly after or before the other. Strictly contemporaneous cospeciation is temporal cospeciation.
duplication	Independent speciation of parasite. Termed a duplication by Page (1994a) by analogy with gene duplication
Fahrenholz's Rule	"Parasite phylogeny mirrors host phylogeny"
horizontal transfer	Generic term for where a lineage associated with one host lineage transfers to another lineage. In the case of hosts and parasites, this is a **host switch**
host switch	Transfer of a parasite from one host to another
incongruence	Discordance between topologies of two or more trees
jungle	Graph displaying all possible reconstructions of the history of association between a pair of host and parasite phylogenies (see chap. 3)
"missing the boat"	Absence of a parasite from a descendant host due to the local absence of the parasite from the ancestral host population (see chap. 12)
pseudocospeciation	A match between host and parasite phylogenies that is not due to **cospeciation**
reconciled tree	A mapping between a parasite and host tree where no **host switching** is permitted
resource tracking	Process where parasite is associated with a set of hosts that share a resource. If the resource is shared by phylogenetically unrelated hosts, then the host and parasite phylogenies will not match.
sorting event	Generic term to cover situation where a parasite lineage is absent from a host, due to extinction, collection failure, or **missing the boat**
temporal cospeciation	Strictly contemporaneous cospeciation
tree mapping	Given a pair of trees a map between them associates every node in one tree with a node in the second tree. Methods such as **reconciled trees** and **jungles** explicitly create such a map.

2

PARSIMONY ANALYSIS OF COEVOLVING SPECIES ASSOCIATIONS

Fredrik Ronquist

Introduction

Closely related organisms not only share traits in their anatomy, morphology, and behavior, they also show extensive similarities in the parasite communities associated with them. Thus, just as morphological and anatomical traits can be used in elucidating phylogenetic relationships, it should be possible to base genealogical conclusions on parasite data. But parasite characters are complicated. If we simply code presence and absence of individual parasite species, we ignore the fact that some parasites are more similar than others. Clearly, hosts that are associated with almost identical, sibling flea species are likely to be more closely related than hosts whose fleas are morphologically divergent and belong to different genera. To fully utilize the parasitological data, we have to take parasite relationships into account.

The first attempt to accomplish this in a quantitative way scored the parasites and their phylogeny as a set of binary characters (Brooks, 1981). Half the characters were used to record the presence or absence of the actual parasite species, that is, the terminal branches in the parasite tree, whereas the remaining characters each represented an interior node in the parasite tree (cf. fig. 2.4). In other words, the parasite data were treated as character-state trees, which were converted into binary characters using additive binary coding. The latter technique was modified to deal with complexities such as parasites occurring on multiple hosts, and was renamed Brooks Parsimony Analysis (BPA) in this new context (Wiley, 1988; Brooks, 1990).

Under the name *component analysis,* similar ideas were being developed in historical biogeography (Platnick and Nelson, 1978; Nelson and Platnick, 1981), where organisms were used to infer area relationships much as parasites were used to infer host relationships in the parasitological context. The similarities between host-parasite studies and cladistic biogeographic analysis soon became evident, and BPA was adapted for use in historical biogeography (Brooks, 1985), just as component analysis was used to tackle the analysis of coevolving species associations (e.g., Humphries et al., 1986).

Much of the methodological discussion that followed confused the distinction between parasitological and biogeographic data on one hand and ordinary character data on the other. As a consequence, heated arguments followed about the meaning of fundamental concepts in character analysis, such as homology and homoplasy, in the coevolutionary and biogeographic context (Sober, 1988). A popular idea at the time was that systematists should focus on elucidating patterns without assuming anything about evolutionary process, making it difficult to resolve methodological controversies. In historical biogeography and coevolutionary analysis, it became evident that homology and homoplasy, as revealed by BPA and component analysis, had no straightforward interpretation in terms of evolutionary events such as dispersals and host shifts (Wiley, 1988).

Other systematists interpreted parsimony analysis as being based on character transformation models, and justifiable to the extent that these models were reasonable representations of the evolutionary process. This view made the relation between method and process explicit, which meant that methodological controversies could at least be accurately understood, if not settled. Ronquist and Nylin (1990) were probably the first to insist that parsimony analysis in historical biogeography and coevolution be based on explicit process models. Such a model should specify a set of events (transformations) that change parasite-host or organism-area associations. Each of these events is assigned a cost, and then the minimum-cost (most parsimonious) explanation of the observed data is sought. A coevolutionary or biogeographic parsimony method that satisfies these criteria is an *event-based method* (Ronquist and Nylin, 1990; Ronquist, 1994, 1998b, 1998c); a method that does not is *pattern-based*.

Event-based methods have many attractive properties. For instance, an event-based analysis specifies a series of events, such as cospeciations and host switches, which could have produced the observed data. There is no need for a posteriori interpretation of mysterious entities such as "reversals" or "items of error" (Ronquist, 1994, 1995, 1997). Furthermore, the process model and the cost assignments of the method reveal its properties. Thus, it is straightforward to compare event-based methods and predict how they will perform when applied to particular problems. Ignoring process in the formulation of a method does not make it more objective—the method's performance is still determined by the nature of the evolutionary process being studied.

Event-based thinking was introduced almost a decade earlier to another field involving the comparison of trees, namely that of inferring species trees from gene trees (Goodman et al., 1979). Genes track species trees much as parasites track the phylogeny of their hosts: the genes codiverge with species just as parasites codiverge with their hosts, and genes infect unrelated species lineages by horizontal transfer just as parasites colonize new hosts through host switching. Page (1988, 1993) was the first to clearly recognize this analogy between the analysis of gene tree–species tree relationships on one hand and parasite-host and organism-area relationships on the other. This brought together the entire field of *parsimony-based tree fitting* (see recent reviews by Page and Charleston, 1998; Ronquist, 1998b).

Under the name *tree reconciliation,* Page (1994a) formalized and developed the method pioneered by Goodman and colleagues; since that time, reconciliation has been widely used in gene tree–species tree comparisons. However, reconciliation is less useful in the coevolutionary context because it is based on a model that does not include host shifting, a process that is clearly relevant in host-parasite systems. Ronquist and Nylin (1990) were the first to successfully incorporate host switches into an explicit event-based method, which will be referred to here as *tracking/switching analysis* (TS). Page later introduced switches into the reconciliation framework and suggested that the problem of finding cost-event assignments could be solved by maximizing the number of inferred host-parasite cospeciations (Page, 1994b). This approach, implemented in the program TREEMAP, may be referred to as *maximum codivergence* (MC). Before discussing these and other event-based methods further, however, we need to cover the basics of the event-based approach in more detail.

The Fundamentals

The kernel in the inference of host-parasite coevolution is the *reconstruction problem.* The given data are (1) a parasite phylogeny, (2) a host phylogeny, (3) data on the host association of each terminal parasite, (4) a model of host-parasite coevolution, and (5) cost assignments to each event in the model. The problem is to find the minimum-cost (most parsimonious) reconstruction of the history of the association. Such a reconstruction specifies a series of past host-parasite associations and a set of events affecting these associations, together producing the associations and the host and parasite phylogenies that can be observed today. The cost of the reconstruction is simply the sum of the cost of the events. The event-cost assignments are essential: they define how the different types of events should be weighted against each other when comparing alternative reconstructions. If each event is associated with a cost that is inversely related to the likelihood of the event (the more likely the event, the smaller the cost), then the most parsimonious reconstruction will also, in some sense, be the most likely explanation of the observed data.

If the host phylogeny is not known, we may be interested in inferring it from the parasite relationships and the parasite-host associations; this is the *estimation problem*. It represents a search through tree space for the best host tree, each host tree being evaluated by fitting the parasite tree(s) onto it, solving the reconstruction problem, and recording the cost. As one would expect of a tree search problem, the estimation problem is computationally complex, considerably more so than the reconstruction problem.

Model Assumptions

The models considered thus far are relatively similar to one another. First, all of them can be described as breaking the host and parasite phylogenies into units corresponding to branches (also known as internodes or edges) (fig. 2.1). The branches (or edges) are often referred to as separate host and parasite *species,* and are connected by nodes (or vertices) representing speciation or splitting events. The inference procedure matches branches and nodes in one phylogeny with branches and nodes in the other. For instance, a particular reconstruction may associate parasite branch p with host branch $h,$ implying that the corresponding lineages were associated in the past.

The host and parasite phylogenies are usually considered known without error. However, uncertainty in the estimate of host and parasite

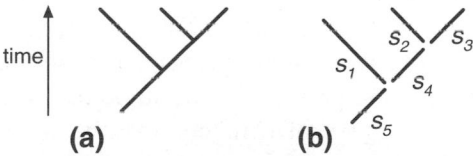

(a)   (b)

FIGURE 2.1. In modeling the evolution of species associations, parasite and host phylogenies *(a)* are usually broken into units corresponding to branches (also called internodes or edges) *(b)*. The branches (s_1 to s_5) are referred to as separate species and treated as homogeneous units.

relationships can easily be taken into account by simply replacing a single input tree with a set of weighted trees expressing the confidence in various host and parasite clades. Such a set may be obtained through bootstrapped parsimony analysis of host or parasite character data, or it may be drawn from a posterior distribution resulting from Bayesian inference of phylogeny (Rannala and Yang, 1996; Yang and Rannala, 1997; Huelsenbeck, Rannala, and Larget, chap. 4).

It is usually assumed that each ancestral parasite lineage was associated with a single host lineage at any particular point in time, the *one-host-per-parasite assumption* (but see Ronquist and Nylin, 1990). This simplification makes the problem easier to treat mathematically but it is also biologically relevant, since it is primarily specific parasites that are likely to retain their host preferences long enough to show phylogenetically constrained association patterns. An individual parasite specimen can only be associated with one individual host at a time, which means that true polyphagy can only be maintained through frequent host shifts within evolutionary lineages. If true polyphagy were common in a group of parasites, we would expect these frequent shifts to extend into phylogenetic time and obscure historical patterns of host usage.

The one-host-per-parasite model remains useful even when several of the terminal parasites are "widespread," that is, associated with multiple hosts. This is particularly true if the widespread parasites represent collections of independently evolving lineages, each associated with a single host. The model is also relevant when wide host ranges consist of one major and a set of more rarely used auxiliary hosts, or a primary and a secondary host used during different life stages of the parasite. Even when none of these cases apply, the one-host-per-parasite model remains a powerful tool for examining a complex world. For instance, it can be used to test whether parasites with wide host ranges evolved recently from more specialized ancestors with phylogenetically conserved host preferences, or if host choice has always been plastic in the group they belong to. Furthermore, if host shifts are common but mostly occur between closely related hosts, historically constrained association patterns may persist and be accurately reconstructed under the one-host-per-parasite assumption.

Another standard assumption is that multiple parasite lineages infecting the same host lineage evolve independently of one another (the *independent-parasites assumption*). There is some evidence from insect-plant associations suggesting that multiple parasite lineages on the same host actually do evolve without interfering with one another (Hougen Eitzman and Rausher, 1994). An alternative model is to assume that a host can carry only a single parasite at a time (Huelsenbeck et al., 1997; Huelsenbeck et al., 2000) (see also chap. 4). This model may under some circumstances be adequate for gene tree–species tree problems but is probably less realistic for most host-parasite systems.

Events and Model Space

There are six types of events that commonly figure in models of host-parasite evolution: *cospeciation* (or codivergence), that is, simultaneous divergence of host and parasite lineages (fig. 2.2a); *duplication* (or independent parasite speciation), whereby the parasite lineage splits into to two daughter lineages associated with the same host (fig. 2.2b); *partial switch* (or simply switch), by which the parasites colonize an unrelated host, speciating in the process (fig. 2.2c); *sorting* (or partial extinction), in which a parasite lineage tracks a host lineage through a host speciation event without speciating itself (fig. 2.2d); *complete switch,* involving a shift to a new host without associated speciation (fig. 2.2e); and *extinction,* wiping out a host or parasite lineage completely (fig. 2.2f). Obviously there are many other possible event types that could be considered, but these six provide a useful starting point in formulating minimal models that can claim some degree of realism.

Several different models and cost assignments have been discussed in the literature. Intuitively it seems that many more, perhaps an infinite number, of varieties may exist. How many models are there that lead to distinct, valid methods? Before examining this question it is necessary to define a set of required properties for a valid event-based tree-fitting method. There will probably never be universal agreement on what those properties should

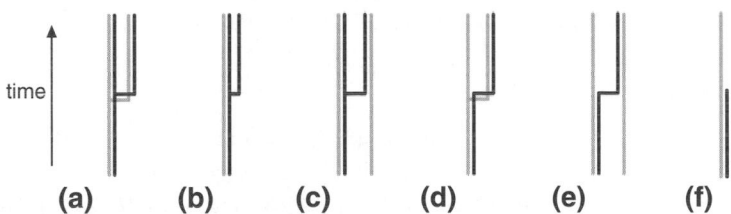

FIGURE 2.2. Six events that have been considered in models of host-parasite coevolution. The host lineage is portrayed in gray, the parasite lineage in black. The events are *(a)* cospeciation; *(b)* duplication; *(c)* partial switch (also referred to simply as a switch); *(d)* sorting event (or partial extinction); *(e)* complete switch; and *(f)* (complete) extinction. The last two events postulate past host-parasite associations (just before the switch or the extinction) that cannot be traced to any living descendants.

be, but let us define three desirable properties here that many workers would probably agree on: *potency, factuality,* and *consistency.* In defining these properties, let us assume we have a host tree H with edges and vertices h, and a parasite tree P with edges and vertices p (see also appendix 2.1).

Potency

A *potent* model should fulfil three basic requirements. First, the method based on the model should be complete in that there should be a solution to all tree-fitting problem instances. Second, there should be problems for which the optimal solution associates internal vertices in the parasite tree with internal vertices or edges in the host tree. If a method always associated parasite ancestors with terminal parasites, it could hardly be called a tree-fitting method. In particular, if the parasite and host trees are identical *(isomorphic)* when the labels of the parasite terminals are exchanged with the labels of their associated hosts, then the method must produce only one optimal solution, fitting the elements in the parasite tree to the corresponding elements in the host tree. Finally, for each type of event in the model on which the method is based, there should be at least some problem for which that event is necessarily implied by the optimal reconstruction. Otherwise that event is redundant in the model.

Factuality

The second desirable property of a tree-fitting method is *factuality.* A factual method will not postulate events requiring ghost ancestral associations that could not be traced in P and H to any of the recent associations. An ancestral association *(p:h)* is traceable if there is a path in P from p to a leaf, along which all associates are either h or descendants of h.

Consistency

Finally, it is desirable that the method is logically *consistent* in that the reconstructions are always biologically possible. For instance, if a reconstruction postulates a switch from h_1 to h_2, then it must be possible that h_1 and h_2 occurred simultaneously. Furthermore, logical consistency in the parsimony framework requires that an event carry a positive cost if it alters an association and zero cost otherwise.

There is some argument about how duplications and cospeciations should be interpreted with respect to cost assignments. Duplications could be viewed as the expected outcome of parasite speciation, since the daughter lineages remain associated with the host of their ancestor, in which case they should be assigned zero cost (Ronquist and Nylin, 1990; Ronquist, 1995). Zero-cost duplications are required in a metric host-distance measure, and metricity is regarded by some workers as a methodological desirability or even necessity (Wheeler, 1993). On the other hand, strict metricity can never be achieved because of the time irreversibility of parasite-host tracking, and it has been argued that duplications should be associated with a positive cost in host-parasite systems (Charleston, 1998; see also chap. 3), as they have been commonly treated in gene tree–species tree analysis.

Cospeciation events could be regarded as the expected outcome of host speciation and assigned zero cost. They have also been associated with negative cost, either explicitly (Charleston, 1998) or implicitly (Page, 1994b; Ronquist, 1998c), the effect being that cospeciation events are maximized within the constraints imposed by the cost induced by other events. However, a negative cost is difficult to defend from a parsimony perspective. Here, I will assume that logical consistency implies a zero or positive cost for cospeciation and duplication events, and strict positive cost for other events.

Now, if factuality is required, then (postulated) extinctions and complete switches can never be included in optimal reconstructions because both of them imply events that cannot be traced in observable descendants (proof in appendix 2.1). This is self-evident for inferred extinctions: extinct lineages do not leave descendants. In a similar way, a complete switch does not leave any descendant on the old host lineage that could reveal the existence of this association. Contrast this with a partial switch, leaving descendants on both the old and new host. In fact, the outcome of a complete switch is indistinguishable from that of a partial switch immediately followed by an extinction event on the old host.

Thus, only four of the six described events can be included in the model of a factual method: duplications, cospeciations, sortings, and (partial) switches. These four event types are intimately related and different combinations of them can explain the same observed pattern. For instance, a cospeciation event (fig. 2.3a) may produce the same outcome as a duplication followed by two sorting events (fig. 2.3b) or a single switching event (fig. 2.3c).

Assume that we assign a constant cost to each of these events: c for cospeciation events, u for duplication events, s for sorting events, and i for switches. If we then examine all models that are based on these four

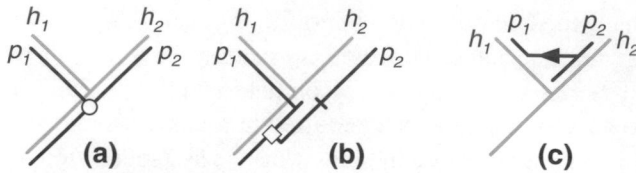

FIGURE 2.3. Interrelation between the events (cospeciations, duplications, sortings, and switches) in the four-event model. The same observed pattern, a pair of sister parasites (p_1 and p_2) being associated with a pair of sister hosts (h_1 and h_2), could be explained by a cospeciation event (*a*), a parasite duplication followed by two sorting events (*b*), or by a host switch (*c*). Host lineage in gray, parasite lineage in black. The following symbols are used for the events: square = duplication, circle = cospeciation, cross line or angle = sorting, arrow = switch.

TABLE 2.1 Model space for consistent, factual, and potent parsimony methods of analyzing host-parasite systems and the associated constraints on cost assignments (the constraints on maximal switch and sorting costs are difficult to define and are not given here; see appendix for details). Each event is assumed to be associated with a fixed constant cost.

Model Type	Name	Cospeciation	Duplication	Sorting	Switch	Cost Constraints
2-event	Two-event (2E)	-	$u \geq 0$	$s > 0$	-	-
3-event	Reconciliation (REC)	$c \geq 0$	$u \geq 0$	$s > 0$	-	$c < u + 2s$
3-event	Tracking/switching (TS)	-	$u \geq 0$	$s > 0$	$i > 0$	$i > u + 2s$
3-event	Cospeciation/switching (CS)	$c \geq 0$	$u \geq 0$	-	$i > 0$	$i > c$
4-event	Four-event (4E)	$c \geq 0$	$u \geq 0$	$s > 0$	$i > 0$	$c < i$
						$c < u + 2s$

events or combinations of them, it turns out that only five of them lead to methods that are potent, factual, and consistent (proofs in appendix 2.1). These models include one two-event, three three-event, and one complete four-event model. Each of these models is associated with a specific set of constraints on the event-cost assignments (table 2.1; appendix 2.1).

All of the models, or close analogues of them, have been discussed in the literature. One of the tree-event models (REC; table 2.1) includes cospeciation, duplication, and sorting events. This is a description of a method that has been frequently discussed in the literature, mostly for fitting gene and species trees, and for which I will be using the name *reconciliation*. According to Page (1994b), "a reconciled tree combines the tree for a host and its associate into a single summary of the historical association between the two entities under the assumption that no horizontal transmission of associates has occurred." I prefer to retain the term *reconciliation* with this meaning rather than using it also for methods that include horizontal transmission (switches) (for a different view, see Charleston and Perkins, chap. 3).

Reconciliation was originally conceived by Goodman and colleagues (1979) and formalized by Page (1994a). Recent papers include empirical applications (Guigó et al., 1996; Slowinski et al., 1997; Page, 2000) as well as methodological advances (Mirkin et al., 1995; Page and Charleston, 1997; Zhang, 1997; Eulenstein et al., 1998; Eulenstein and Vingron, 1998; Ma et al., 2000). From its original conception, reconciliation has been a method for minimizing the number of duplications, the number of losses (sorting events), or the total number of duplications and losses. In the parsimony framework, this corresponds to assigning an arbitrary unit cost to duplications, losses, or both, and zero cost to the other event(s) in the model. As noted above, however, assigning zero cost to losses leads to a logically inconsistent method with undesirable properties. Assume, for instance, that we have a gene tree with only two terminals, with both of these gene copies occurring in the same species. The optimal solution requires one gene duplication, but unless losses have a positive cost, that duplication could be placed in any ancestor of the species instead of in the species itself, without increasing the cost. As long as zero-cost losses are avoided, however, many different cost assignments to the events in the reconciliation model are possible (table 2.1).

The method based on the two-event model (2E; table 2.1) is a close analogue of reconciliation, the only difference being that it models cospeciations as the result of one duplication and two sortings instead of as a true codivergence event. It still has the characteristic property of reconciliation, namely that switches are not considered. Of the other three-event models, one is the basis for tracking-switching analysis (TS; table 2.1) (Ronquist and Nylin, 1990; Ronquist, 1995, 1998b); it includes switches, but cospeciations are represented by their duplication-sorting equivalent rather than as true cospeciations. The final three-event model is identical to one formulated by Maddison (1997) for folding gene trees into species trees, except that he did not recognize the necessity to include duplication events. The method based on this model fits trees together assuming that only switches break the historical pattern created by duplications and cospeciations; it may be referred to as cospeciation-switching analysis (CS; table 2.1). Finally, the full four-event model represents the superset of all the others; it was first discussed by Page (1994b) (4E; table 2.1). For each model, the cost assignments must be chosen such that they fulfil certain conditions (table 2.1). If these are not met, then some events will never occur in optimal reconstructions or the method becomes inconsistent (proofs in appendix 2.1).

Previous discussions of models that include switches have not fully recognized the distinction between partial and complete switches. One reason for this is that a partial switch is equivalent to a combination of a duplication event followed by a complete switch. Thus, the initial formulation of tracking-switching analysis clearly models switches as complete rather than partial, even though most (but not all; see proof in appendix 2.1) inferred switches in optimal reconstructions can be reinterpreted as being partial rather than complete by simply moving the switch from an edge in the parasite tree to the vertex beneath it, negating the duplication event at that vertex. Similarly, Page (1994b, fig. 5) allowed the possibility of both immediate descendants of a splitting event in the parasite tree switching from their original host, a scenario that requires at least one complete switch or an extinction event. If allowed, this nonfactual combination of two switches will occur in the optimal reconstructions for some problems (proof in appendix 2.1), and complete switches are therefore best avoided in models for parsimony-based tree fitting.

Maximum Codivergence and Parsimony

Parsimony methods are minimization procedures: they find the most likely explanation of some observed data by minimizing the cost of the implied events. In this sense, maximization is foreign to parsimony. Nevertheless, the event-based parsimony framework and its algorithms can be used for maximization methods by simply assigning a negative cost to the events that should be maximized. Minimization of a negative cost is equivalent to maximization of a benefit value (Ronquist, 1998c).

The method proposed by Page (1994b) and subsequently referred to as maximum codivergence (MC; alternatively maximum cospeciation or maximum vicariance) (Ronquist, 1995, 1998b, 1998c) fits a host and parasite tree together by maximizing the number of cospeciations. MC can be described as a maximization variant of the four-event parsimony method described above; it is equivalent to assigning an arbitrary negative cost to cospeciations and zero cost to all other events (Ronquist, 1998c) (table 2.2). If it were possible to postulate nonfactual events, then MC wouldn't work because it would be possible to have an infinite number of cospeciations for any data set. However, the four-event model guarantees factuality.

MC has two important advantages: it is intuitively appealing to focus on cospeciations in studying the history of host-parasite associations, and

TABLE 2.2 Cost assignments of some event-based parsimony methods marginally touching (jungles) or falling outside (other methods) the space described in table 2.1. In modified BPA, n is the number of vertices (nodes) in H between the two host edges involved in the shift, plus one. Fitch optimization cannot be used for tree fitting but is powerful in detecting duplication-switching patterns (table 2.3).

Model		Event			
Type	Name	Cospeciation	Duplication	Sorting	Switch
4-event	Maximum codivergence (MC)	$c = -1$	$u = 0$	$s = 0$	$i = 0$
4-event	Jungles	$c \leq 0$	$u > 0$	$s > 0$	$i > 0$
3-event	Modified BPA	-	$u = 0$	$s = 1$	$i - n$
2-event	Fitch optimization	-	$u = 0$	-	$i = 1$

MC avoids the difficulty of specifying the exact cost of all the events in the model. But these advantages come with some cost. For instance, MC can detect cospeciation patterns but not other types of historically structured association patterns. More important, MC is less powerful than a true parsimony method in identifying cospeciating lineages and in detecting significant cospeciation patterns. The reason is that MC reconstructions may mix cospeciation events that are well supported by the data with more uncertain ones, while a method with a positive cost associated with sorting and switching events, and a zero or positive cost associated with cospeciation events, tends to restrict the inferred cospeciation events to the better-substantiated cases. The latter is more satisfactory in the context of testing cospeciation hypotheses. MC is particularly sensitive to duplication events, as will be shown below.

A perhaps less important problem with MC is that the event-cost assignments lead to some problems having no solution. Assume, for instance, that all parasites are associated with the same host. Then there are an infinite number of optimal zero-cost (or zero-benefit) solutions. In other cases, MC may produce a large number of equally optimal reconstructions, some of which postulate a ridiculously large number of switches and sorting events.

Charleston (1998) suggested a more refined protocol by modifying the event-cost assignments of MC. He used a negative or zero cost for cospeciations and a positive cost for the other events (table 2.2). This gives a hybrid between MC and a true parsimony method based on the four-event model. Because of the problems with maximization of codivergence, Charleston's method will perform best when the cost of cospeciations is set to zero, in which case the method falls within the definition of a true parsimony method given above (table 2.1).

Modified Brooks Parsimony Analysis

I made a clear distinction above between event-based and pattern-based parsimony methods. Despite the important differences in the conceptual framework, however, pattern-based analytical protocols can sometimes be easily translated into event-based equivalents. I will illustrate this with Brooks Parsimony Analysis, one of the most commonly used pattern-based methods in coevolutionary and biogeographic inference. It is abundantly clear that BPA is a pattern-based protocol (Brooks, 1981; Wiley, 1988; Brooks, 1990). It is true that the results of BPA can be interpreted a posteriori in terms of events (Wiley, 1988), but it has never been possible to formalize this translation. Even if such a formalization were developed, BPA would still not be equivalent to an event-based protocol because the analysis is not based on a model with associated event-cost assignments.

BPA is usually understood as a method that translates the host (or parasite) tree into a set of binary characters (fig. 2.4a–b). For our purposes,

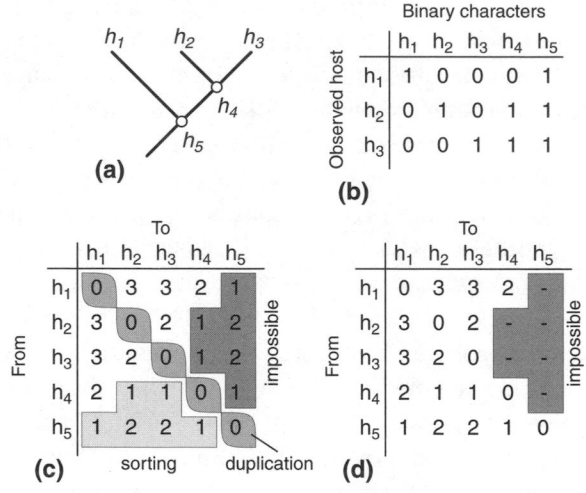

FIGURE 2.4. Brooks Parsimony Analysis and an event-based modification of it. *(a)* A host tree. *(b)* A set of binary characters, derived by additive binary coding, describing the host tree. *(c)* A cost matrix equivalent to the set of binary characters. There are four different types of cells in the matrix. The uncolored cells can be interpreted as switches, the light gray cells in the lower triangle as sorting events, and the intermediate gray cells in the diagonal as duplications. The dark cells, however, cannot be interpreted in terms of events, since they imply a reversal in time. If these cells are associated with an arbitrary large value (symbolized by the dash), ensuring that such transitions are not postulated in optimal host-parasite mappings, a true event-based method is obtained *(d)*.

however, it is more convenient to describe BPA as a translation of the host tree into a cost or step matrix (fig. 2.4c), which is then optimized onto the parasite tree. If there are no parasite terminals associated with more than one host terminal, the cost matrix gives results identical to those obtained with the binary character set both in terms of ancestral state assignments and total cost. This follows from the equivalence of character state trees with their additive binary representation.

Let h be an edge (species) in the host tree. Then each cell in the BPA cost matrix can be understood as defining the cost of a parasite lineage moving from an initial association with host h_i to an association with host h_j. Depending on the relation between h_i and h_j, the cells fall into four different types (fig. 2.4c), three of which can be translated into events. The first type is found in the diagonal, has $h_i = h_j$ and a cost of zero. If the parasite lineage undergoes speciation between the initial and final associations, this type requires one or more duplication events, each with zero cost. In the second type, found in the lower triangle of the cost matrix, h_j is a descendant of h_i. This transition involves one or more sorting events and the cost is equivalent to the number of nodes between h_i and h_j, which is the same as the number of implied sorting events. The total cost in the BPA matrix thus corresponds to a cost of 1.0 for each sorting event. The third type comprises those cells where h_j is not a descendant or ancestor of h_i. In this case, h_j and h_i could have existed simultaneously as independent host lineages, and these transitions can therefore be interpreted as switches. The cost is equivalent to n, which is one plus the number of nodes in the host tree on the path between h_i and h_j. Thus, in contrast with the methods discussed above, switches do not have a constant cost. Instead, they are weighted according to the distance in the host tree between the involved hosts (see also Maddison, 1997). Finally, the fourth type of cells are those in which h_j is an ancestor of h_i. Such a transition implies a reversal in time and cannot be interpreted as a biological event. To convert BPA into an event based method, it is necessary and sufficient to associate these cells with an arbitrary large cost, ensuring that this type of transition will never be postulated in optimal reconstructions (fig. 2.4d).

In summary, the event-based method that results from modification of BPA in the manner described above models cospeciation events in terms of their duplication-sorting equivalent, has a unit cost for sorting events, and associates switches with a cost, which is a function of the node distance between the involved hosts (table 2.2). The method may appropriately be called modified BPA. An interesting difference between modified BPA and the methods discussed previously is that, although switches are modeled as complete, the switches in optimal reconstructions appear always to be reinterpretable a posteriori as partial switches. At least, the proof for theorem 7 (appendix 2.1) clearly does not apply to modified BPA. Nevertheless, it may be advantageous to model switches as partial rather than complete in modified BPA to guarantee strict factuality.

Previous criticisms of BPA (Page, 1990; Ronquist and Nylin, 1990) have focused on the representation of switches when using additive binary coding of the host tree. Each switch is then associated with one or more reversals in characters representing ancestors of the hosts involved in the switch, an apparent contradiction if reversals are interpreted as extinctions. However, if these reversals are not seen as extinction events but as the binary expression of the switching cost, then this criticism becomes inappropriate. The original formulation of BPA is still logically inconsistent, however, because of the finite cost associated with impossible host transitions. This will lead to inconsistent optimal reconstructions for problems where it is possible that the parasites have first tracked a host phylogeny and then backcolonized an ancestral host and tracked the host phylogeny once again (fig. 2.5). Since such patterns are likely to be rare, BPA can be viewed as a close approximation of modified BPA.

The event-based derivation of BPA nicely shows the advantages of the event-based framework. Once the event-cost assignments have been defined (table 2.2), the properties of BPA can easily be understood. Thus, BPA is an appropriate method when it is desirable to weight switches according to the distance in nodes between the hosts in the host phylogeny. It

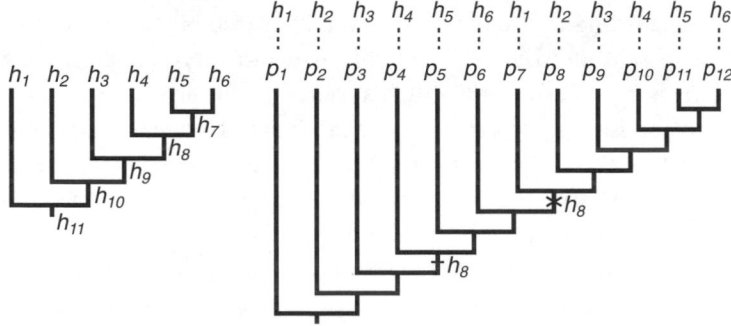

FIGURE 2.5. A case in which Brooks Parsimony Analysis (BPA), as originally formulated, will produce a set of optimal solutions, none of which can be interpreted in terms of biologically possible events. The optimal solutions for this problem will suggest that the parasites first tracked the host tree to h_8 or its descendants (cross line) and then tracked the tree backward in time (cross) to h_9 or h_{10}.

is probably true for many associations that switches between closely related hosts are more likely than switches between more distantly related hosts. However, the cost function used by BPA weights so heavily against distant switches that they are unlikely to occur at all in optimal reconstructions. A weaker weighting function is probably more realistic for most associations. The event-based parsimony framework allows us to use both the BPA and weaker cost functions and test them against each other with respect to their ability to find significant historical host-parasite association patterns (see below).

Significance Tests

Parsimony analysis always produces one or more most-parsimonious reconstructions given some data, a model, and suitable event-cost assignments. This is true even for random data. Thus, we need some method for testing the significance of the results of a parsimony reconstruction or, in other words, evaluating the evidential support or robustness of the conclusions.

One technique for assessing significance in parsimony-based tree fitting is to compare the observed cost with a distribution of cost values expected by chance if there had been no historical association between the trees. The most common way of estimating this random distribution is to calculate the cost for a number of data sets in which one or both trees have been replaced with random trees drawn from some specified tree distribution (Page, 1988). The two standard distributions used in these tests are the model where all labeled trees are equally probable (equiprobable trees: Rosen, 1978; Nelson, 1979) and the tree universe generated by a random birth-death process (often referred to as a Markov or Yule process) (Markov trees: Simberloff et al., 1981; Simberloff, 1987; Page, 1988). The significance of the observed fit is determined by comparing it with the distribution of cost values expected by chance. Say that the observed parasite tree could be fitted with the observed host tree with a total cost of 5.0, and that 960 of 1,000 random parasite and host trees had a higher cost when fitted together. Then the probability of the two trees fitting as well or better by chance would be 4%, or $p = 0.04$, a significant result at the $\alpha = 0.05$ level.

Instead of replacing observed parasite or host trees with random trees, we can generate random data sets by permuting (shuffling) the parasite or host terminals, and thus the associations between them (Siddall, 1996; Ronquist, 1998b). There are two advantages of using this technique: (1) we do not need to choose an appropriate tree distribution; and (2) we can control for the effect of tree topology. Choosing an appropriate tree distribution is problematic because tree shape significantly affects the cost of tree fitting. The more balanced two trees are, the easier it is to fit them together. The reason is that there are far fewer distinct balanced trees, so the chance that they will fit well by chance is much higher. Thus, the significance test will be affected by differences in symmetry between the true and the random trees. If the random trees are more unbalanced than true trees, then p will be underestimated and the significance test will suffer from an inflated type I error rate. If the random trees are more balanced, on the other hand, p will be overestimated and the test will become unnecessarily conservative. The effect is well illustrated by comparing Markov and equiprobable trees. Markov trees are on average more symmetric and the cost of fitting two random trees together is thus lower (fig. 2.6). These considerations suggest that significance tests should rely on random trees from distributions that are likely to overestimate the proportion of balanced trees, such as the Markovian universe (Mooers and Heard, 1997).

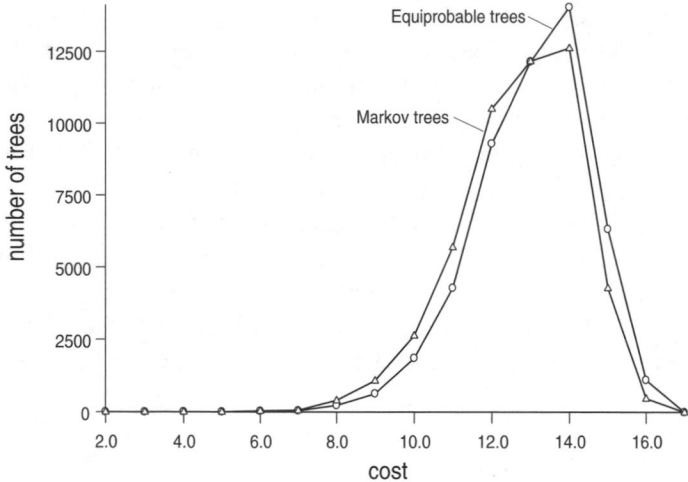

FIGURE 2.6. Tree shape significantly affects the expected cost of fitting two randomly chosen trees together. Thus, two trees picked randomly from a Markov distribution, in which balanced trees are relatively more common, will tend to have a lower cost than trees picked from a distribution where all distinct labeled trees are equiprobable (equiprobable distribution). If the true tree distribution contains more balanced trees than the one used for significance testing, p values will be underestimated, with an inflated type I error rate as a consequence. The distributions shown were calculated from 50,000 pairs of random trees with 10 taxa in each.

However, even if the random universe overestimates the proportion of balanced trees, significance tests based on picking new tree topologies would still have an inflated type I error rate for balanced trees. Consider for instance two perfectly symmetric 4-taxon trees. The probability that they will fit perfectly by chance is $p = .33$, but random Markovian trees would give $p = .074$ and random equiprobable trees would give $p = .067$. The problem is particularly cumbersome because balanced trees will more often have reasonably good fit values by chance and may therefore be preferentially chosen for significance testing. Such problems are avoided by permutation tests, which control for tree balance.

Whether significance testing is done by drawing new trees at random or by permuting terminals, they can involve the host tree, the parasite tree, or both. The most general test involves randomization of both parasite and host trees, but other variants are also useful. For instance, if many parasites share a host, then conserved host-usage patterns (duplication patterns) among the parasites can be tested by randomization of only the parasite tree, whereas randomization of only the host tree tests for historical constraints determined solely by host relationships (cospeciation patterns).

If the host tree were inferred from the parasite data rather than from independent evidence, it would not be appropriate to test its significance by comparing it with random host trees. Since it has been inferred from the parasite data, it *should* have lower cost than all other possible host trees. In these cases, an appropriate significance test would have to involve randomization of the parasite data and inference of new host trees from each of these data sets (Siddall, 1996). The host tree inferred from the original parasite data would be deemed as well supported if it had lower cost than expected in trees inferred from randomized parasite data.

Finding Optimal Event Costs

A serious impediment for the use of parsimony methods has been the problem of finding optimal event costs. In an early attempt to attack the problem, Ronquist (1995) proposed a technique that sets the switching cost just below the value, producing a solution completely without switches. This technique is difficult to justify and produces reasonable solutions only for some simple problems. Goloboff (1997) developed a general method for adjusting transformation weights of binary and multistate characters, but the weighting function he uses includes a constant, the value of which cannot be estimated from the data, and the function has an arbitrarily set switch between a linear and a concave part. It is obvious that this weighting function cannot be used for the four-event model by just counting all events, since the event costs would then be outside of the possible cost space (table 2.1) for some problems. For instance, Goloboff's weighting function would suggest that a pure switching explanation is equally parsimonious to a pure cospeciation explanation for two congruent parasite and host trees. The method will probably do better if some events, such as cospeciations and possibly duplications, are not counted, but the fundamental problem with the arbitrary concavity constant remains.

It is first here that I can present a simple, easily justifiable, and computationally feasible parsimony-based solution to the cost-optimization problem that does not rely on arbitrary ad hoc decisions on the shape of weighting functions. Suppose that we test the existence of phylogenetically constrained host-parasite association patterns by some randomization test. Then it would make sense to choose event-cost assignments that maximize our chances of detecting historical constraints in the data. In this way, it is possible to obtain optimal event-cost assignments for each analyzed problem through exploring the space of possible cost-event assignments. To some extent, these optimal cost assignments can be interpreted as estimating the relative frequency of different types of events: the larger the optimal cost, the less common the events are in the studied system. To illustrate this approach, let us examine some hypothetical historically constrained patterns that may arise in parasite-host associations.

Constrained Patterns

It may be heretical to claim this in a book about cospeciation, but simultaneous host-parasite speciation is not the only process generating phylogenetically conserved association patterns detectable with parsimony analysis; such historical patterns are also generated by duplication. This book gives plenty of examples of cospeciation patterns, but duplication patterns are probably at least as common in host-parasite associations, for instance in insect-plant associations. The host-plant association of gall wasps (Liljeblad and Ronquist, 1998) provide an extreme example (fig. 2.7). When host-plant families are mapped onto the phylogeny of this group with more than 1,300 members, only 20 switches are evident and there is no detectable pattern of cospeciation with the host plants. More than 98% of the speciation events in the family represent duplications with respect to host-plant family.

Sorting and switching occur on top of the constrained cospeciation and duplication processes. We can thus imagine a series of possible patterns mixing constrained and unconstrained events. Each of these patterns should

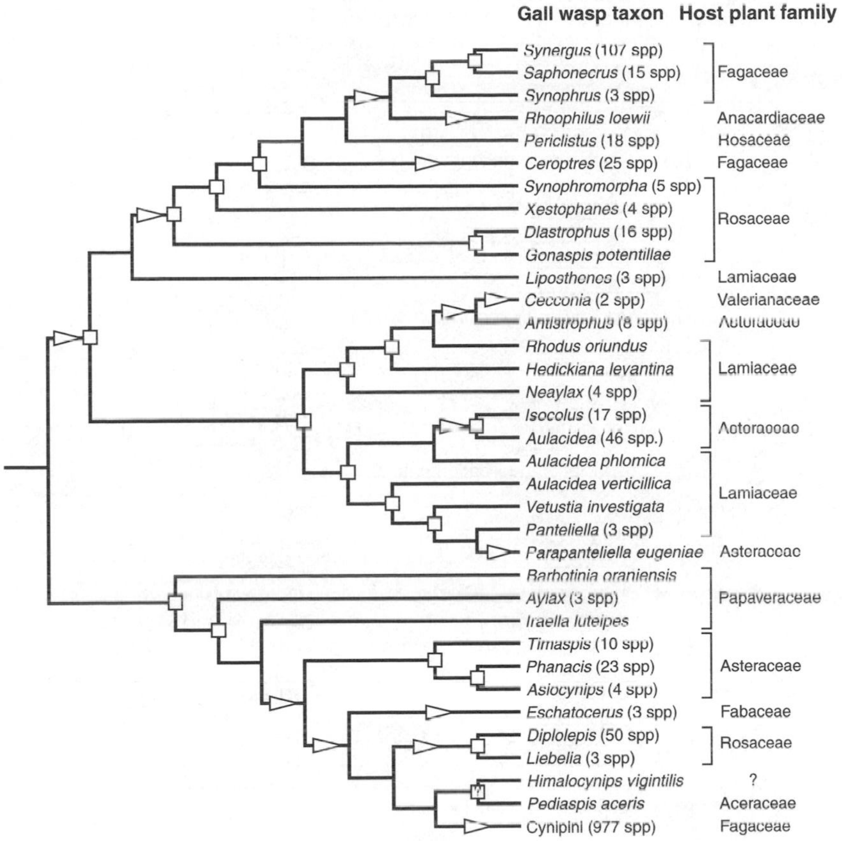

Gall wasp taxon   Host plant family

Synergus (107 spp)
Saphonecrus (15 spp)   Fagaceae
Synophrus (3 spp)
Rhoophilus loewii   Anacardiaceae
Periclistus (18 spp)   Rosaceae
Ceroptres (25 spp)   Fagaceae
Synophromorpha (5 spp)
Xestophanes (4 spp)
Diastrophus (16 spp)   Rosaceae
Gonaspis potentillae
Liposthenes (3 spp)   Lamiaceae
Cecconia (2 spp)   Valerianaceae
Antistrophus (8 spp)   Asteraceae
Rhodus oriundus
Hedickiana levantina   Lamiaceae
Neaylax (4 spp)
Isocolus (17 spp)
Aulacidea (46 spp.)   Asteraceae
Aulacidea phlomica
Aulacidea verticillica
Vetustia investigata   Lamiaceae
Panteliella (3 spp)
Parapanteliella eugeniae   Asteraceae
Barbotinia oraniensis
Aylax (3 spp)   Papaveraceae
Iraella luteipes
Timaspis (10 spp)
Phanacis (23 spp)   Asteraceae
Asiocynips (4 spp)
Eschatocerus (3 spp)   Fabaceae
Diplolepis (50 spp)   Rosaceae
Liebelia (3 spp)
Himalocynips vigintilis   ?
Pediaspis aceris   Aceraceae
Cynipini (977 spp)   Fagaceae

FIGURE 2.7. Example of a host-parasite system characterized by duplications with occasional switches rather than by cospeciation: the evolution of host-plant family preferences of gall wasps (Cynipidae). One of several optimal reconstructions is mapped onto a recent hypothesis of higher gall-wasp relationships (Liljeblad and Ronquist, 1998). The event symbols are as in figure 2.3. Most of the 20 switches between host-plant families in the Cynipidae occur on the basal branches in the phylogeny (14 are contained in the given tree); the remaining more than 1,300 cynipid speciation events are almost all duplications with respect to host-plant family. In addition to the given affiliations, there is one species each in *Periclistus* and *Diastrophus* associated with Smilacaceae, one species each in *Aulacidea, Timaspis,* and *Phanacis* associated with Apiaceae, and one species in *Phanacis* associated with Lamiaceae.

be detectable with a suitable method. We may consider the series to include pure cospeciation patterns (fig. 2.8a), duplication-switching patterns (fig. 2.8b), cospeciation-duplication patterns (fig. 2.8c), cospeciation-sorting patterns (fig. 2.8d), cospeciation-switching patterns (fig. 2.8e), and patterns mixing more than two types of events (fig. 2.8f). This series includes all

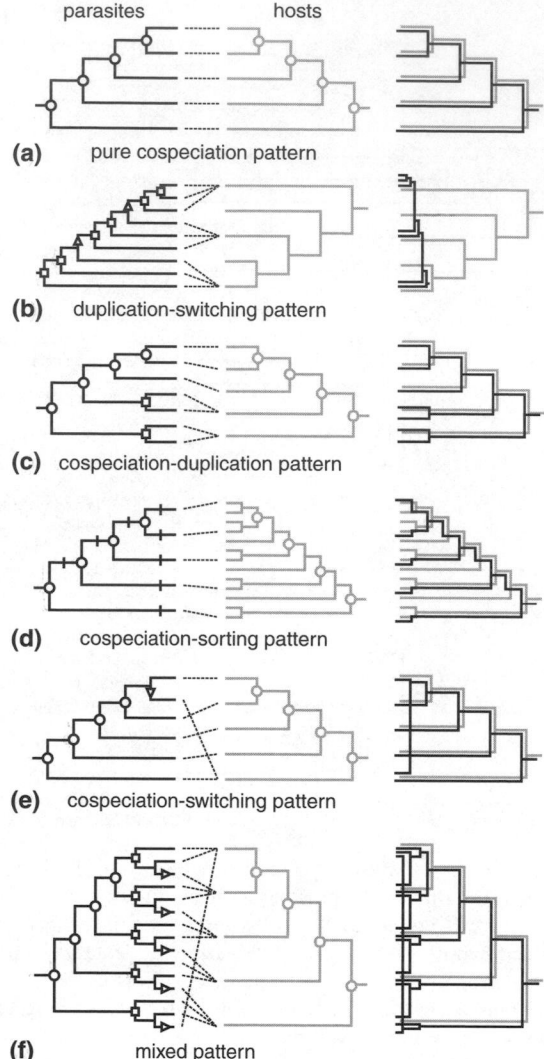

parasites   hosts

(a)   pure cospeciation pattern

(b)   duplication-switching pattern

(c)   cospeciation-duplication pattern

(d)   cospeciation-sorting pattern

(e)   cospeciation-switching pattern

(f)   mixed pattern

FIGURE 2.8. Six different hypothetical patterns in the evolution of host-parasite associations. Each pattern is given as a tanglegram (to the left) and a trackogram (to the right), with the host tree in gray and the parasite tree in black. The trackograms and the indicated events in the tanglegrams are based on the best fit under optimal event costs (fig. 2.9). Symbols for events are as in figure 2.3. *(a)* Pure cospeciation pattern. *(b)* Duplication-switching pattern. *(c)* Cospeciation-duplication pattern. *(d)* Cospeciation-sorting pattern. *(e)* Cospeciation-switching pattern. *(f)* Mixed pattern with cospeciations, duplications, and switches.

possible combinations of one constrained and one unconstrained process, since duplication-sorting patterns are indistinguishable from cospeciation-sorting patterns.

For a general cost optimization it would be necessary to vary the cost of three different events (the cost of the fourth being set to some arbitrary value) and optimize the p value of the significance test over this parameter space. To avoid four-dimensional diagrams, let us accept a zero cost for cospeciations, set sorting events to unit cost, and vary the costs of switches and duplications. When the cost-optimization procedure is applied to the set of hypothetical patterns under these conditions, two important empirical observations emerge (fig. 2.9). First, the chance of detecting historically constrained association patterns is not always maximized in a single point in parameter space but may obtain the same maximal value throughout a broad range of parameter values. Throughout this range, the optimal reconstruction is the same. Second, the parameter values maximizing the chances of detecting constrained patterns are similar for most problems: the important thing is to assign a low cost to duplications and cospeciations and a higher, roughly equal cost to switches and sorting events. The only type of pattern that cannot be adequately analyzed under these conditions is the cospeciation-sorting pattern (fig. 2.8d, 2.9d), which requires a switch cost that is considerably higher than the sorting cost.

The power of the parsimony approach is well illustrated when its performance is compared with that of two alternative methods for analyzing coevolving associations: maximum codivergence (MC) and Fitch optimization. The latter technique maps hosts onto the parasite phylogeny by minimizing the number of switches against a background of duplications. Thus, it may be described as based on a subset of the four-event model in which only duplications and switches are allowed (cf. table 2.2). Consider the six patterns discussed above (fig. 2.8). MC detects the pure cospeciation, cospeciation-sorting, and cospeciation-switching patterns as significant but fails with the duplication-switching, cospeciation-duplication, and mixed patterns (table 2.3). Fitch optimization only detects the duplication-switching pattern. Parsimony analysis under the four-event model detects all patterns as significant except the cospeciation-sorting pattern, even when cost assignments are set to standard values (zero for duplications and cospeciations, 1.0 for sortings, and 2.0 for switches) for all problems. When costs are optimized for each problem separately, there is not a single problem for which the parsimony method is beaten by any of the other two methods (table 2.3).

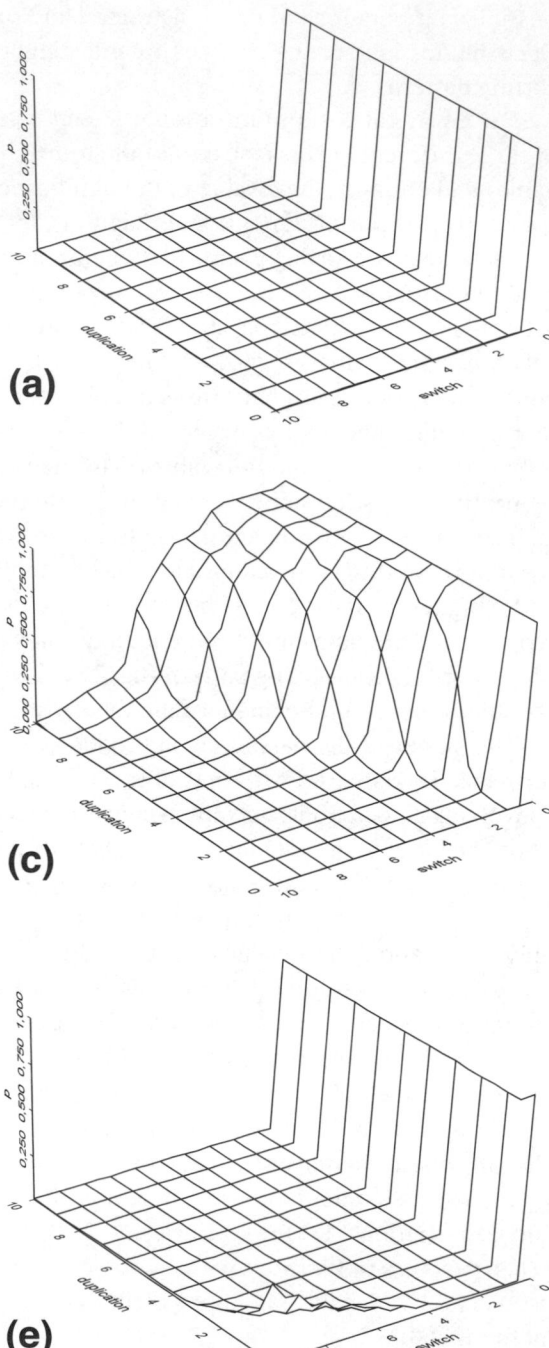

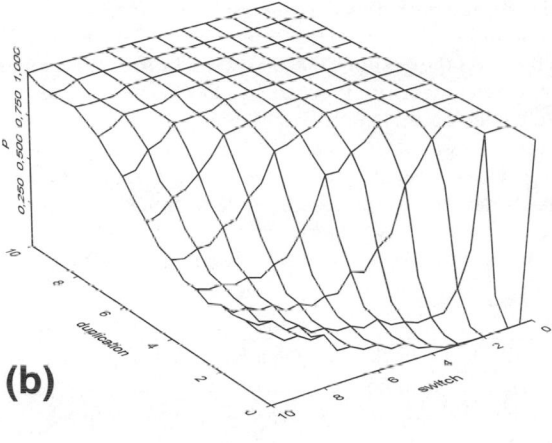

(b)

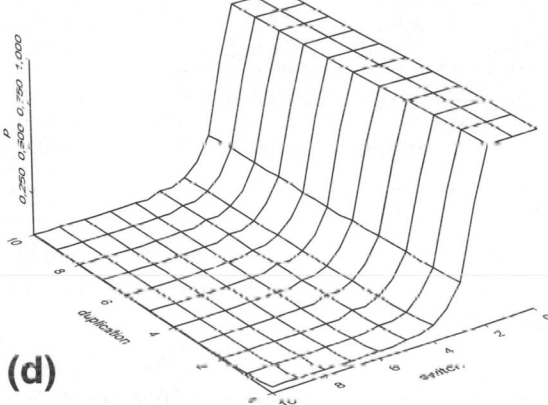

(d)

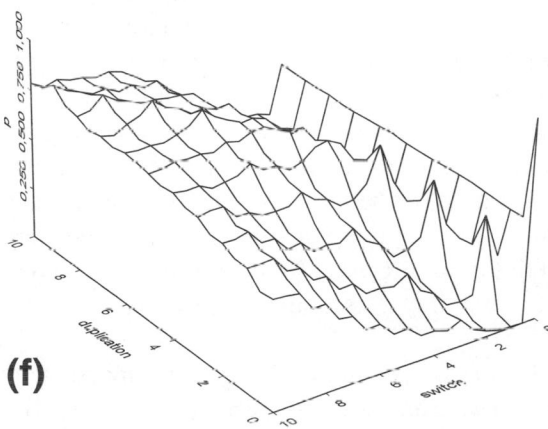

(f)

FIGURE 2.9. Exploration of the cost space for six hypothetical evolutionary patterns, corresponding to figure 2.8a–f. Cospeciation events were assigned zero cost, sorting events unit cost (1.0), and the switch and duplication costs were varied between 0.0 and 10.0 in increments of 0.5. The height in the cost landscape is determined by the p value, which is the percentage of random trees with a cost equal to or lower than that observed; the lower the value, the better. Each p value was calculated by comparing the observed cost with that of 10,000 tree pairs in which the parasite and host terminals had been randomly permuted. Observe that the optimal cost assignment for most problems is to have zero duplication cost and a positive but moderately low switching cost. The only exception is the cospeciation-sorting pattern, for which it is important to have a relatively high switching cost.

TABLE 2.3 Performance of parsimony and parsimonylike methods in detecting some hypothetical patterns that could occur in host-parasite associations (fig. 2.8). The values in the table are p-values based on the cost of 1,000,000 permutations of host and parasite terminals calculated using exact algorithms (Ronquist, 2000). The methods are maximum codivergence (MC), Fitch optimization (Fitch), and four-event parsimony analysis with general cost assignments ($c = u = 0.0, s = 1.0$, and $i = 2.0$) (General 4E) or with optimal cost values for each pattern (found by setting $s = 1.0$ and varying the other costs between 0.0 and 10.0) (Optimized 4E). Parsimony analysis based on the four-event model is by far the most powerful method of detecting historical patterns, particularly for associations characterized by cospeciation mixed with duplications or switches.

	Method			
Pattern (fig. 2.8)	MC	Fitch	General 4E	Optimized 4E
a. Pure cospeciation	0.017	1.000	0.017	0.017
b. Duplication-switching	1.000	0.004	0.006	0.004
c. Cospeciation-duplication	0.206	0.333	0.0002	0.0002
d. Cospeciation-sorting	0.012	1.000	1.000	0.012
e. Cospeciation-switching	0.027	1.000	0.028	0.016
f. Mixed	0.982	0.326	0.042	0.009

The hypothetical patterns discussed above illustrate how sensitive the maximization of cospeciation is to noise arising from other processes, particularly duplication and switching, in detecting cospeciation patterns. The problem occurs because maximization will allow poorly supported cospeciation events in randomized data, decreasing the difference between observed and randomized data.

In some cases, the performance of cospeciation-maximization methods can be improved considerably by randomizing only the host tree and not the parasite tree. Thus, the cospeciation-duplication pattern (fig. 2.8c) will be detected as significant by MC if only the host tree is randomized. However, this approach does not always work. For instance, all randomized data sets have the same number of cospeciation events as observed for the mixed pattern (fig. 2.8f) even when the parasite tree is kept intact. It has been stated that randomization of the host tree is a more conservative test of cospeciation than randomization of the parasite tree under MC (cf. Farrell and Mitter, 1998) because the host tree is generally smaller than the parasite tree. The cospeciation-duplication pattern considered here (fig. 2.8c) clearly shows that this is not the case. Actually, randomization of the host tree may be the only possibility of finding significant patterns with MC for some associations.

The problems associated with maximization of cospeciation are easily avoided by using a true parsimony method. For instance, an optimized parsimony analysis of the problematic pattern discussed above (fig. 2.8f) will reveal significant historical constraints both when the host and parasite terminals are permuted (table 2.3) and when only the host terminals are permuted. In the latter case, the analysis will detect more cospeciation events than expected by chance ($p = .05$), in contrast with maximum cospeciation, which will not report a significant amount of codivergence ($p = 1$).

Widespread Terminals

A recurring problem in tree fitting is that of "widespread" terminals. In the coevolutionary context, a parasite terminal is widespread when it is associated with multiple host terminals. If the one-host-per-parasite assumption is accepted for ancestors, then the existence of widespread terminals will have to be reconciled with this notion. The most obvious approach is perhaps to treat each widespread taxon as an unresolved clade consisting of one lineage for each host in its repertoire (Page, 1994b). The ancestral host of this terminal clade can then be determined according to one of three separate methods: the *ancient, recent,* and *free* options (Sanmartin, Enghoff, and Ronquist, 2001). According to the ancient option, the ancestral host of the hypothetical terminal parasite clade is the host that is the most recent common ancestor of all the hosts attacked by the parasite (see also Page, 1994b). The recent option instead assumes that the ancestral host is identical to one of the species attacked by the parasite today, the other hosts having been acquired by recent colonization. Finally, the free option considers every possible ancestral host of the terminal parasite clade in turn, assigning a minimal cost to each host assignment by resolving the terminal polytomy in the most favorable way for that particular ancestral host assignment.

Algorithms

For biologists, algorithms are mainly interesting in terms of the complexity of the problems they can analyze, an ability determined by the accuracy and speed of the algorithms. With respect to accuracy, computer scientists talk about *exact* and *heuristic* algorithms. Exact algorithms are guaranteed to find the optimal solution to a problem; heuristic algorithms will find solutions that may or may not be optimal. The time complexity of an algorithm is given as a function of some parameter or parameters that determine the size of the problem. Constants are often omitted because the general type of the function is more important in determining the feasibility of analyzing large problems. The function may be linear, $O(n)$; quadratic, $O(n^2)$;

exponential, $O(e^n)$; or even more complex, like $O(n!)$. In general terms, only polynomial-time algorithms (functions of the type $O(n^x)$, where x is an integer constant) are fast enough to be applicable to problems of realistic size. The smaller the value of x, the faster the algorithm: a linear function ($O(n)$) is better than a quadratic function ($O(n^2)$), a quadratic function is better than a cubic ($O(n^3)$), and so on. The difference in time consumption between a linear and quadratic algorithm can be considerable for large problems.

The reconstruction problem (fitting a given H tree to a given P tree) is easily solved given that switches are prohibited (that is, under the reconciliation model); exact solutions can be found in $O(m + l)$ time, where m is the number of parasite species and l is the number of sorting events (losses) (e.g., Zhang, 1997; Ma et al., 2000). The estimation problem is considerably more complex, since it involves a tree search; but ordinary heuristic tree-search algorithms can be combined with optimization shortcuts (Goloboff, 1996; Ronquist, 1998a) such that each alternative H tree can be evaluated in $O(m/n)$ time, which is exceedingly fast. Even more interesting, if it is possible to specify a maximum number k of parasites simultaneously infecting a single host lineage, then the estimation problem can be solved exactly in polynomial time (M. T. Hallett and J. Lagergren, pers. comm.).

As soon as switches are introduced, however, it becomes significantly more difficult to find fast exact algorithms, even for the reconstruction problem. The complexity arises because of the difficulty of separating possible from impossible combinations of host switches (Ronquist, 1995). One reason for switch incompatibility is that switches may require different, incompatible sequences of speciation events in the host tree. Consider, for instance, two potential switches on a host tree for which there are two possible splitting sequences (fig. 2.10a–b). Switch 1 is possible given one sequence (fig. 2.10a) but not the other (fig. 2.10b), whereas the reverse is true for switch 2. Thus, switch 1 and switch 2 are incompatible; they cannot both occur among parasites associated with this group of hosts. The problem is easy to solve if the order of the nodes in the host tree is known (Ronquist, 1995). When the order of the nodes is known, the tree is referred to as a *(labeled) history* (Brown, 1994).

Even if the host history is specified, however, switch incompatibility may arise because of temporal constraints. Consider two potential switches on a specified host history (fig. 2.10c). The switches are compatible only if switch 4 occurred before switch 3. A reconstruction requiring the reverse switching order in the parasites is impossible. To avoid this problem, an algorithm needs to keep track of where the parasites are on long host

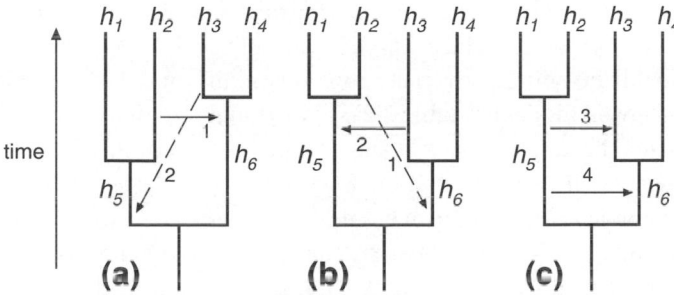

FIGURE 2.10. Switch incompatibility causes problems in developing fast algorithms for host-parasite tree fitting. The reason is that switches may imply different sequences of splitting events in the host tree, and certain switch combinations may therefore be impossible. Switch 1 is possible, and switch 2 is impossible given one sequence of splitting events in the host tree *(a)*, whereas the reverse is true for the other possible sequence of splitting events *(b)*. Thus, the switches are incompatible. Even if the order of splitting events is known, switches must occur in the right order to be compatible. For instance, switches 3 and 4 are both possible given a particular sequence of splitting events in the host phylogeny *(c)*, but they are only compatible if switch 3 occurs after switch 4.

branches that span one or more splitting events occurring in other parts of the host tree (Ronquist, 1995).

The first algorithm to solve the reconstruction problem with switches was a nonpolynomial explicit enumeration algorithm that would be impossible to use for problems of any realistic size (Ronquist and Nylin, 1990). Now it is known that the reconstruction problem can be solved with switches in polynomial time if the host history is known (Ronquist, 1995, 1998c). The fastest algorithms of this type known to date (Ronquist, 2000) have time complexity $O(mn^2)$, m still being the number of parasites and n the number of hosts. If the host tree is given but the history is unknown, then the optimal splitting order must be searched for by explicit enumeration, and the worst-case time complexity is no longer a polynomial. However, the exact $O(mn^2)$ algorithm can easily be modified to provide a heuristic solution, which represents an upper bound on the cost of the most optimal reconstruction, for instance by arbitrarily picking one of the possible host histories. A lower bound is easily obtained by ignoring incompatibility among switches. The fastest such lower-bound algorithm is similar to the exact algorithm but computes in $O(mn)$ time (Ronquist, 2000).

Page (1994b) and Charleston (1998) presented other algorithms that deal with switches. Page provided a nonpolynomial exact algorithm and a heuristic algorithm, none of which correctly account for switches

(Ronquist, 1995; Charleston, 1998). Charleston's algorithm attempts to construct a complex graph, a *jungle,* containing all reconstructions that could possibly be optimal given certain constraints on the cost assignments of the four-event model (cf. table 2.2). Unfortunately, Charleston's rules do not construct the jungle correctly, since some possibly optimal solutions are omitted (proof in appendix 2.1). Furthermore, Charleston's dynamic-programming algorithm for traversing the jungle does not accurately account for the complexity of host switches even when the host history is known (proof in appendix 2.1). With suitable corrections, Charleston's algorithms can compute a lower bound on solutions in $O(mn^3)$ time, but this is considerably slower than the fastest known algorithms. In addition, many of the techniques for pruning nonoptimal solutions from the jungle are likely to be of limited use for large and complex or randomized data sets.

Current attempts to solve the estimation problem (searching for the best H tree) when switches are considered rely on the $O(mn)$ and $O(mn^2)$ lower and upper bound tree-fitting algorithms mentioned above (Ronquist, 2000). Algorithms of the k-width type, similar to the ones developed for the reconciliation approach, are more difficult to apply when switches are allowed, and current optimization shortcuts cannot be easily adapted to the estimation problem with switches included. Even so, good heuristic solutions may be obtained for large problems reasonably fast using techniques such as stepwise addition and branch swapping, familiar from heuristic parsimony inference of phylogeny (Ronquist, 2000).

Conclusions

Rather than concentrating on cospeciations, I have attempted in this chapter to provide a broad overview of parsimony-based tree fitting as applied to host-parasite problems. The fundamental principles of event-based methods have been understood for some time. With the recent addition of the powerful cost-optimization technique described herein and faster algorithms, it is time to fully benefit from the power of parsimony analysis by working with more refined cost assignments, larger problems, and more thorough hypothesis testing than has been possible previously. By looking beyond cospeciation we will be able to discover more about other processes generating historical patterns in host-parasite systems. At the same time, we will learn more about cospeciation itself. Despite the recent addition of new techniques to the toolbox of host-parasite biologists, parsimony analysis still has much to offer, not the least because of its ability to handle more realistic host-parasite evolution models than other methods. When these lines are written, parsimony is still the only technique that can analyze host-parasite systems under the full four-event model.

Computer Implementation

The algorithms and analytical procedures described herein are implemented in a computer program (TreeFitter) that can be downloaded from the author's home page, <www.ebc.uu.se/systzoo/staff/ronquist.html> (Ronquist, 2000). Executables for Macintosh and Windows platforms are available, as well as the C++ source code.

Acknowledgments

Rod Page, Mark Siddall, and an anonymous reviewer provided useful criticism on a first version of the chapter. The research reported here was supported by the Swedish Natural Science Research Council.

References

Brooks, D. R. 1981. Hennig's parasitological method: A proposed solution. *Systematic Zoology* 30:229–49.

———. 1985. Historical ecology: A new approach to studying the evolution of ecological variation. *Annals of the Missouri Botanical Garden* 72:660–80.

———. 1990. Parsimony analysis in historical biogeography and coevolution: Methodological and theoretical update. *Systematic Zoology* 39:14–30.

Brown, J. K. M. 1994. Probabilities of evolutionary trees. *Systematic Biology* 43:78–91.

Charleston, M. A. 1998. Jungles: A new solution to the host/parasite phylogeny reconciliation problem. *Mathematical Biosciences* 149:191–223.

Eulenstein, O., B. Mirkin, and M. Vingron. 1998. Duplication-based measures of difference between gene and species trees. *Journal of Computational Biology* 5:135–48.

Eulenstein, O., and M. Vingron. 1998. On the equivalence of two tree mapping measures. *Discrete Applied Mathematics* 88:103–28.

Farrell, B. D., and C. Mitter. 1998. The timing of insect/plant diversification: Might *Tetraopes* (Coleoptera: Cerambycidae) and *Asclepias* (Asclepiadaceae) have co-evolved? *Biological Journal of the Linnean Society* 63:553–77.

Goloboff, P. A. 1996. Methods for faster parsimony analysis. *Cladistics* 12:199–200.

———. 1997. Self-weighted optimization: Tree searches and character state reconstructions under implied transformation costs. *Cladistics* 13:225–45.

Goodman, M., J. Czelusniak, G. W. Moore, A. E. Romero-Herrera, and G. Matsuda. 1979. Fitting the gene lineage into its species lineage: A parsimony strategy illustrated by cladograms constructed from globin sequences. *Systematic Zoology* 28:132–68.

Guigó, R., I. Muchnik, and T. F. Smith. 1996. Reconstruction of ancient molecular phylogeny. *Molecular Phylogenetics and Evolution* 6:189–213.

Hougen Eitzman, D., and M. D. Rausher. 1994. Interactions between herbivorous insects and plant-insect coevolution. *American Naturalist* 143:677–97.

Huelsenbeck, J. P., B. Rannala, and B. Larget. 2000. A Bayesian framework for the analysis of cospeciation. *Evolution* 54:352–64.

Huelsenbeck, J. P., B. Rannala, and Z. Yang. 1997. Statistical tests of host-parasite cospeciation. *Evolution* 51:410–19.

Humphries, C. J., J. M. Cox, and E. S. Nielsen. 1986. *Nothofagus* and its parasites: A cladistic approach to coevolution. In *Coevolution and systematics,* edited by A. R. Stone and D. L. Hawksworth, 55–76. Oxford: Clarendon Press.

Liljeblad, J., and F. Ronquist. 1998. A phylogenetic analysis of higher-level gall wasp relationships. *Systematic Entomology* 23:229–52.

Ma, B., M. Li, and L. Zhang. 2000. From gene trees to species trees. *SIAM Journal of Computation* 30:729–52.

Maddison, W. P. 1997. Gene trees in species trees. *Systematic Biology* 46:523–36.

Mirkin, B., I. Muchnik, and T. F. Smith. 1995. A biologically consistent model for comparing molecular phylogenies. *Journal of Computational Biology* 2:493–507.

Mooers, A. Ø., and S. B. Heard. 1997. Inferring evolutionary processes from phylogenetic tree shape. *Quarterly Review of Biology* 72:31–54.

Nelson, G. 1979. Cladistic analysis and synthesis: Principles and definitions, with a historical note on Adanson's *Famille des Plantes* (1763–1764). *Systematic Zoology* 28:1–21.

Nelson, G., and N. I. Platnick. 1981. *Systematics and biogeography: Cladistics and vicariance*. New York: Columbia University Press.

Page, R. D. M. 1988. Quantitative cladistic biogeography: Constructing and comparing area cladograms. *Systematic Zoology* 37:254–70.

———. 1990. Component analysis: A valiant failure? *Cladistics* 6:119–36.

———. 1993. Genes, organisms, and areas: The problem of multiple lineages. *Systematic Biology* 42:77–84.

———. 1994a. Maps between trees and cladistic analysis of historical associations among genes, organisms, and areas. *Systematic Biology* 43:58–77.

———. 1994b. Parallel phylogenies: Reconstructing the history of host-parasite assemblages. *Cladistics* 10:155–73.

———. 1995. TREEMAP computer program, distributed by the author. Glasgow: University of Glasgow.

———. 2000. Extracting species trees from complex gene trees: Reconciled trees and vertebrate phylogeny. *Molecular Phylogenetics and Evolution* 14:89–106.

Page, R. D. M., and M. A. Charleston. 1997. From gene to organismal phylogeny: Reconciled trees and the gene tree/species tree problem. *Molecular Phylogenetics and Evolution* 7:231–40.

———. 1998. Trees within trees: Phylogeny and historical associations. *Trends in Ecology and Evolution* 13:356–59.

Platnick, N. I., and G. Nelson. 1978. A method of analysis for historical biogeography. *Systematic Zoology* 27:1–16.

Rannala, B., and Z. Yang. 1996. Probability distribution of molecular evolutionary trees: A new method of phylogenetic inference. *Journal of Molecular Evolution* 43:304–11.

Ronquist, F. 1994. Ancestral areas and parsimony. *Systematic Biology* 43:267–74.

———. 1995. Reconstructing the history of host-parasite associations using generalised parsimony. *Cladistics* 11:73–89.

———. 1997. Dispersal-vicariance analysis: A new approach to the quantification of historical biogeography. *Systematic Biology* 46:195–203.

———. 1998a. Fast Fitch-parsimony algorithms for large data sets. *Cladistics* 14:387–400.

———. 1998b. Phylogenetic approaches in coevolution and biogeography. *Zoologica Scripta* 26:313–22.

———. 1998c. Three-dimensional cost matrix optimisation and maximum cospeciation. *Cladistics* 14:167–72.

———. 2000. TreeFitter. Computer program distributed by the author. Uppsala: University of Uppsala.

Ronquist, F., and S. Nylin. 1990. Process and pattern in the evolution of species associations. *Systematic Zoology* 39:323–44.

Rosen, D. E. 1978. Vicariant patterns and historical explanation in biogeography. *Systematic Zoology* 27:159–88.

Sanmartín, I., H. Enghoff, and F. Ronquist. 2001. Patterns of animal dispersal, vicariance and diversification in the Holoarctic. *Biological Journal of the Linnean Society* 73:345–90.

Siddall, M. E. 1996. Phylogenetic covariance probability: Confidence and historical constraints. *Systematic Biology* 45:48–66.

Simberloff, D. 1987. Calculating the probabilities that cladograms match: A method of biogeographic inference. *Systematic Zoology* 36:175–95.

Simberloff, D., K. L. Heck, E. D. McCoy, and E. F. Connor. 1981. There have been no statistical tests of cladistic biogeographical hypotheses. In *Vicariance biogeography: A critique,* edited by G. Nelson and D. E. Rosen, 40–63. New York: Columbia University Press.

Slowinski, J., A. Knight, and A. P. Rooney. 1997. Inferring species trees from gene trees: A phylogenetic analysis of the Elapidae (Serpentes) based on the amino acid sequences of venom proteins. *Molecular Phylogenetics and Evolution* 8:349–62.

Sober, E. 1988. The conceptual relationship of cladistic phylogenetics and vicariance biogeography. *Systematic Zoology* 37:245–53.

Wheeler, W. C. 1993. The triangle inequality and character analysis. *Molecular Biology and Evolution* 10:707–12.

Wiley, E. O. 1988. Parsimony analysis and vicariance biogeography. *Systematic Zoology* 37:271–90.

Yang, Z., and B. Rannala. 1997. Bayesian phylogenetic inference using DNA sequences: A Markov chain Monte Carlo method. *Molecular Biology and Evolution* 14:717–24.

Zhang, L. 1997. On a Mirkin-Muchnik-Smith conjecture for comparing molecular phylogenies. *Journal of Computational Biology* 4:177–87.

APPENDIX 2.1
Notation

Consider a set S of n taxa with phylogenetic relationships described by a rooted binary tree, T, which consists of a set of vertices, $V(T)$, connected by a set of edges, $E(T)$. The set of vertices can be divided into the leaf set, $V_L(T)$, with n leaves (vertices of degree one) uniquely labeled by a taxon in S, the set $V_I(T)$ with $n-1$ internal vertices of degree three, and the singleton set $V_R(T)$ containing the unique root vertex of degree one. The edge set can be similarly divided into $E_L(T)$, $E_I(T)$, and $E_R(T)$.

Denote the immediate ancestral (father) edge of an edge e as $f(e)$, and the immediate children edges $c(e)$, arbitrarily separated as the left child $c_l(e)$ and the right child $c_r(e)$. Consider an edge e as directed from the vertex $v_f(e)$ and directed into the vertex $v_c(e)$. If an edge e_1 is a descendant of another edge e_2, we write $e_1 \subset e_2$ or $e_2 \supset e_1$ (as if edges were sets of descendant leaves). Denote by $\mathrm{lca}(e_1, e_2)$ the least common ancestral edge of e_1 and e_2 and by $\mathrm{dist}(e_1, e_2)$ the number of vertices separating e_1 and e_2. Assume that the time for the splitting event corresponding to a vertex v_i is given by $t(v_i)$. Then two edges $e_1, e_2, \in E(T)$ are *simultaneous* (the corresponding lineages were concurrent at some point in time) if $t(v_c(e_1)) > t(v_f(e_2))$ and $t(v_c(e_2)) > t(v_f(e_1))$; this is denoted $e_1 \mid e_2$. If the edges could be simultaneous given some possible function $t(v)$, this is denoted e_1 / e_2.

Now the reconstruction problem can be formulated as follows. We have a rooted binary tree of host relationships H, a similar tree of parasite relationships P, and a mapping φ of leaves in P to leaves in H. Assume that an edge $p \in E(P)$ is associated with a single edge $h \in E(H)$ at any single point in time (the one-host-per-parasite assumption) and denote this association $(p{:}h)$. Each edge $p \in E(P)$ starts out being associated with the edge $i(p) \in E(H)$ (the *in*-state), and it ends being associated with the edge $o(p) \in E(H)$ (the *out*-state). A *solution* Φ is a set of associations $(p{:}h)$ specifying in- and out-states in H for all edges $p_i \in E(P)$; $\Phi = \{(p_i{:}i(p_i)), (p_i{:}o(p_i))\}$ where $o(p_i) = \varphi(v_c(p_i))$ for all p_i directed into leaves in P.

The cost of a solution $K(\Phi)$ is found by translating it into events, each of which has a cost. Consider an internal vertex $v \in V_i(P)$ and the three edges adjacent to it: the ancestral edge p_a, and the children edges p_l and p_r (fig. 2.11a). A solution associates each internal vertex v in P with three edges in H: the out-state of the ancestral edge, $o(p_a)$, and the in-states of the two children edges, $i(p_l)$ and $i(p_r)$. Similarly, each edge $p \in E(P)$ is associated with two edges in H: the in-state, $i(p)$, and the out-state, $o(p)$. The solution is evaluated using a biological *model* $M = \{(\varepsilon_i, k_i)\}$ specifying a set of different types of allowed events, ε_i, and a cost function, k_i, for each event type. An event type can either relate to an internal vertex $v \in V_i(P)$, in which case it specifies a set of allowed combinations $(o(p_a), i(p_l), i(p_r))$ consistent with the type, or it can map to an edge $p \in E(P)$, in which case it specifies a set of allowed combinations $(i(p), o(p))$. If all elements in a solution Φ relating to edges and vertices in P correspond to event types in the model, then the solution

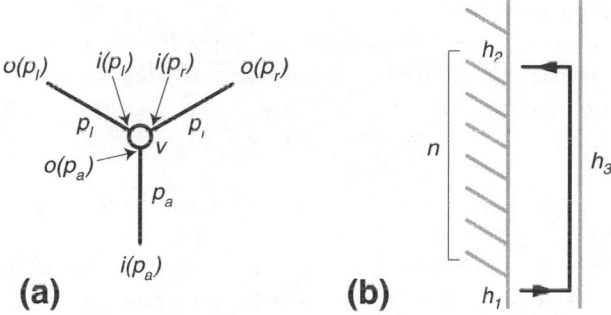

FIGURE 2.11. (*a*) Notation used in the exploration of model space. We consider a vertex v in the parasite tree and the three edges connected to it: the ancestral edge p_a and the two descendant edges p_l and p_r. Each edge p is associated with an initial host, $i(p)$, and a final host, $o(p)$. (*b*) If complete switches are allowed in the model, they cannot always be reinterpreted a posteriori in terms of factual events. For instance, when two sibling parasite edges are associated with widely separated ancestral (h_1) and descendant (h_2) hosts, the optimal solution may postulate an ancestral association with, or a double switch through, an unrelated host lineage (h_3).

is possible under the model and its cost, $K(\Phi)$, can be calculated as the sum of the cost of the events,

$$\sum_i k_i(o(p_i), i(c_l(p_i)), i(c_r(p_i))) + \sum_i k_i(i(p_i), o(p_i)), \quad p_i \in E_i(P),$$

where each k_i is chosen appropriately according to M. The associations around a vertex or edge in P may translate to several possible event types, in which case calculating the cost of a given solution becomes an optimization problem.

A *method* is a procedure that finds the solution with minimal total cost using a particular model and its cost assignments. In finding this solution it is useful to consider partial solutions. Denote by $\kappa(p{:}h)$ the cost of a partial solution pertaining to a subtree of P rooted at p and containing the association $(p{:}h)$, and by $\kappa_{min}(p{:}h)$ the minimum cost of all such solutions.

Desirable Properties

Potency of a method can be formalized in the following rules:
1. Completeness: For all problem instances there must exist a finite number of solutions Φ such that $K(\Phi)$ is finite.
2. Recognition of congruence: If H and P are isomorphic when the leaves are exchanged according to φ, then there must be only one optimal solution Φ with each $o(p)$ being the image of p in H.
3. Nonredundancy: For each of the event types in the model on which the method is based, there must exist a problem instance with an optimal solution Φ which translates to at least one instance of this event type.

Factuality requires that

4. All past associations $(p{:}h)$ postulated in a solution should be traceable to a recent association (a member of φ). An ancestral association $(p{:}h)$ is traceable if, in the solution, there is a path in P from p to a leaf, along which all associates are either h or descendants in H of h.

Finally, logical consistency requires that

5. Events should carry a positive cost if they alter associations and have zero cost otherwise.

6. Events implied at vertices in P must be instantaneous and involve edges in H that could have been simultaneous; events implied on the edges of P must not require a reversal in time flow.

Method Space

We consider a supermodel with six different types of events (cf. fig. 2.3). Co-divergence events, duplications, and partial switches map to vertices $v \in V_I(P)$ (fig. 2.11a), with codivergence events requiring that $o(p_a) = f(i(p_l)) = f(i(p_r))$ and $i(p_l) \neq i(p_r)$, duplication events requiring that $o(p_a) = i(p_l) = i(p_r)$, and partial switching events requiring that $i(p_l) \mid i(p_r)$ and either $o(p_a) = i(p_l)$ or $o(p_a) = i(p_r)$. Sorting events, complete switches, and extinction events map to edges $p \in E(P)$. Sorting events require that $i(p) \subset o(p)$, the number of switching events being determined by $\mathrm{dist}(i(p),o(p))$. A complete switch requires that there is a point along the edge p where there could have been a switch from $(p{:}h_1)$ to $(p{:}h_2)$ such that $h_1 \mid h_2$. A postulated (as opposed to an observed) extinction event requires that there is a leaf in P or in H, which has not been observed. In all cases we consider $i(p) = o(p)$ to be permissible, and this relation carries no cost (it is a nonevent).

A fixed cost is applied to each event. The cost must be positive for all events except codivergence events or duplications, for which zero cost is also permissible (cf. table 2.1). The proofs below will repeatedly refer to a triplet of edges (p_a, p_l, and p_r) around a node $v \in V_I(P)$ (fig. 2.11a), and this will alternately be considered as the full P tree or only part of it.

Lemma 1. A method based on a model including complete extinction events or complete switches either produces nonfactual solutions or the model is redundant. A model based on the remaining four events or a subset of them is factual and consistent, and there is only one way of translating associations around vertices and edges in P to events.

Proof. A postulated (as opposed to observed) extinction of a parasite or host lineage involved in an association $(p{:}h)$ necessarily leads to that association not being traceable among the observed associations. Postulation of a complete switch from $(p{:}h_1)$ to $(p{:}h_2)$ necessarily implies that the association $(p{:}h_1)$ cannot be traced. The factuality and consistency of solutions translating to duplications, sorting events, or codivergence events follow trivially from their definitions and cost assignments. Now consider a partial switch from $(p_a{:}h_1)$ to $(p_l{:}h_1)$ and $(p_r{:}h_2)$, $h_1 \mid h_2$. The associations

$(p_a{:}h_1)$ and $(p_l{:}h_1)$ will be traceable in the descendants of p_l, and $(p_r{:}h_2)$ will be traceable in the descendants of p_r, guaranteeing factuality. The unique relation between events and combinations of associations follows from the event definitions, which are mutually exclusive.

Lemma 2. Any potent, factual, and consistent method must include duplication events in its model.

Proof. By lemma 1 we have that only sorting events are allowed on edges of P. Consider a parasite tree with only two leaves (fig. 2.11a), both mapping to the same host leaf; thus $o(p_l) = o(p_r)$. Depending on the number of sorting events introduced on p_l and p_r, we have either $i(p_l) = i(p_r), i(p_l) \subset i(p_r)$, or $i(p_r) \subset i(p_l)$. The only combination that translates to an event is $i(p_l) = i(p_r)$, and we have that duplications must be included in the model of a complete method.

Lemma 3. A factual method that does not include switching events must include sorting events in its model to be complete.

Proof. Consider a parasite tree with two leaves mapping to different host leaves, which are not siblings (fig. 2.11a). A method based on a model without sortings cannot have any events mapping to edges of P, since sorting events are the only factual events associated with edges in P. Thus, we have $i(p_l) = o(p_l)$ and $i(p_r) = o(p_r)$. Since $i(p_l) \neq i(p_r)$, duplication at v is not possible, and since $i(p_l)$ and $i(p_r)$ are not siblings, codivergence at v is not possible. Thus, the method is incomplete unless sorting events are included.

Lemma 4. A method based on a model with only duplication and switching events does not recognize congruent H and P.

Proof. Consider a parasite tree with two leaves (fig. 2.11a). Since sorting events are not allowed, we have $i(pl) = o(pl)$ and $i(pr) = o(pr)$. If $o(p_l)$ and $o(p_r)$ are siblings in H with immediate ancestor $h_a = f(o(p_l)) = f(o(p_r))$, we have that $o(p_a) = h_a$ by condition (2). However, if only duplications and switches are allowed, $o(p_a)$ cannot be h_a and the method violates condition (2).

Lemma 5. Consider a triplet of edges in P (fig. 2.11a). If a solution is factual, then it must be true that either $o(p_a) \subseteq o(p_l)$ or $o(p_a) \subseteq o(p_r)$. If switches are not included, then it must be true that $o(p_a) \subseteq o(p_l)$ and $o(p_a) \subseteq o(p_r)$.

Proof. This follows directly from the definition of factuality.

Lemma 6. Let $h_a = \mathrm{lca}(o(p_l), o(p_r))$ (fig. 2.11a). Then an optimal solution must have $o(p_a) \subseteq h_a$.

Proof. By lemma 5 any solution where $o(p_a) \not\subseteq h_a$ must have $o(p_a) \supseteq h_a$. A solution with $o(p_a) \supseteq h_a$ cannot have node v associated with a switch, because there is no $i(p_l) \subseteq o(p_l)$ such that $i(p_l) \mid o(p_a)$, and the same applies to $i(p_r)$. Assume that the solution with $o(p_a) = h_a$ has cost K. Then setting $o(p_a)$ to some $h \supset h_a$ can maximally decrease the cost by $s\,\mathrm{dist}(h, h_a)$ (by removing sorting events from p_a) but will increase the solution by at least $2s\,\mathrm{dist}(h, h_a)$ (the sorting events must be accounted for independently on p_l and p_r, since v does not correspond to a switching event). Thus, the lemma is proven. See also Charleston (1998, theorem 1), but note that the proof presented there is not complete since it does not consider the possibility that changing $o(p_a)$ from h_a to some $h \supset h_a$ may save events in other parts of the solution.

Theorem 1. A two-event method based on a model with duplications of cost $u \geq 0$ and sorting events of cost $s > 0$ is potent, factual, and consistent. This is the only valid tree-fitting method based on a model with one or two events.

Proof. Examine an internal vertex $v \in V_I(P)$ (fig. 2.11a). From lemma 5 and 6 we have that there is only one possible optimal solution, namely $o(p_a) =$ lca($o(p_l), o(p_r)$). Since this solution always exists, and is associated with a finite number of sorting events on the edges p_l and p_r, we have fulfilment of condition (1). Assume that $o(p_l)$ and $o(p_r)$ are siblings; then the solution has $o(p_a) =$ lca($o(p_l), o(p_r)$) $= f(o(p_l)) = f(o(p_r))$ and condition (2) is satisfied. Since optimal solutions must include sorting events when $o(p_l) \neq o(p_r)$, condition (3) is also satisfied. Factuality and consistency follows from lemma 1. That this is the only valid method based on a one-event or two-event model follows from lemmas 2, 3, and 4.

Lemma 7. If a model includes duplication and sorting events, then the model is redundant with respect to codivergence events unless $c < u + 2s$.

Proof. A codivergence event at v is only possible if $o(p_l)/o(p_r)$. Let $h_a =$ lca($o(p_l), o(p_r)$), and set $h_l = c(h_a)$ and $h_r = c(h_a)$ such that $h_l \subseteq o(p_l)$ and $h_r \subseteq o(p_r)$. Then setting $i(p_l) = h_l, i(p_r) = hr$, and $o(p_a) = h_a$ will translate to a codivergence event with cost c. If we instead set $i(p_l) = i(p_r) = h_a$, then the event at node v must be a duplication. With two extra sorting events on edges p_l and p_r, from h_a to h_l and h_r, respectively, we get p_l and p_r associations identical to those produced by the codivergence, but the cost is now $u + 2s$. The codivergence and duplication-sorting sets of associations around vertex v are equivalent with respect to the rest of the solution (cf. fig. 2.3). Thus, codivergence events are redundant unless $c < u + 2s$.

Theorem 2. A three-event method based on a model with duplications of cost $u \geq 0$, codivergence events of cost $c \geq 0$, and sorting events of cost $s > 0$ is potent, factual, and consistent if $c < u + 2s$.

Proof. Factuality and consistency follow from lemma 1 and completeness from theorem 1. The model is not redundant with respect to duplications and sortings by lemma 3 and 4, and by lemma 7 it is not redundant with respect to codivergence events provided that $c < u + 2s$. Condition (2) is satisfied by lemmas 5 and 6.

Lemma 8. Any method based on a model including switches is complete.

Proof. Assume a post-order traversal of P. For each triplet of edges (fig. 2.11a) set $i(p_l) = o(pl)$ and $i(p_r) = o(p_r)$. Then arbitrarily set $o(p_a) = i(p_l)$ or $o(p_a) = i(p_r)$, corresponding to a switch. This will always provide a solution with finite cost $K = ni$, where $n = |VI(P)|$.

Theorem 3. A method based on a three-event model with duplications of cost $u \geq 0$, codivergence events of cost $c \geq 0$, and partial switches of cost $i > 0$ is potent, factual, and consistent if $i > c$.

Proof. We have fulfilment of condition (1) from lemma 8. Since sorting events are disallowed, we have $i(p_l) = o(p_l)$ and $i(p_r) = o(p_r)$. Assume that $o(p_l)$ and $o(p_r)$ are siblings. Then $o(p_a)$ must be either $o(p_l)$, $o(p_r)$, or $f(o(p_l)) = f(o(p_r))$. If $o(p_a) = f(o(p_l)) = f(o(p_r))$, we get the cost c (for a codivergence event); otherwise we get the cost i (for a switch). Provided that $i > c$, the codivergence solution wins and fulfilment of condition (2) is guaranteed for isomorphic H and P (but not for isomorphic subtrees of anisomorphic H and P). Assume that $o(p_l)$ and $o(p_r)$ are not siblings. Then if $o(p_l) = o(p_r)$ a duplication event is required and if $o(p_l) \mid o(p_r)$ a switch is required, and hence condition (3) is satisfied. Factuality and consistency follow from lemma 1.

Lemma 9. Assume that M is the maximum number of nodes separating leaves in H. Then for a model with sortings of cost $s > 0$ and partial switches of cost i, there are conceivable problems requiring switches if $i < u + s(M + 1)$ (codivergence events not included) or $i < c + s(M - 1)$ (codivergence events included).

Proof. Consider a two-taxon tree P (fig. 2.11a). Assume that $o(p_l)$ and $o(p_r)$ are leaves in H, that these leaves are not siblings or identical, and that $\text{dist}(o(p_l), o(p_r)) = M$. If v translates to a switch, then there are two equally optimal solutions, one with $o(p_a) = o(p_l)$ and one with $o(p_a) = o(p_r)$, both with cost i. If v does not translate to a switch, then there is one optimal solution, with $o(p_a) = \text{lca}(o(p_l), o(p_r))$. If codivergence events are included in the model and are not redundant, then v translates to a codivergence event and the cost is $c + s(M - 1)$; otherwise v translates to a duplication event and the cost is $u + s(M + 1)$. Thus, switches are not redundant given the conditions in the lemma.

Theorem 4. A three-event method with duplications of cost $u \geq 0$, sorting events of cost $s > 0$, and partial switches of cost $i > 0$ is potent, factual, and consistent if $u + s(M + 1) > i > u + 2s$.

Proof. Completeness, that is, fulfilment of condition (1), follows from theorem 1. Assume that $o(p_l)$ and $o(p_r)$ are siblings. From lemma 5 and 6 we have that $o(p_a)$ is either $o(p_l)$, $o(p_r)$, or $f(o(p_l)) = f(o(p_r))$. If $o(p_a) = f(o(p_l)) = f(o(p_r))$, then we must have a duplication at v and one sorting event on each of p_l and p_r, total cost $u + 2s$. Otherwise there must have been a switch at v and the cost is i. The codivergence solution is locally optimal if $i > u + 2s$, and this guarantees fulfilment of condition (2) for isomorphic H and P (but not necessarily for isomorphic subtrees of anisomorphic H and P). This also shows that there are optimal solutions requiring duplications and sorting events. Lemma 9 demonstrates that switches are required for some problems and thus condition (3) is fulfilled if $i < u + s(M + 1)$. Factuality and consistency follow from lemma 1.

Theorem 5. A method based on a four-event model with codivergence events of cost $c \geq 0$, duplications of cost $u \geq 0$, sortings of cost $s > 0$, and partial switches of cost $i > 0$ is potent, factual, and consistent if $c < u + 2s$, $c + s(M - 1) > i > c$, and $s < (M - 1)(i - c)$, where M is the largest number of nodes in H and P trees.

Proof. By theorem 1 we have that completeness is satisfied (condition (1)). Consider a triplet of nodes in P (fig. 2.11a) and assume that $o(p_l)$ and $o(p_r)$ are siblings. Then we have by lemmas 5 and 6 that $o(p_r)$ equals $o(p_l)$, $o(p_r)$, or $f(o(p_l)) = f(o(p_r))$. Consider the last case, in which switching is not possible. By lemma 7 we have that $c < u + 2s$ and the optimal solution will translate v to a codivergence event rather than a duplication event. Then by lemma 9 condition (2) is satisfied for isomorphic H and P (but not necessarily for isomorphic subtrees of anisomorphic H and P) if $i > c$. This also demonstrates that codivergence events are not redundant. Duplications are not redundant by lemma 2 and switches by lemma 9 if $c + s(M - 1) > i > c$. Now assume that H and P are completely asymmetric and isomorphic after relabeling and contain maximally M nodes. Remove one of the two leaves in P separated from the root by the largest number of vertices. Then the optimal solution without sortings will have cost $(M - 1)i$, since only switches are now possible at nodes in P. The optimal solution with sortings will have cost $(M - 1)c + s$. It follows that there are problems requiring sorting events as long as $s < (M - 1)(i - c)$ and condition (3) (nonredundancy) is fulfilled. Factuality and consistency follow from lemma 1.

Theorem 6. The five methods shown above to be potent, factual, and consistent are the only such methods based on subsets of the six-event model and having fixed event costs.

Proof. This follows directly from lemma 1 and theorems 1–5.

Functional Factuality

Theorem 7. There are problems for which a method based on a model with nonfactual events (complete switches or extinctions) will produce optimal solutions that cannot be reinterpreted a posteriori in terms of factual events.

Proof. If we allow extinctions of cost $e > 0$, then we can introduce complete switches on edges in P with cost $i + e$ (a partial switch followed by extinction of one of the P lineages), which would correspond to allowing complete switches of cost $w = i + e$. Consider a triplet of edges in P, where the ancestral edge $p_a \in R(P)$. Assume further that p_l supports a subtree of P isomorphic with the H subtree rooted at h_1, that p_r supports a subtree of P isomorphic with the H subtree rooted at h_2, that $h_1 \subseteq h_2$, and that dist$(h_1, h_2) = n$. Finally, assume that there is an edge $h_3 \in E(H)$ such that $h_3 \mid h_1$ and $h_3 \mid h_2$ (fig. 2.11b). To be able to reinterpret a solution with $o(p_l) = h_1$ and $o(p_r) = h_2$ in terms of factual events we must have $o(p_a) = h_1$. This does not allow any switches and the cost is $\kappa_{min}(p_l{:}o(p_l) = h_1) + \kappa_{min}(p_r{:}o(p_r) = h_2) + u + ns$. Consider setting $o(p_l)$ to h_2 instead of h_1. This will alter the cost of the solution by $n(i - c)$ or $n(w + u - c)$ because it will remove n cospeciation events and replace them with partial switches or combinations of duplications and complete switches. This means that the minimal cost of a solution interpretable a posteriori as factual is $\kappa_{min}(o(p_l) = h_1) + \kappa_{min}(o(p_r) = h_2) + u + n(\min(s, i - c, w + u - c)$. Consider the cost of instead associating the edge p, with two complete switches, $h_1 \to h_3$ and $h_3 \to h_2$ (associating v with two switches from h_3 will give a similar result). The cost of this solution is $\kappa_{min}(p_l{:}o(p_l) = h_1) + \kappa_{min}(p_r{:}o(p_r) = h_2) + u + 2w$, which means that it will win if $2w < n(\min(s, i - c, w + u - c)$. Since we have $s > 0$, $i > c$, and $w + u > c$ (from lemma 1 and theorem 3, noting that a duplication and a complete switch are equivalent to a partial switch), it follows that there are values of n (barring limits to the total size of H and P) such that optimal solutions will postulate complete switches that cannot be reinterpreted in terms of factual events.

Jungle Excursions

Definitions. The notation of this section follows Charleston (1998) and is summarized here to the extent it differs from the notation used previously. The immediate ancestor vertex of a tree vertex or edge v is denoted v'. A jungle $J(H, P, \varphi)$ is a directed graph which contains all the potentially optimal mappings of the vertices of P into H. A jungle vertex u associates a vertex $p \in V(P)$ with a vertex or an edge $h \in H$. If h is a vertex in H, then u corresponds to a codivergence event and is said to be of type 1, which is denoted $u = (p{:}h)^*$. If h is an edge in H, then u is said to be of type 2 and is denoted $u = (p{:}h)$ (without an asterisk). This corresponds to a duplication event or a partial switch. The set of vertices in jungle J which are directed into vertices v_x and v_y are called the feasible parent set of v_x and v_y and is denoted $N\{v_x, v_y\}$. A solution Φ is a set of connected jungle vertices with each vertex in P being represented once. The cost of a solution is the total cost of the implied events. The event costs can take any values under the following constraints: codivergence events $c \leq 0$, duplication events $u > 0$, sorting events $s > 0$, and (partial) switches $i > 0$ (cf. table 2.2).

Proposition 1. Charleston's lemma 3 is incorrect. If it is applied, the resulting jungles do not include all the conceivably optimal solutions.

Proof. Charleston's lemma 3 reads as follows: "Suppose that p_x, $p_y \in P$ are siblings and the jungle J associating P with host tree H contains vertices $v_r = (p_r{:}h_r)$ and $v_y = (p_y{:}h_y)$, where h_x, h_y are also siblings. Then with the cost constraints of table 2.1 $[c \leq 0, u > 0, s > 0, i > 0]$, $N\{v_r, v_y\} = \{(p_x' = p_y'{:}h_x' = h_y')\}$." As is obvious from the proof given in Charleston's appendix A, there is an asterisk missing: the feasible parent set of v_x and v_y should have been given as $N\{v_x, v_y\} = \{(p_x' = p_y'{:}h_x' = h_y')^*\}$.

Consider a simple problem with two leaves in H and three leaves in P (fig. 2.12a). Label the P vertices p_1 through p_5, the H vertices h_1 to h_3, and the corresponding H edges h_{1e} to h_{3e}. Using Charleston's theorem 1, which is correct, we easily find that the jungle contains five possible solutions, and the associated costs are easily obtained (table 2.4). Solutions 1 and 2 can never be optimal because $2i > c + i$ under the given cost constraints. Similarly, solution 4 can never be optimal because

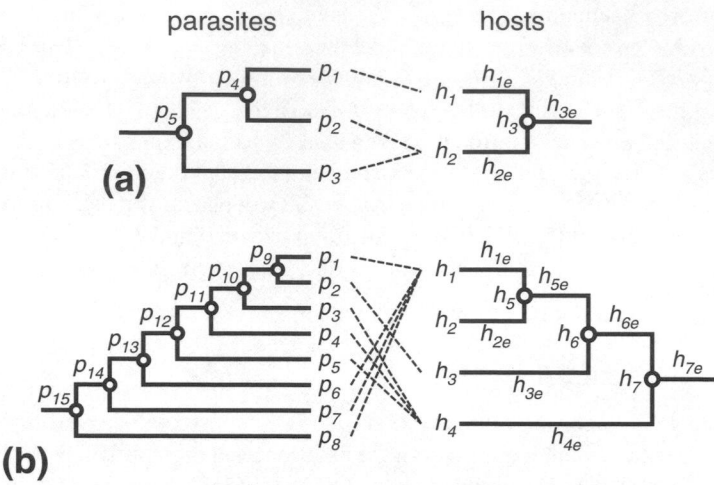

FIGURE 2.12. Two cases in which the algorithms described by Charleston (1998) will not provide exact solutions. See text and tables 2.4 and 2.5 for further details.

TABLE 2.4 All possible solutions to the problem given in figure 2.12a. Both solutions 3 and 5 are conceivably optimal under different cost assignments allowed by Charleston (1998). Yet, Charleston's lemma 3 incorrectly excludes solution 3 as guaranteeably nonoptimal.

Solution	J Vertex 1	J Vertex 2	Events	Cost
1	$(p_4{:}h_{1e})$	$(p_5{:}h_{1e})$	2 switches	$2i$
2	$(p_4{:}h_{1e})$	$(p_5{:}h_{2e})$	2 switches	$2i$
3	$(p_4{:}h_{1e})$	$(p_5{:}h_3)^*$	1 switch, 1 codivergence	$c+i$
4	$(p_4{:}h_{2e})$	$(p_5{:}h_{2e})$	1 switch, 1 duplication	$u+i$
5	$(p_4{:}h_3)^*$	$(p_5{:}h_{3e})$	1 duplication, 1 sorting, 1 codiv.	$c+s+u$

$u + i > c + i$. However, both solutions 3 and 5 are conceivably optimal. Solution 3 is optimal when $i < u + s$ and solution 5 when $i > u + s$; both these conditions can occur under the given cost constraints. All solutions Φ must contain the jungle vertices $v_x = (p_1{:}h_1)$ and $v_y = (p_2{:}h_2)$, since these are given by φ. Because p_1 and p_2 are siblings and h_1 and h_2 are also siblings, we can apply Charleston's lemma 3 and then obtain the feasible parent set as $N\{v_x, v_y\} = \{(p_4{:}h_3)^*\}$. Lemma 3 thus excludes solution 3 and the proof is complete.

Comments. The feasible parent set should have been given as $N\{v_x, v_y\} = \{(p_x'{:}h_x')^*, (p_x'{:}h_x), (p_x'{:}h_y)\}$. But this result is given already by Charleston's theorem 1, making lemma 3 superfluous. Charleston's proof of lemma 3 fails because it is not sufficient to retain the jungle vertices that are conceivably optimal when we look locally at them and their immediate descendants. Locally suboptimal vertices may well be included in globally optimal solutions, as shown by solution 3 (table 2.4) in the example discussed above.

The failure of local decision rules also means that it is impossible to construct jungles containing only the conceivably optimal solutions until we have traversed the jungle and calculated the cost of the possible solutions. The jungle can be weeded somewhat before traversing it, notably by using Charleston's theorems 1 and 2, but will still contain a lot of solutions that are demonstrably nonoptimal after the jungle has been traversed.

Proposition 2. Dynamic-programming traversal of a jungle using Charleston's method for dealing with switches gives erroneous costs for some solutions.

Proof. Consider a problem (fig. 2.12b) and a possible solution (table 2.5). This solution is globally optimal under some cost assignments allowed by Charleston, for instance $c = -1.0$, $u = 1.0$, $s = 1.0$, and $i = 2.0$, which is easily shown with exact algorithms. Global optimality is not crucial for the proof, however, so it will not be demonstrated here. We need only show that Charleston's method of jungle traversal does not allow correct calculation of the total cost of this solution under the given event costs.

Denote by $\kappa_{min}(v)$ the minimal raw cost (the cost of all implied events, without considering possible incompatibility among switches) of solutions rooted at jungle vertex v. According to Charleston (1998), the true minimal cost (considering switch incompatibility) can be obtained by simply adding $s\,\mathrm{SIM}_{min}$, where s is the sorting cost and SIM_{min} is the minimum Switch Incompatibility Measure, that is, the minimum number of extra sorting events induced by any set of host switches required by solutions rooted at v and having raw cost $\kappa_{min}(v)$. The set of sets of host switches in solutions rooted at v and having minimal raw cost $\kappa_{min}(v)$ is called the feasible switch set of v and is denoted $\Gamma(v)$.

A true dynamic-programming algorithm must be able to calculate $\kappa_{min}(v)$ recursively from all possible pairs of immediate descendant vertices, v_i and v_j. Now consider $v = (p_{10}{:}h_4)$ (fig. 2.12b). One of the descendant vertices is $(p_3{:}h_4)$.

TABLE 2.5 A possible solution to the problem given in figure 2.12b. The described mechanism for jungle traversal (Charleston, 1998) does not allow calculation of the cost of this solution (see text), which is globally optimal under some permissible event costs.

J Vertex	Switch Added	Event Added	Cumulative Cost
$(p_9{:}h_{1e})$	$h_{1e} \to h_{3e}$	1 switch	i
$(p_{10}{:}h_{4e})$	$h_{4e} \to h_{1e}$	1 switch	$2i$
$(p_{11}{:}h_{4e})$		1 duplication	$u + 2i$
$(p_{12}{:}h_{4e})$		1 duplication	$2u + 2i$
$(p_{13}{:}h_{1e})$	$h_{1e} \to h_{4e}$	1 switch	$2u + 3i$
$(p_{14}{:}h_{1e})$		1 duplication	$3u + 3i$
$(p_{15}{:}h_{1e})$		1 duplication	$4u + 3i$

The other descendant is either one of $(p_9:h_{1e})$, $(p_9:h_{3e})$, or $(p_9:h_6)^*$. We find that the minimum raw cost is obtained if we choose $(p_9:h_6)^*$, in which case we get $\kappa_{min}(p_{10}:h_4) = c + s + i = -1.0 + 1.0 + 2.0 = 2.0$ with associated switch set $\Gamma(p_{10}:h_4) = \{(h_4 \to h_6)\}$. But to obtain the correct cost of solution A (table 2.4) we should have chosen $(p_9:h_{1e})$ instead to give us $\kappa(p_{10}:h_4) = 2i = 4.0$ with associated switch set $\Gamma(p_{10}:h_4) = \{(h_4 \to h_1, h_1 \to h_3), (h_4 \to h_3, h_3 \to h_1)\}$ (two possible switch sets are included because we could have obtained the same minimal raw cost by choosing $(p_9:h_{3e})$ instead of $(p_9:h_{1e})$). At vertex $(p_{13}:h_1)$ we have to consider adding the switch $h_1 \to h_4$ to the existing set. If we chose the optimal solution at $v = (p_{10}:h_4)$, then we must reach the conclusion at $v = (p_{13}:h_1)$ that it is impossible to add the switch $h_1 \to h_4$, and all solutions that include the jungle vertices $(p_{10}:h_4)$, $(p_{11}:h_4)$, $(p_{12}:h_4)$, and $(p_{13}:h_1)$ will erroneously be determined to have an infinite (arbitrarily large) cost. Thus, we cannot obtain the correct cost of the considered solution (table 2.4) using Charleston's approach, and the proof is complete.

Comments. Note that this type of problem occurs in the jungle approach even in this simple example, where H is completely asymmetric and only one H history is possible. When H is larger and more balanced, and many alternative H histories are possible, the problem becomes more serious. If the H history is known, however, the problem can easily be solved by subdividing each edge in H stretching over one or more splitting events occurring in other parts of H (Ronquist, 1995).

3

LIZARDS, MALARIA, AND JUNGLES
IN THE CARIBBEAN

Michael A. Charleston and Susan L. Perkins

Introduction

In this chapter we describe the jungles method for comparing host and parasite phylogenies, and use it to analyze the relationships between the *Anolis* lizards and the malaria varieties infecting them in the eastern Caribbean. We begin with a description of the general problem of reconciling related phylogenies, and then show how it applies to the lizard and malaria phylogenies.

The *jungle* is a method devised by Charleston (1998) to solve the problem of mapping a "parasite" phylogeny into a "host" phylogeny. The "parasite" may be anything from a louse to a retrotransposon, and the "host" could be a gene, or genome, whale, or island chain. Phylogeny reconciliation is the process of estimating the historical associations between two phylogenies, that is, mapping the "parasite" phylogeny into the "host" phylogeny, reconciling their differences (Goodman et al., 1979; Page and Charleston, 1997; Eulenstein et al., 1998). Page uses *reconciliation* in a more restrictive way to mean a mapping that precludes host switching (Page and Charleston, 1997), but here we allow host switches. The problem is mathematically identical over a large set of biological problems as has been discussed by Page (1993a) and others. The data relate to two varieties of *Plasmodium:* one *P. azurophilum* red, the other *P. azurophilum* white, the names based on the types of blood cells they prefer to inhabit in their hosts. For the *azurophilum* red it transpires that there are three possible optimal solutions, and that of these we choose the one involving one better supported host switch, as opposed to three less supported ones. There is a lower degree of cospeciation (contemporaneous speciation of host and parasite) between the *azurophilum* white and the lizard hosts, though it remains significant.

Jungles

We have already seen the desirability of reconstructing the joint history of related phylogenies. If we can with confidence state that cospeciation events took place between, say, a host and its parasites, then the two lineage bifurcation events must perforce have occurred at the same moment in evolutionary time. We can use such hypotheses to relate evolutionary rates, since the same amount of time can be hypothesized to have taken place from the cospeciation events to the present day. Also, suppose that we have good evidence supporting a close correspondence between the phylogeny of a geographical region and its endemic species—then we may even use one history to date the other.

All consistent methods (see later) of reconstructing these joint histories amount to the creation and evaluation of maps from the vertices in the parasite tree into the host tree. In other words, not every point in the host tree need be mapped to, though every extant taxon in the parasites should map to its known host(s). Since the internal vertices (also called nodes) in the parasite tree correspond to speciation events, we are interested in whether they map to nodes in the host tree. We naturally regard places when internal parasite vertices map to internal host vertices as cospeciation events (see fig. 3.1). We have said "consistent" above, which we shall define a

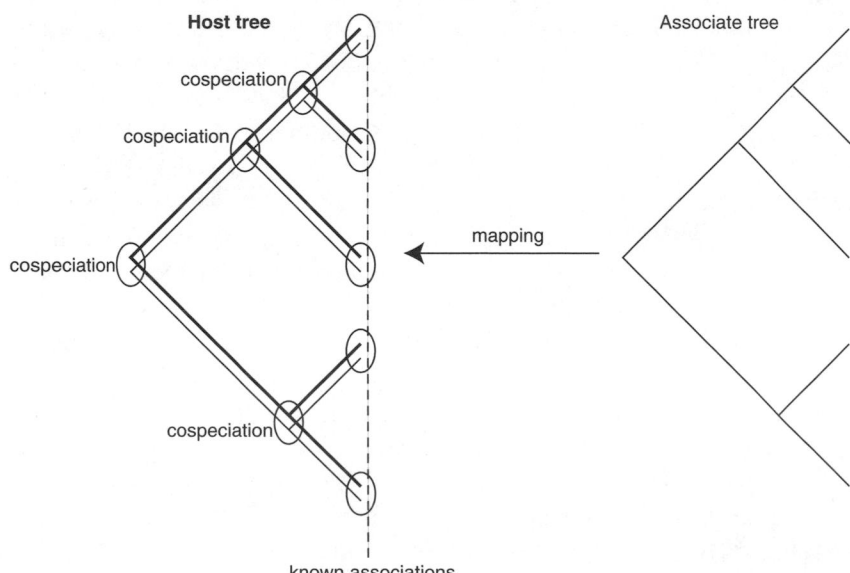

FIGURE 3.1. Mapping an associate tree into a matching host tree. These two trees match perfectly, and we reconstruct four cospeciation events.

little more formally now. For the purposes of this discussion, a method of phylogeny reconciliation (into which category a number of methods fall: Brooks, 1988; Page, 1988, 1993a; Charleston, 1998) is consistent if (1) it can account for a perfect match between phylogenies, and (2) any reconstruction output by the method has a biological interpretation (c.f. Ronquist, chap. 2). Clearly all methods discussed thus far satisfy the first requirement: given a perfect match between the host and associate trees, such that renaming the associate tips with the name of the host which they infect yields the host tree, then there can only be one optimal solution, which is that the trees are perfectly congruent. The second requirement is less easy to satisfy, and this difficulty has led to the eventual failure of Brooks Parsimony Analysis (BPA) (Page, 1990; Ronquist and Nylin, 1990).

We also require of our methods that they be *complete*—that is, that any triplet of host and parasite phylogenies and the set of known associations between their tips, must be admissible as input to our method. We would certainly regard a method as incomplete if we could not apply it in all situations! Though all the methods mentioned here are complete, they do have different computational complexities, and there exist combinations of host and associate trees, and their extant associations, such that the computational requirements are too great to render their solution feasible. In these rather problematic instances the computational complexity is exacerbated by a very low level of agreement between the trees. Fortunately we may take the reasonable stand that in situations in which there is very low agreement between the trees, we would not be hypothesizing coevolutionary events anyway, so would not be attempting to find the optimal match between the trees. Thus the computational complexity of reconciling two badly matched phylogenies is not generally a restriction to our investigations.

In figure 3.1 it is clear that the simplest explanation for the relationship between the two trees is one of perfect cospeciation: every time the host diverges, so does the parasite, and no parasite lineages are lost (which would be possible if there were extinction or if we simply failed to find a parasite on the host). In a case like this it is easy to find the best reconstruction of what went on in the past for these two trees, and we can avoid having to choose some arbitrary cost function. But what can we do if the trees don't match up so nicely? We need to find some kind of optimality criterion by which we can judge among a set of possible mappings of the parasite tree into the host tree, which is (are) the best one(s).

It is usually the choice of optimality criterion (cost function) that forms the main source of dissent among proponents of different methods. This is to be expected because, while it is generally true that optimality criteria—such as the number of parsimony steps in Brooks Parsimony Analysis method of reconciling host and associate tree—are definitely minimized when there is a perfect fit of data to the model, when the data deviate from this perfect match, then the optimality criteria disagree in just "how bad" the deviation is.

Brooks uses a property of the parsimony reconstruction method for phylogenetic trees, which is that a data set constructed with binary characters corresponding to inclusion in clades in the tree will fit that tree perfectly, and will require the minimum number of character state changes on the tree to account for the "observed" binary data (Brooks, 1988). The method is very simple, and is easy to interpret in the case of a perfect match between host and associate trees. The parasite tree is coded as a set of binary characters such that all taxa descendant from a given internal node are assigned one character state, and the others are assigned the other state. Thus a set of statements of monophyly is constructed for the parasite tree. These "data" are then fed into a parsimony-based tree finding program (such as PAUP* or Hennig86), and the most parsimonious reconstruction of those character states on the host tree is found.

If the trees match up, then the number of character-state changes required is minimal for strict cospeciation, but it is not clear how to interpret the change of "states" when the match is not perfect. Page (1990) has investigated this quite thoroughly, as have Ronquist and Nylin (1990). One of the indirect consequences of this failure is that a posited host switch of an associate taxon between host lineages in a sense causes host switches in all the ancestors of that associate, right back to the common ancestor of the host lineages—a clear problem!

A more statistically sound optimality criterion for selection of optimal maps from the parasite to the host tree would be in terms of the likelihood of the reconstruction, which is proportional to the probability of observing the extant host/parasite associations and trees given some stochastic model of cospeciation between host and associated phylogeny. However, Maximum Likelihood (ML) methods are notoriously computationally intensive and are not for the fainthearted, nor those without huge computing power at their fingertips. The other difficulty is of choosing a model upon which to base any likelihood analysis: while it is true that hypothesis testing is only possible within a sound statistical framework, we do not believe that for the most part there exists the information in the data sets to estimate the parameters of a maximum likelihood model for host-parasite interactions with any surety. For further discussion of maximum likelihood and Bayesian approaches, the interested reader is directed to the chapters by Huelsenbeck, Rannala, and Larget (chap. 4), and by Rannala and Michalakis (chap. 5).

We turn to a slightly different model, in which we assign a cost to each kind of coevolutionary event that we can observe for a pair of related phylogenies (for recent reviews of cost-based methods see Ronquist and Nylin, 1990; Page and Charleston, 1998). This cost can be thought of as the log likelihood of each kind of event occurring, so the sum of a set of log likelihoods for events A, B, C, and so on is the log likelihood of the combined event, that all of A, B, C, and so on happen. However, it will be simpler for the moment to put aside this interpretation, as it does not readily extend to homogeneous stochastic models (that is, probabilistic models with parameters that remain constant through time). Note that minimizing an overall cost is equivalent to maximizing parsimony—hence such cost-based approaches can be thought of both in terms of parsimony and of likelihood. Later we shall discuss how the costs can be extended to be more realistic with little extra computational complexity.

In this model we identify four kinds of events, as with Page (1996), and these are described in box 3.1. We use the convention that cospeciations are represented by solid dots and duplications by hollow dots. Since host switches can be thought of as a duplication and subsequent relocation of one descendant lineage, they too are represented by hollow dots. Lineage sorting events have no symbol—they should be clear from the context. These event types cover all the events that we can logically distinguish from two associated phylogenies (see Ronquist, chap. 2). For instance we cannot tell just how far a parasite lineage "makes it" down a host lineage if eventually it dies out, as shown in figure 3.2.

Problem Input

The input data to an instance of the phylogeny reconstruction problem are often represented as a tanglegram, sensu Page (1993b), as shown in figures 3.3 and 3.12. In these figures the host phylogeny is on the left and the associate on the right. Associations between the two sets of extant taxa are indicated by dotted lines between the tips of the trees.

In figure 3.4 we see one way in which the trees in the tanglegram might be reconciled. The reconstruction, which may or may not be optimal, has cospeciation events posited at the root and further cospeciation events down the tree, and two host switches. Another explanation could be given, in which a duplication occurred in the associate tree prior to the root of the host tree, followed by some lineage sorting events and cospeciations, with no host switches. If the biological system that gave rise to these trees is

Box 3.1. The coevolutionary events counted by jungles

Cospeciation: Both host and associate speciate at the same time. We attempt to maximize the frequency of these events.	
Duplication: The associate speciates independently of the host.	
Lineage sorting: At a divergence in the host phylogency, the associate lineage does not apparently diverge. It either fails to go down the new lineage at all, "missing the boat" (Paterson, et al. 1997), it becomes extinct on the new lineage, or we simply fail to find it.	
Host switching: At a duplication event, one new lineage moves to a different host lineage. This does not differentiate between the cases where the duplication occurs first and the new parasite lineage subsequently moves, or the switch occurs and the newly isolated populations diverge and "speciate."	

such that host switching events are very unlikely—for instance in gene tree–species tree systems—then we would clearly favor the latter explanation. In other circumstances, when host switching is frequent, then we would favor the former. Other explanations to this solution were presented by Ronquist (1995).

As we can see, there can be a large number of solutions for any given instance of the problem; and as has been mentioned, the difficulty worsens when there is less agreement between the trees (see fig. 3.5). In the worst

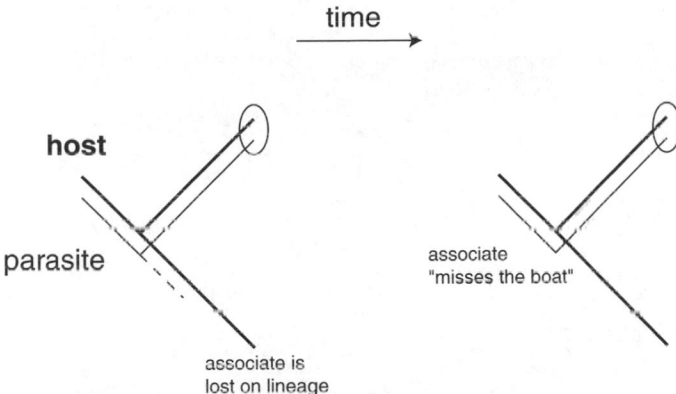

FIGURE 3.2. Two possible scenarios of lineage loss, which we cannot logically distinguish from each other. In the first, the associate cospeciates with the host but later dies out on one of the new descendant host lineages, whereas in the second case the associate "misses the boat" altogether at the formation of a new host lineage. It is not possible to distinguish these scenarios based solely on tree topology: we also need to invoke biological and/or temporal arguments.

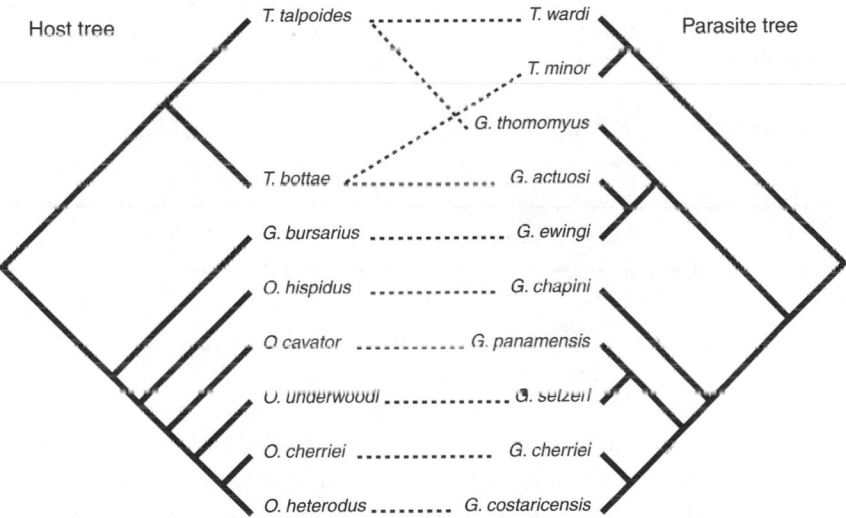

FIGURE 3.3. An example tanglegram, showing the familiar gopher and louse phylogenies and their associations, from Hafner and Nadler (1988), reproduced in the form used by Charleston (1998).

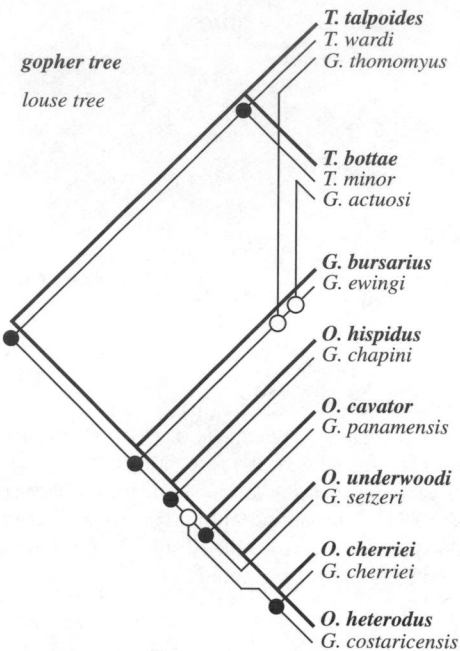

gopher tree

louse tree

T. talpoides
T. wardi
G. thomomyus

T. bottae
T. minor
G. actuosi

G. bursarius
G. ewingi

O. hispidus
G. chapini

O. cavator
G. panamensis

O. underwoodi
G. setzeri

O. cherriei
G. cherriei

O. heterodus
G. costaricensis

FIGURE 3.4. A possible reconciliation of the tanglegram in figure 3.3 (after Charleston, 1998). With nicely matched trees there are relatively few optimal solutions. With less well-matched trees there are vastly more possible solutions (see fig. 3.5).

case the number of conceivably optimal mappings from parasite tree P into host tree H grows more than exponentially with the number of taxa in either tree, though with very similar trees the rate of growth is small. The computational complexity is mainly due to the admissibility of host switches into our possible explanations of the relationship between the trees. If no host switches are allowed, then the problem is of quite a different character and can be solved in polynomial time, that is, an amount of time that grows as a polynomial function of the number of taxa involved. The no-host-switches-allowed problem is referred to by Page as "reconciled trees" and is solved in his program GENETREE. We extend the term here to include explanations allowing host switches; the distinction of usage should be borne in mind. One of the consequences of allowing host switches in a solution is that for any given set (usually a pair) of siblings which have been assigned positions on the host tree H, there can be more than one possible location for their immediate ancestor on H.

Consider a pair of parasite taxa, p_1 and p_2, which are hypothesized to be on nodes h_1 and h_2, respectively, and suppose that the immediate ancestor

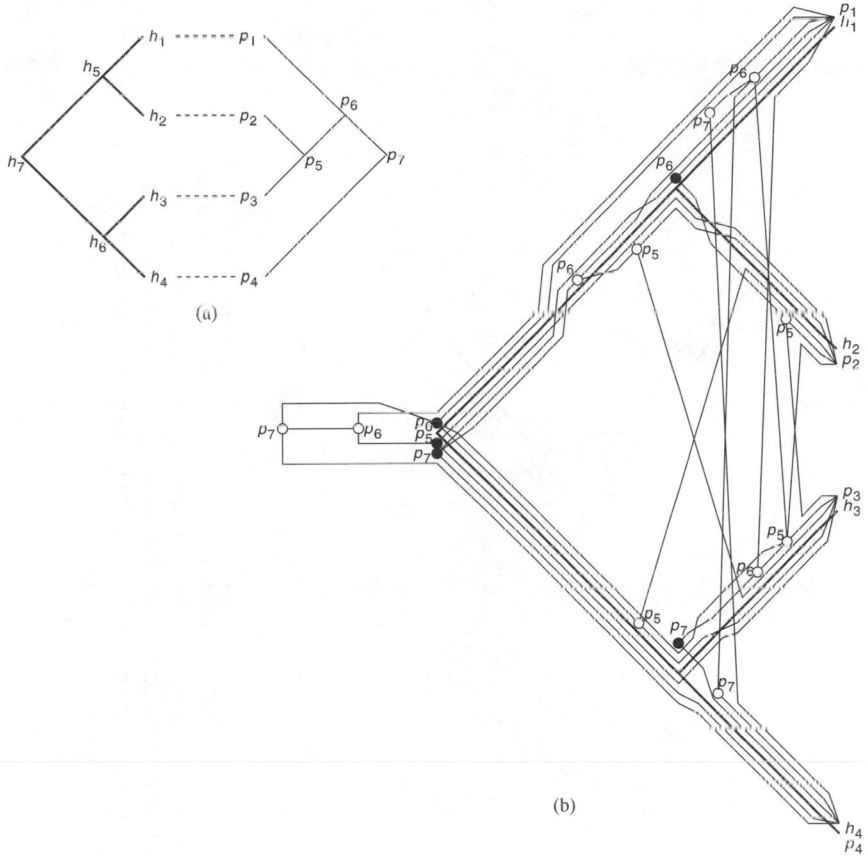

FIGURE 3.5. The tanglegram *(a)* and complete jungle *(b)* for a pair of badly matched trees. The complexity of the jungle means that there are many potential solutions embedded in it even for just four host and parasite taxa. Note the complexity due to host switches.

of p_1 and p_2 is p', and that of h_1 and h_2 is h'. Under the no-host-switches rule the only position we could place p' is at h' or on one of its ancestors (see fig. 3.6a). If we were to allow host switches as well, then we might also place p' anywhere on the immediate ancestors of h_1 or of h_2, since we could then explain the location of p' with some hypothesized lineage sorting events and a host switch (fig. 3.6b). Obviously if h_1 and h_2 are far apart in H, then there will be more locations available, and so more solutions possible.

Another issue arising directly from the permission of host switches is the "order" in which they are hypothesized to occur. This is a more subtle concern that requires some care. Consider the host switches shown in

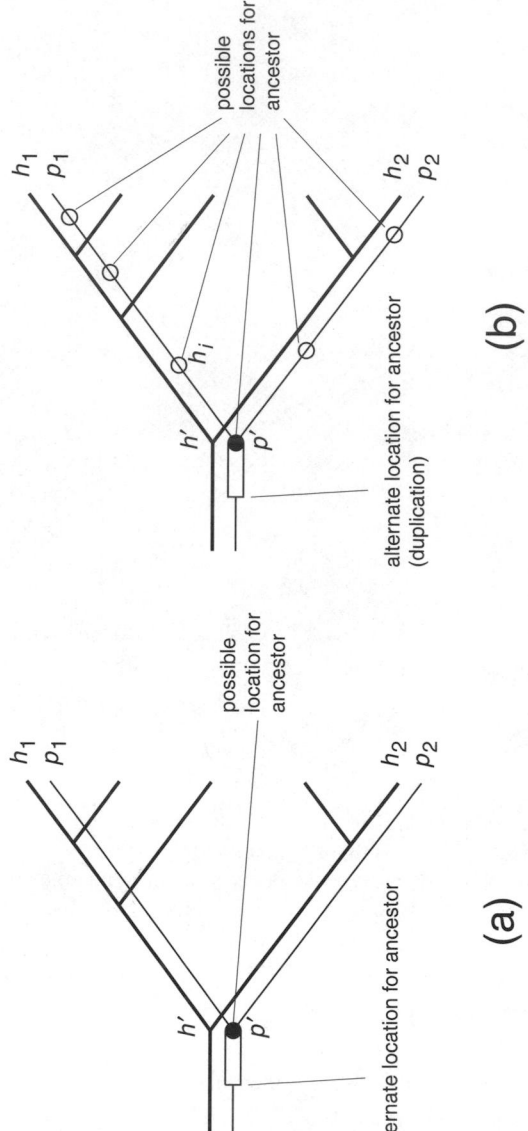

FIGURE 3.6. Consequences of prohibiting or allowing host switches. If host switches are prohibited, then the only possible location for p', the most recent common ancestor (MRCA) of p_1 and p_2, is on the MRCA of h_1 and h_2, at h' in the figures. This can either be at the host vertex itself (implying a cospeciation event) or immediately beforehand (implying a duplication and subsequent lineage sorting events). If on the other hand host switches are permitted, then all the branches on the host tree are permitted as locations for p', with one descendant associate lineage having to jump the gap between the two host lineages (h', h_1) and (h', h_2).

figure 3.7a. There, the switch labeled s starts from h_s in the host tree H and arrives eventually at h_t. Clearly there are a number of possible locations for the "landing site" of this switch, from the final location at h_t right back to the root of the host tree. In this case, h_e is the earliest possible location for the landing site of s, since with only two host switches we would only unnecessarily add extra lineage sorting events if we had s land earlier. With more host switches in a solution, we may find other weak switch incompatibilities with other host switches, necessitating moving the landing site of s further back on the host tree. We do not consider moving the "takeoff" point of the switch, since according to our model host switching only takes place after a duplication, and we cannot expect to trace host-switching lineages that do not leave a sibling lineage on their current host.

In this sense a posited host switch induces a partial ordering relation of the host tree H, that is, it makes a statement about the relative positions in time of some of the host nodes. Because each host switch potentially imposes such ordering constraints on the host tree, we may need to consider all the possible sequences of host switches through time, since if we consider one switch to have occurred before another, we may arrive in quite different conclusions about the coevolution of the two taxonomic groups. For a set of k host switches there are $k!$ orderings, so this plainly aggravates the complexity of the problem.

Clearly there are computational difficulties arising from the large number of solutions, so if we were to list all the potential solutions and assess them individually for a given instance of the problem, the problem could become intractable. Fortunately things aren't quite this bad, because we can make the observation that the associate lineages (parasite, gene, etc.) can be treated as behaving completely independently after they diverge from each other. That is, once parasites have speciated, they no longer interact with each other. This might seem like a strong assumption to make, but in fact the same assumption has been made in all the previous attempts at the solution of this problem. If we were to allow in our model, interactions between the associate lineages, then the complexity introduced is likely to make things essentially insoluble.

Once we have accepted that for the sake of our approach we shall ignore between-lineage interactions, we can treat them independently also. Think for a moment about the solutions to a given instance of the host/parasite phylogeny reconciliation problem. The solution is a mapping from the associate tree P into the host tree H. That is, it is a collection of associations between parasite vertices (nodes) and host vertices or edges (branches). The

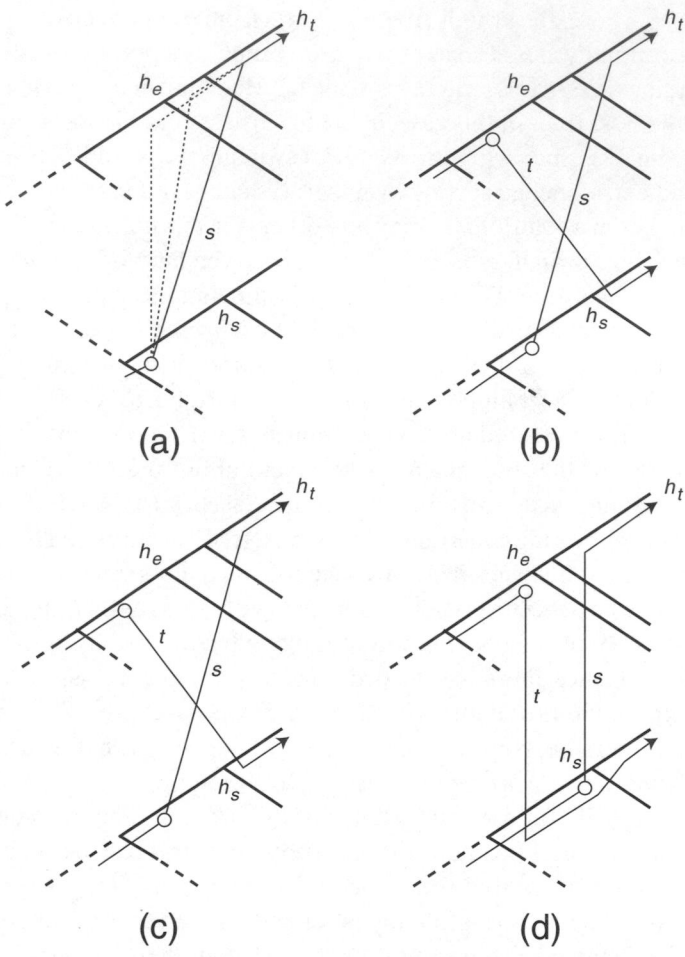

FIGURE 3.7. Host switches place partial ordering constraints on the host tree. In *(a)* the existence of the host switch *s* constrains the host tree *H* because the takeoff branch and the landing site branch must overlap in time at some point. In this example this means h_s', the immediate ancestor of h_s, must precede h_t, and the landing site can be anywhere back in time up to and including h_e. In *(b)* we introduce a second host switch *t*, which puts further constraints on the host tree, and which means we must move the landing sites of *s* or *t* to accommodate both switches. The two possible solutions are shown in *(c)* and *(d)*.

collection of associations between these two sets of vertices, representing the parasite and host taxa, completely describes the posited relationship between the trees. So as shown in figure 3.8, all the solutions are collections of these vertex associations. Note that many of the same associations are common to many of the solutions to the jungle. By listing all the solutions

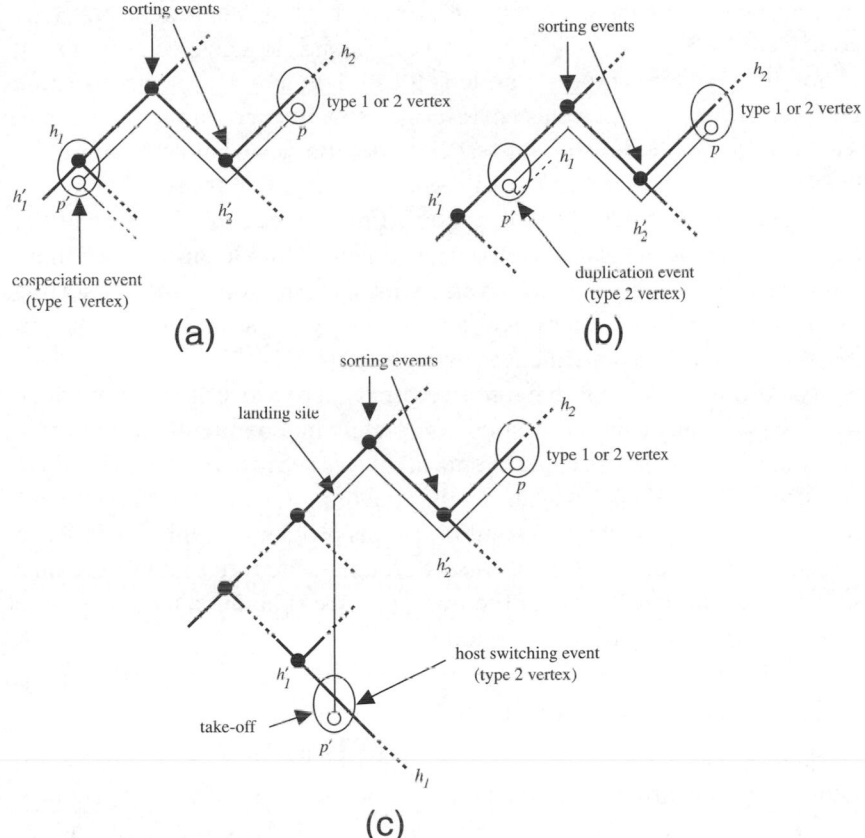

FIGURE 3.8. The evolutionary history of an associate branch is completely determined by the location of its beginning h' and ending h in the host tree. In *(a)* we see a cospeciation, in *(b)* a duplication, and in *(c)* a host switch. Any subsequent lineage sorting events are determined by the host nodes between h' and h, or in the case of a host switch, between the landing site of the switch and h. In each of these cases the total cost for the arc is independent of the other arcs (ignoring for the moment the question of host switch incompatibility), and therefore is determined completely by the individual j-arc events.

we are creating extra work for ourselves, since the components of a single solution to the problem are identical in different solutions. It would certainly be beneficial if we could somehow list each hypothesized association only once.

Now if the associate lineages are independent of each other, then the evolutionary history of a branch of the associate tree can be completely described by the location of the ancestral and descendant nodes on the host tree, as shown in figure 3.8. If the ancestral node of a given branch is exactly coincident with a host node, then this corresponds to a cospeciation event. If not, that is, if the associate node is located on an edge of the host tree between two nodes, then this corresponds to a lineage duplication event. The location of the descendant associate determines what coevolutionary events are hypothesized to have taken place after the ancestral associate node, being host switches and lineage sorting events. If the location of the descendant associate is not itself a descendant of that location, then there must have been a host switch event, with perhaps some lineage sorting events thrown in. The other possibility is that there was no host switch, but again there may be some lineage sorting events.

The astute reader will have noticed that there are situations in which the lineages really can't be treated completely independently, when there are two or more host switching events involved, because the order in which conflicting switches occur can affect the coevolutionary events that we have to propose to explain a given solution to this problem. In this case we can only treat as independent those parts of the parasite tree that involve host switches on unrelated parts of the host tree. We shall need a couple more items of terminology to help explain.

Host Switch Incompatibility

Since as has been mentioned previously the existence of a host switch induces a partial ordering on the nodes of the host tree, there may of course be situations in which two posited host switches are in conflict: they induce partial orderings of H that cannot both exist at the same time. There are two cases to worry about here, one being a case of true incompatibility and the other being only an apparent incompatibility, which we can solve with the judicious inclusion of some lineage sorting events. The former case, illustrated in figure 3.9, is called strong incompatibility. The latter case is called weak incompatibility, because the problem is quite soluble (as in fig. 3.7a–d). Ronquist (1995) discussed issues pertaining to host switch (in-)compatibility earlier.

Jungles

The jungle is a representation of all the possible locations of all the extant and ancestral associate taxa at once. In this sense it contains all the mappings from P into H, and we sift through it to find the "best" mapping. As we have mentioned, there are any number of ways of scoring a mapping between parasite and host trees, and in this method the allowable scoring

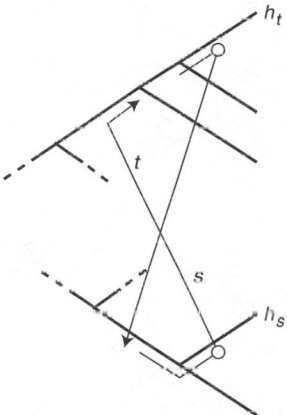

FIGURE 3.9. The host switches *s* and *t* are strongly incompatible: no matter how far back in time we move the landing site of either, we cannot prevent them from overlapping. Thus, we cannot reconcile the two without time-traveling parasites!

systems represent a continuous region in 4-dimensional space. Since we are hoping to maximize the agreement between the phylogenies of interest, we assign a low cost to each point where the trees match, that is, to cospeciation events. In fact all we require is that the cospeciation cost c be strictly less than the other event costs, so we can assign $c = 0$ and to the other events we assign a positive cost, corresponding to the fact that we are attempting to find a maximal match between the host and associate trees, in accordance with Fahrenholz's 1913 rule that parasite phylogeny mirrors host phylogeny (see chap. 2 for further discussion of cost schemes).

Figure 3.10 shows a simple tanglegram with just four × four taxa, and the jungle that shows all the optimal solutions and some of the nonoptimal ones for this instance of the phylogeny reconciliation problem.

Caribbean *Anolis* and *Plasmodium*

The Caribbean tropics are home to several species of *Anolis* lizards and lizard malaria (fig. 3.11). This study focuses primarily on the small islands of the Lesser Antilles, where two described species are found. One of these, *P. floridense,* was not used, as it occurred on only four islands sampled. The other species, *P. azurophilum,* was originally described as being capable of infecting both erythrocytes and white blood cells of the lizard host (Telford Jr., 1975). Subsequent genetic analysis of these forms (here called *P. azurophilum* red and *P. azurophilum* white to refer to the blood cell class where they were found), however, suggests that they are distinct

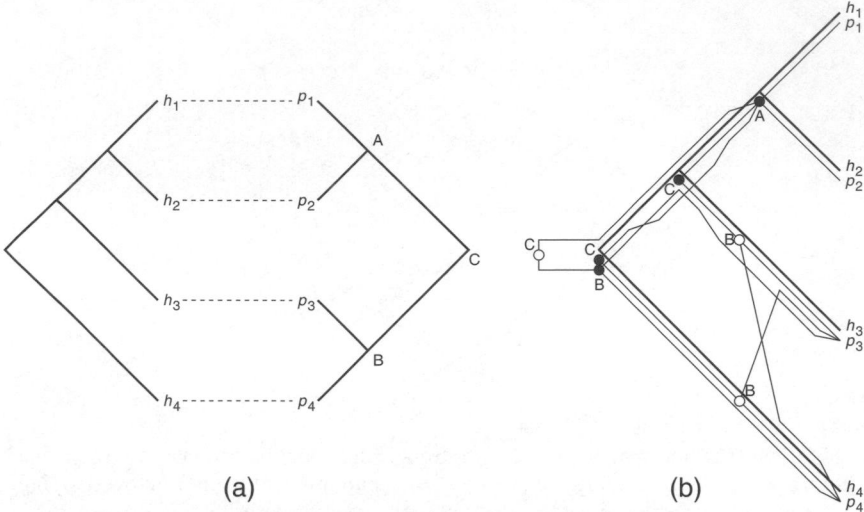

FIGURE 3.10. Simple tangle and the jungle for four × two taxa. The jungle shows all the possibly optimal solutions. In this case there are only three, and each contains the same cospeciation event at A. The solutions are then distinguished by the location of the other cospeciation event, and the corresponding host switches.

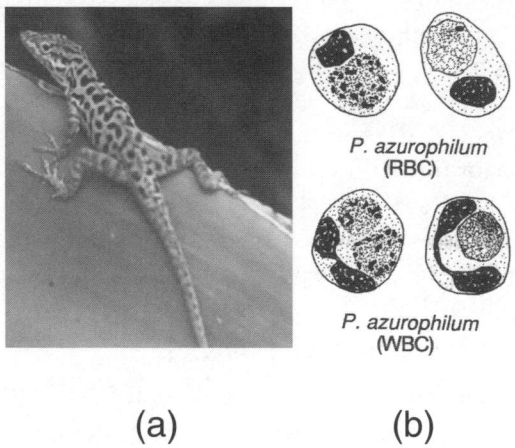

(a)   (b)

FIGURE 3.11. *Anolis* and *Plasmodium azurophilum* (photograph of *A. sabanus* by S. L. Perkins).

lineages (Perkins, 2001). We analyzed these forms separately. The differing solutions of the two forms lends additional support to the idea that these represent separate evolutionary lineages.

Note that these results are highly preliminary, and as further data accumulate we expect the relationships of the *P. azurophilum* to change. They are presented mostly as illustrative of the jungles approach and should not be regarded as definitive of the *P. azurophilum* phylogeny.

P. azurophilum Red

To construct the phylogenetic tree of the parasites, a portion of the cytochrome *b* gene (554 bp) was sequenced for each of three infections of each form (red and white) from each island, with the exception of Guadeloupe, where only two infected lizards were collected, one with *P. azurophilum* red and one with *P. azurophilum* white (data collected by SLP). No differences were observed within island samples except for two distinct haplotypes (here called *A* and *B*) observed for *P. azurophilum* red from the island of St. Kitts. Neighbor-joining, maximum parsimony, and maximum likelihood analyses (under F81 model, found with MODELTEST: Posada and Crandall, 1998) all produced identical trees, which differed only in their branch lengths. Support for these trees was not strong, however, due to the small number of variable sites in this sequence data. Enforcing a molecular clock did provide slightly better resolution of the trees, with no statistically significant difference in the log-likelihood scores (*P. azurophilum* red: −804.24390 vs. −807.97434, df = 7, *P. azurophilum* white. −780.78305 vs. −783.95509, df = 5).

The host tree was adapted from Roughgarden (1995). This tree was constructed from combined data of morphological (squamation), karyotypic, immunological analysis of blood albumins and protein electrophoretic data from over 20 loci from a variety of other published studies. The tree supports the differentiation of anoles into two main groups, the *bimaculatus* anoles and the *roquet* anoles, which occur above and below an imaginary line separating Dominica and Martinique. (The Puerto Rican anole [*A. gundlachi*] is part of the *cristatellus* subgroup of Greater Antillean anoles.) Only *Anolis* species which were observed to be infected with malaria parasites were included in our host tree. The tanglegram that illustrates the associations of these taxa is shown in figure 3.12.

The jungle for this tanglegram contains 1,643 solutions embedded in it—not shown, for obvious reasons. The reconstructions shown in figure 3.13 are optimal under the costing scheme used, with zero cospeciation cost and duplication, lineage sorting, and host switching costs at 1, 2, and 3, respectively.

The second and third reconstructions remain optimal over a large region of cost values, and we find that no matter what is the set of event costs, we cannot do better than five cospeciation events. To test the number of cospeciating parasite nodes for significance we randomize the parasite tree. This is because we are testing the degree to which the parasite tree follows

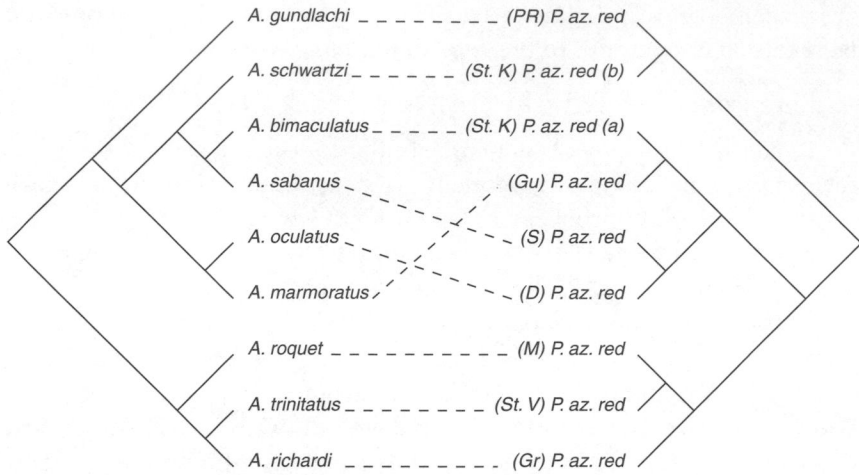

FIGURE 3.12. Relationships between Caribbean *Anolis* and *Plasmodium azurophilum* red (host tree adapted from Roughgarden, 1995, with parasite tree and host/parasite associations from Perkins, 2000).

the host tree. It is therefore inappropriate to randomize the associations themselves, or the host tree. The parasite trees were randomized according to the distribution produced by the usual Yule model of tree growth, that is, that each tip is equally likely to split at a given speciation event. We found that five cospeciation events was significant with this method ($p \leq .03$).

The solutions present interesting interpretations, which follow.

Solution 1

There is a single host switch from Martinique in the south, to the common ancestor of the St. Kitts, Saba, Dominique, and Guadelupe anoles. The large number of lineage sorting events seems unlikely: malaria is endemic in the anoles in the region and the chance of so many incipient *Plasmodium* lineages "missing the boat" (10 in all) makes this reconstruction implausible. Indeed, we found that increasing the cost s of lineage sorting events leads to this solution falling from the optimal set, superseded by solutions 2 and 3, below.

Solution 2

Again, we require a host switch northward from the Martinique group, followed later by two host switches from St. Kitts and Saba to Guadeloupe and Dominica, respectively. These latter two switches are against the direction of prevailing ocean currents. For this reason we rate this as a less likely solution to the problem of reconciling the two phylogenies.

Solution 3

The third solution is very similar to the second, though the two more recent host switches hypothesized are from south to north, in the same direction as the prevailing ocean currents, which lend them credence.

Note, however, that direction of the host switches posited in solutions 1 and 2 supports movement of parasites within the lizard hosts as they are in agreement with Gorman and Kim (1976), whereas solution 3 is more in accordance with the parasites traveling in windblown vectors.

P. azurophilum White

The *P. azurophilum* white shows a quite different pattern from *P. azurophilum* red: the distribution and the best estimate of the phylogeny are quite different, and this leads to quite disparate jungles and optimal reconstructions.

The jungle deriving from the above tanglegram is a little more manageable within the confines of a page, containing only 101 solutions. However, the greatest number of cospeciations that can be hypothesized is only 4 of a possible 7, which, while statistically significant ($p < .05$ by randomization of parasite tree, using TreeMap 1.0), allows a great many potentially optimal solutions. The jungle is shown in two parts. From the tanglegram in figure 3.14 we can see that ultimately we need to find the feasible locations in the host tree of the ancestral parasite nodes labelled b and e in the malaria tree. Figure 3.15 shows the jungle as constructed up to the assignment of locations to these parasites, and figure 3.16 depicts the feasible parent sets of all those pairs of locations. This breaking up of the figures serves joint purposes: both in making the complete jungle a great deal less confusing, and in showing a little more how the jungle is constructed.

The unique optimal solution shown in figure 3.17a contains four cospeciation events, one host switch, one duplication, and five lineage-sorting events. This solution remains optimal even when the cospeciation cost is increased to zero, and over a range of other cost values, though of course if the host switch cost is raised sufficiently there will be no switches in the optimal solution. This is shown in 3.17b. Note that there are still four cospeciation events, and the solutions do not differ very much from each other.

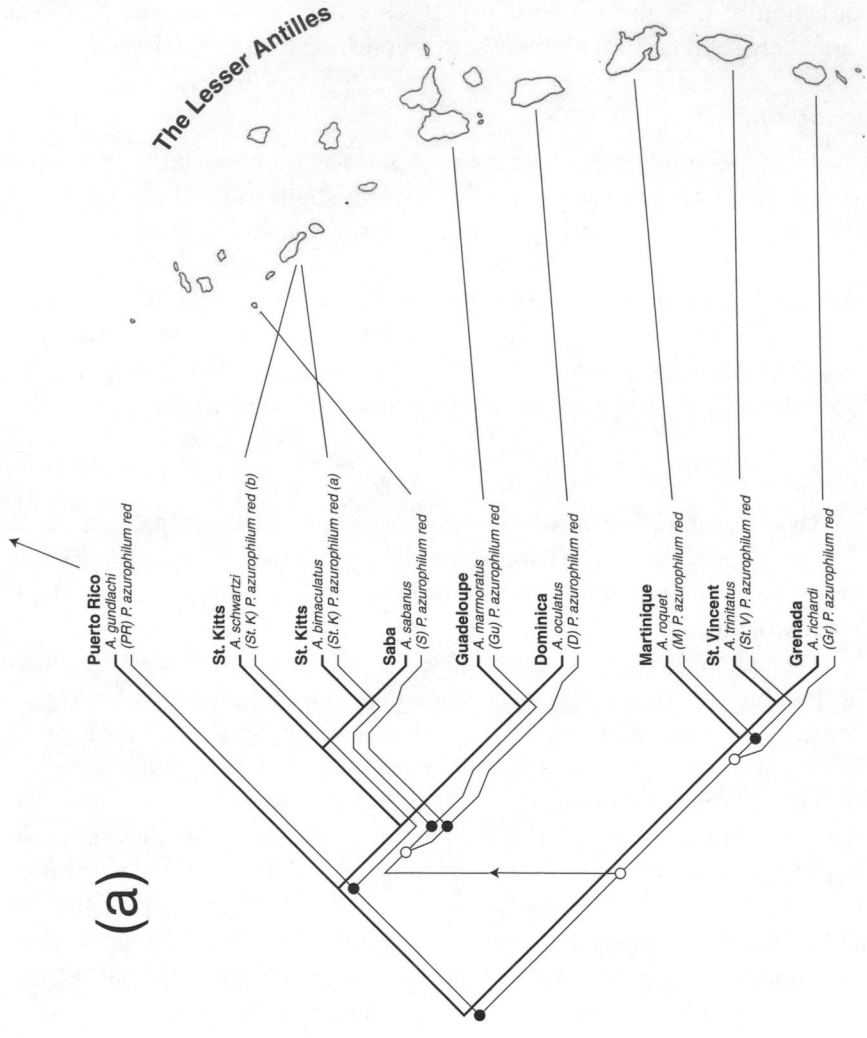

The Lesser Antilles

(a)

Puerto Rico
A. gundlachi
(PR) P. azurophilum red

St. Kitts
A. schwartzi
(St. K) P. azurophilum red (b)

St. Kitts
A. bimaculatus
(St. K) P. azurophilum red (a)

Saba
A. sabanus
(S) P. azurophilum red

Guadeloupe
A. marmoratus
(Gu) P. azurophilum red

Dominica
A. oculatus
(D) P. azurophilum red

Martinique
A. roquet
(M) P. azurophilum red

St. Vincent
A. trinitatus
(St. V) P. azurophilum red

Grenada
A. richardi
(Gr) P. azurophilum red

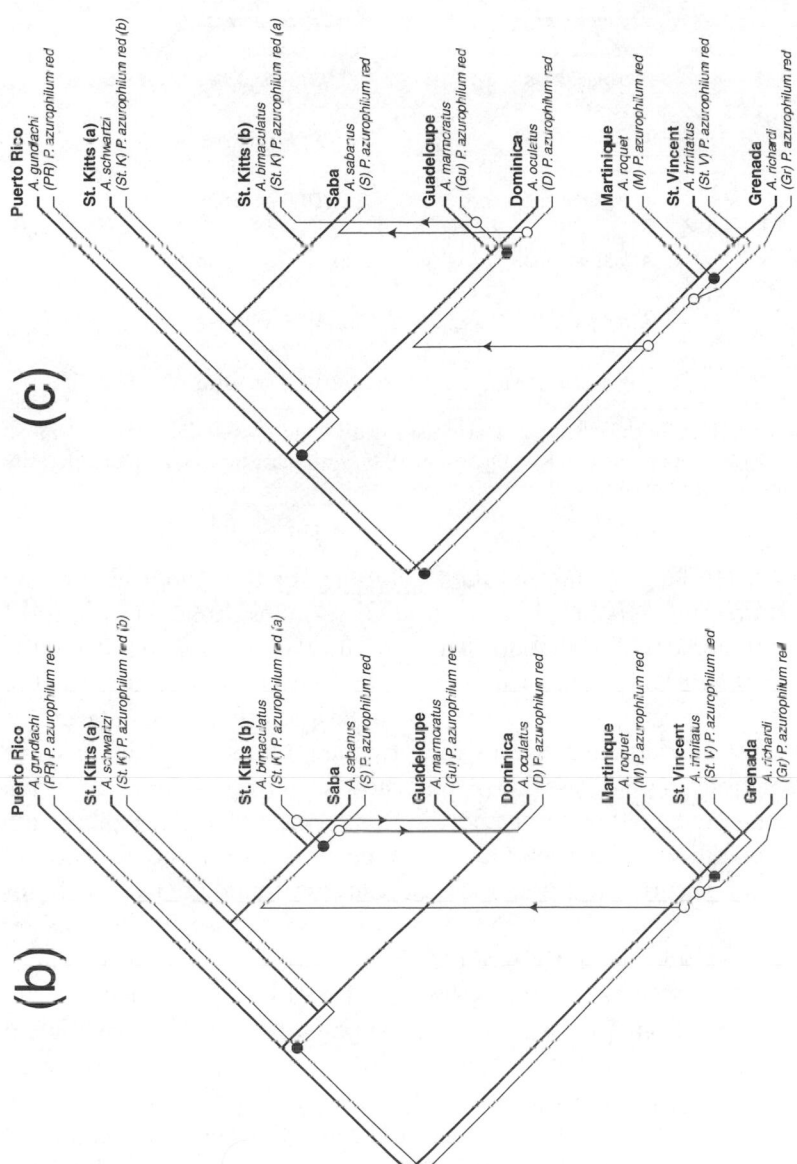

FIGURE 3.13. Possible reconstructions of shared history between *Anolis* and *Plasmodium azurophilum* red.

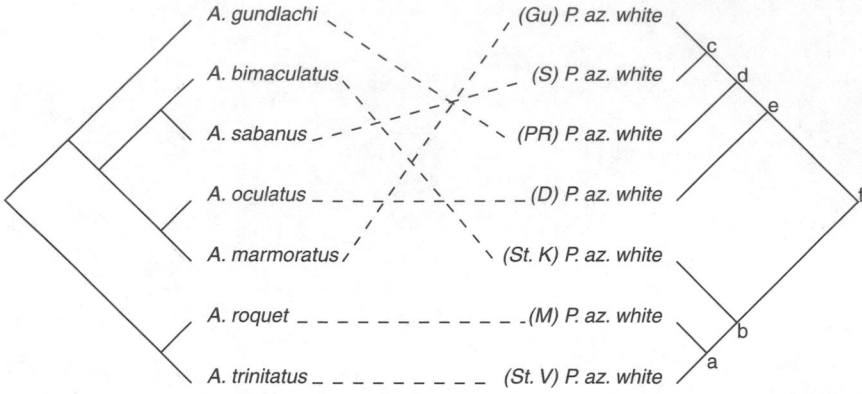

FIGURE 3.14. Relationships between Caribbean *Anolis* and *Plasmodium azurophilum* white (host tree adapted from Roughgarden, 1995, with parasite tree and host/parasite associations from Perkins, 2000).

In all, the tanglegrams and their solutions for the *Anolis* lizards and their *P. azurophilum* parasites for the two varieties do not show a great deal of agreement, though there is in both the *P. azurophilum* red and the *P. azurophilum* white significant degree of cospeciation. It is important to note that these areas *do* share a history of cospeciation: its significance is the subject of continuing investigation by the authors. Note also that the best solutions in both cases involved a host switch from the clade in the southern islands to those in the north. This seems a plausible explanation of the relationships between the trees, since it is also in the direction of the prevailing currents. If both the best solutions found are true, then this suggests a mechanism by which the malaria migrated among the species: by hitching a ride in a host lizard (c.f. Censky et al., 1998) as it rafts from one island to another, or perhaps within an insect vector blown along with wind currents from one island to another. Clearly there remains a great deal more to be learned about this collection of hosts, parasites, and their areas of endemism. Data are at present somewhat limited, but the system is large enough to permit interesting and important new elucidations of biological cospeciation and biogeography through studies of this kind.

The Next Steps

We have taken the reader through the basics of the jungle, a mathematical construct that comprises the entire solution space to the highly complex problem of mapping a parasite phylogeny into its host phylogeny, in order

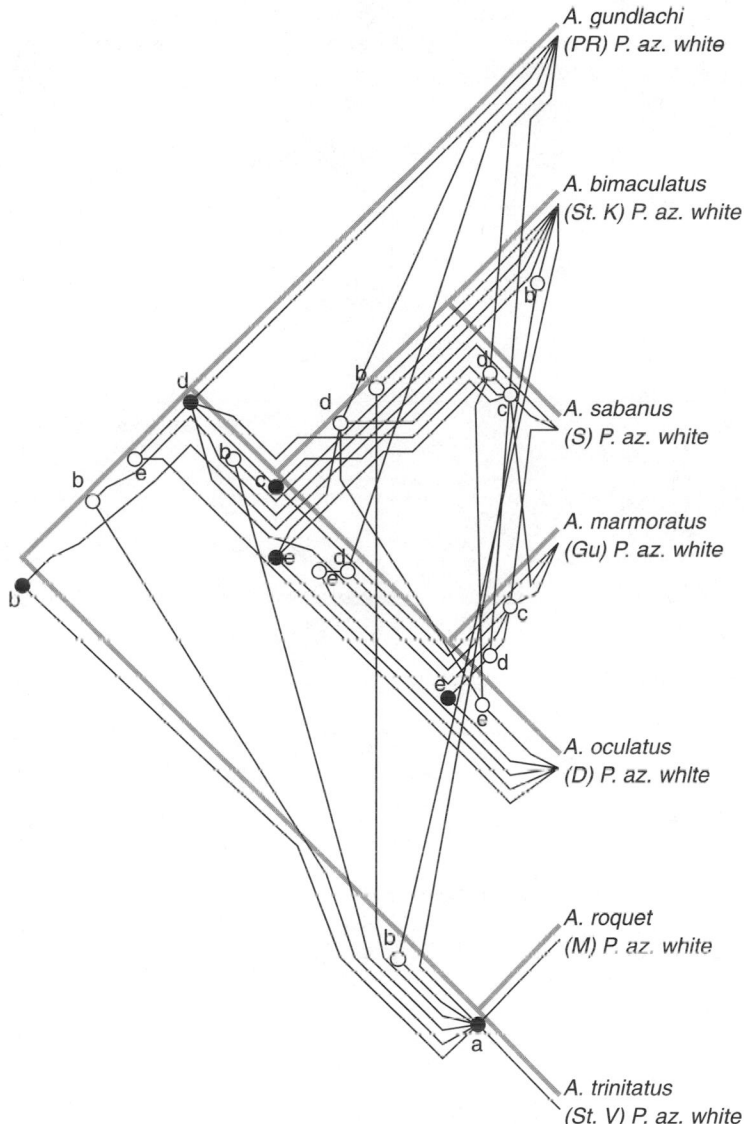

FIGURE 3.15. Partial jungle (part 1) for *Anolis* and *Plasmodium azurophilum* white. This shows the jungle as constructed up to the assignment of possible host locations for the hypothetical parasite nodes labeled *b* and *e*.

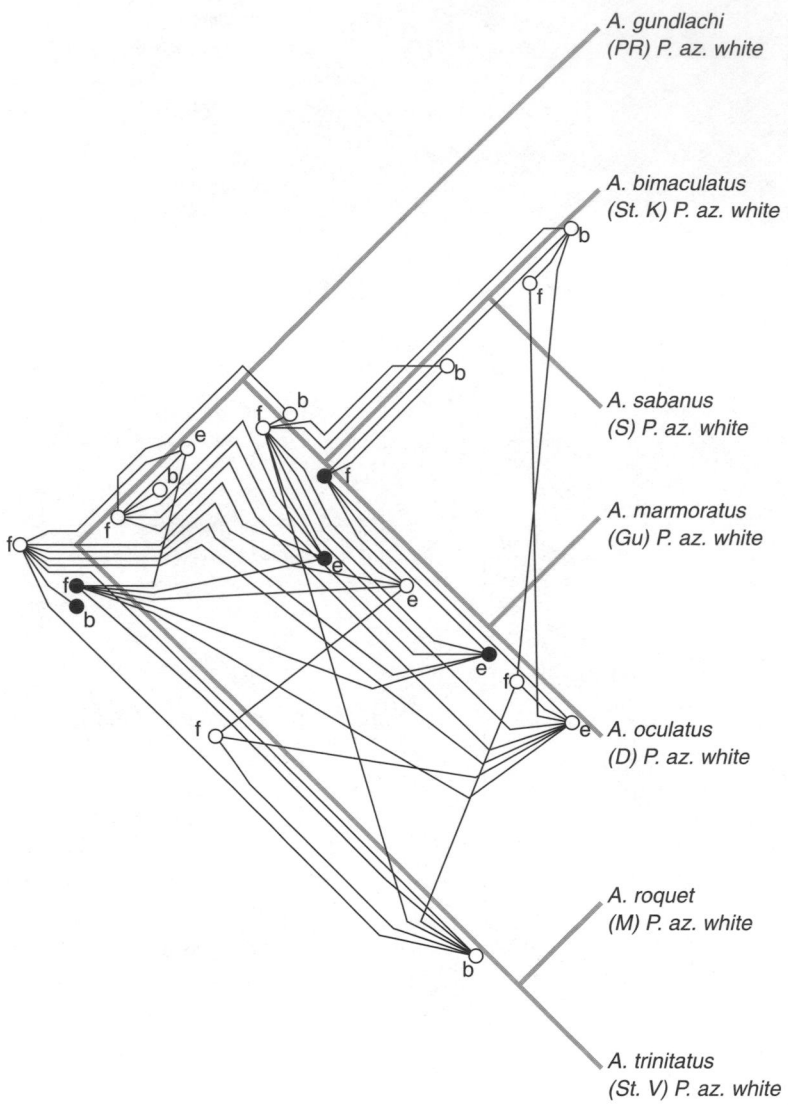

FIGURE 3.16. Partial jungle (part 2) for *Anolis* and *Plasmodium azurophilum* white. This shows the completion of the jungle, having constructed all the possible locations for parasite nodes *b* and *e*, amounting to the set of all possible locations for the root *f* of the *P. azurophilum* tree in the *Anolis* tree.

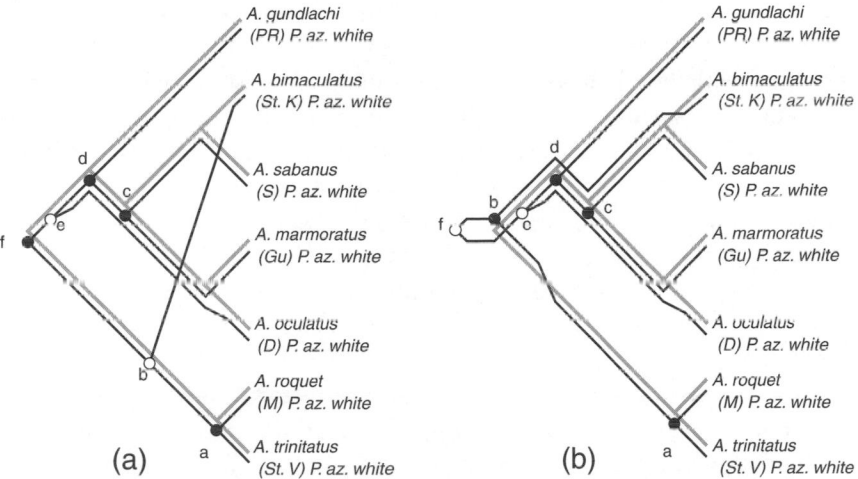

FIGURE 3.17. The optimal solutions for *Anolis* and *Plasmodium azurophilum* white. *(a)* This solution remains optimal over a large range of event costs. *(b)* This solution is optimal when host switches are given a very high cost, effectively prohibiting them altogether. There remain four cospeciation events (eight *lineage* cospeciations), two duplication events (four *lineage* duplications), and nine lineage sorting events.

to estimate best the phylogenetic relationships between their histories. (More detail can be found in Charleston, 1998.) The method takes advantage of the assumptions of previous solutions to this problem, to wit, that parasite lineages are hypothesized to act independently of each other after lineage divergence. This allows separate lineages to be assessed independently and costs assigned without reference to the other lineages: a huge saving in computing time and space, when one considers the number of potential solutions that exist for even moderately sized instances of this problem. The fact that a globally optimal solution can be obtained is clearly appealing: most problems in phylogenetic inference require the use of heuristics, which cannot guarantee ever finding the best solution.

There is, however, an obvious information gap: that of assigning event costs. We have mentioned above that there is often, perhaps generally, a broad range of event costs over which the same solutions tend to remain optimal. In a sense this allows us to estimate these costs: if the solution that might be the most biologically reasonable to us remains optimal over a very small range of event cost values, then this might make us considerably less in favor of a solution than if the range of permitted costs was very large. In the extreme case in which the host and associate trees are identical, we of course come back to the optimal solution being perfect cospeciation, and that solution will remain optimal over *all* event costs, provided we continue with the stipulation that the cospeciation cost c is strictly less than the other costs. But there remains the problem of estimating these costs biologically: the a posteriori assessment of event costs is not as satisfactory as having good grounds for estimates of the costs in advance (but see also chap. 2). Further research must be directed toward not only development of theoretical tools such as jungles and the other methods described in this text, but also biological (parasitological, ecological, genetic) research experimentation in order to find real rates at which the events described here occur.

The theoretical tools must continue to advance. Jungles as they are stated here make use of some assumptions that may not always apply. For instance, genetic parasites such as retrotransposons may not ever be able to exist outside certain locations in the host genome. Such a restriction on the niches available to the parasite markedly changes the nature of the jungle because it effectively removes the independence among parasite lineages. In this case the jungle becomes much simpler, but there are other cases possible in which parasites may actively compete with each other, or provide niches of their own. For instance, most parasites occupy different niches at different stages in their life cycles, and as yet no tools exist to cope with such complexities. Much anecdotal evidence indicates that host switching between more distantly related species is less likely than between more closely related ones, but rigorous experimentation remains rare. This may well be one of the rare cases in which the biological information and the theoretical tools available are at about the same level—and both need to advance from these initial steps. It may be that as we obtain more sophisticated tools for modeling such complex behaviors we lose resolving power when testing among models. It may be that we are overestimating the degree to which phylogenies should match—how much does Fahrenholz's Rule really apply to nature? How likely is host switching relative to lineage sorting?

Since all biological systems interact, there clearly must be phylogenetic relationships within them. There are more parasites than nonparasites; there are more genes than genomes; there are more species than areas, and each interaction among genes, species, parasites, and ecological areas generates a wealth of potential information for study. Now that we are beginning to understand the way things interact, we shall be much better able to deal with the complexities of life on earth.

Software

Jungles are in the process of being implemented in TREEMAP 2 by R. D. M. Page and M. A. Charleston. The interested reader should contact either of those people for a copy of that program.

Acknowledgments

The authors are very grateful for helpful comments and guidance provided by Jos. J. Schall, Roderic D. M. Page, and F. Ronquist. M.A.C. was funded by BBSRC Bioinformatics grant BIF 05332, and S.L.P. was funded by grants from the NSF (USA) and the National Geographic Society to Jos. Schall and a Graduate Training Grant from NSF to S.L.P.

REFERENCES

Brooks, D. R. 1988. Macroevolutionary comparisons of host and parasite phyloge-nies. *Annual Review of Ecology and Systematics* 19:235–59.
Censky, E. J., K. Hodge, and J. Dudley. 1998. Over-water dispersal of lizards due to hurricanes. *Nature* 395:556.
Charleston, M. A. 1998. Jungles: A new solution to the host/parasite phylogeny reconciliation problem. *Mathematical Biosciences* 149:191–223.
Eulenstein, O., B. Mirkin, and M. Vingron. 1998. Duplication-based measures of difference between gene and species trees. *Journal of Computational Biology* 5:135–48.
Goodman, M., J. Czelusniak, G. W. Moore, A. E. Romero-Herrera, and G. Matsuda. 1979. Fitting the gene lineage into its species lineage: A parsimony strategy illus-trated by cladograms constructed from globin sequences. *Systematic Zoology* 28:132–68.
Gorman, G. C., and Y. J. Kim. 1976. Anolis lizards of the eastern Caribbean: A case study in evolution. Part 2, Genetic relationships and genetic variation of the *bimaculatus* group. *Systematic Zoology* 25:62–77.
Hafner, M. S. and S. A. Nadler. 1988. Phylogenetic trees support the coevolution of parasites and their hosts. *Nature* 332:258–59.
Page, R. D. M. 1988. Quantitative cladistic biogeography: Constructing and com-paring area cladograms. *Syst Zool* 37:254–70.
———. 1990. Component analysis: A valiant failure? *Cladistics* 6:119–36.
———. 1993a. Genes, organisms, and areas: The problem of multiple lineages. *Systematic Biology* 42:77–84.
———. 1993b. Parasites, phylogeny and cospeciation. *International Journal for Par-asitology* 23:499–506.
———. 1996. Temporal congruence revisited: Comparison of mitochondrial DNA sequence divergence in cospeciating pocket gophers and their chewing lice. *Systematic Biology* 45:151–67.

Page, R. D. M., and M. A. Charleston. 1997. Reconciled trees and incongruent gene and species trees. In *Mathematical hierarchies in biology,* edited by B. Mirkin, F. R. McMorris, F. S. Roberts, and A. Rzhetsky, 57–70. Providence, R.I.: American Mathematical Society.

———. 1998. Trees within trees: Phylogeny and historical associations. *Trends in Ecology and Evolution* 13:356–59.

Paterson, A. M., and R. D. Gray. 1997. Host-parasite cospeciation, host switching, and missing the boat. In *Host-parasite evolution: General principles and avian models,* edited by D. H. Clayton and J. Moore, 236–50. Oxford: Oxford University Press.

Perkins, S. L. 2000. Species concepts and malaria parasites: Detecting a cryptic species of Plasmodium. *Proceedings of the Royal Society of London,* ser. B, 267:2345–50.

———. 2001. Phylogeography of Caribbean lizard malaria: Tracing the history of vector borne parasites. *Journal of Evolutionary Biology* 14:34–45.

Posada, D., and K. A. Crandall. 1998. MODELTEST: Testing the model of DNA substitution. *Bioinformatics* 14:817–18.

Ronquist, F. 1995. Reconstructing the history of host-parasite associations using generalized parsimony. *Cladistics* 11:73–89.

Ronquist, F., and S. Nylin. 1990. Process and pattern in the evolution of species associations. *Systematic Zoology* 39:323–44.

Roughgarden, J. 1995. *Anolis lizards of the Caribbean: Ecology, evolution, and plate tectonics.* Oxford: Oxford University Press.

Telford Jr., R. S. 1975. Saurian malaria in the Caribbean: *Plasmodium azurophilum* sp. nov., a malarial parasite with schizogony and gametogony in both red and white blood cells. *International Journal for Parasitology* 5:383–94.

4

A STATISTICAL PERSPECTIVE
FOR RECONSTRUCTING THE HISTORY
OF HOST-PARASITE ASSOCIATIONS

John P. Huelsenbeck, Bruce Rannala, and Bret Larget

Introduction

The study of host-parasite associations benefits from two very different, but complementary, approaches. The ecological approach examines the ecology and population genetics of host-parasite associations with the goal of understanding the factors that determine the host range of a parasite, the prevalence and intensity of parasites among hosts, host susceptibility, and the cost of infection (e.g., Cheng, 1964; Esch et al., 1977). Host-parasite assemblages are often thought of as coevolving; an adaptive change in the host to avoid infection by a parasite is countered by adaptation by the parasite. Often, the ecological approach aims to understand the details of coevolution (see Price, 1980). The second approach takes a phylogenetic perspective and looks for similarities in the phylogenies of the hosts and parasites. Such an approach can provide an indication of whether the association between a particular host and parasite is ancient or relatively recent. A phylogenetic perspective can also allow the evolutionary biologist to infer cases of cospeciation (a speciation event in the host causes a speciation in the associated parasite lineage[s] through allopatric speciation) and the importance of processes such as host switching.

Superficially, at least, the phylogenetic approach to the problem of inferring the history of host-parasite association seems simple: take phylogenies for the hosts and parasites and compare them. This approach appears simple because it avoids much of the labor that is involved with collecting ecological and genetic data on hosts and parasites. However, the comparison of host-parasite phylogenies presents a number of difficult biological and statistical problems, not all of which have been solved. For example, what do similarities or differences between the host and parasite phylogenies reveal about the biology and evolutionary history of the association?

Also, is it reasonable to assume that host and parasite phylogenies are known without error, as is common in previous analyses of host-parasite association, and if it is not, how can uncertainty about inferences of host-parasite association be accommodated? Finally, how can different models of host-parasite association be compared?

In this chapter, we discuss some recent developments in the study of host-parasite associations using phylogenies. In particular, we concentrate on statistical approaches that examine these questions: (1) are the histories of hosts and parasites phylogenetically independent? (2) are the histories of hosts and parasites identical? and (3) how can the history of cospeciation, host switching, and lineage sorting be inferred?

Statistical Tests of Host-Parasite Association

Explicit methods for examining host-parasite phylogenies date back at least 20 years (e.g., Brooks, 1979, 1981). Page (1988) was among the first to perform phylogenetic hypothesis tests of host-parasite association, and since the late 1980s several additional hypothesis tests have been formulated. Figure 4.1 outlines the questions that can currently be tested in a phylogenetic framework with the interpretation that can be drawn if the specific hypothesis is rejected or tentatively accepted. Questions that can be addressed include, (1) are host and parasite trees independent? (2) are host and parasite trees identical? and (3) are the speciation times for host and parasite identical?

How Similar Are Phylogenies for Hosts and Parasites?

If processes such as host switching are rare relative to speciation by the hosts and parasites, then the phylogenies of the hosts and parasites should be similar. Figure 4.2 summarizes the results of a simulation that illustrates this point (Huelsenbeck et al., 1997). Here, phylogenies of 13 host species were generated under a birth-death process. Parasites were assumed to evolve on this phylogeny and cospeciate (whenever a host speciates, so too does the associated parasite). The only process that causes the host and parasite trees to differ is host switching by the parasite. Here, we use a very simple model of host switching, first discussed by Huelsenbeck et al. (1997). Parasites colonize new hosts at a constant rate. When a host switch occurs, one of the other host lineages in existence at that time (and which leaves descendants) is colonized, with each host lineage having equal probability of being colonized. The colonizing parasite drives the parasite species previously associated with the host to extinction. The x-axis

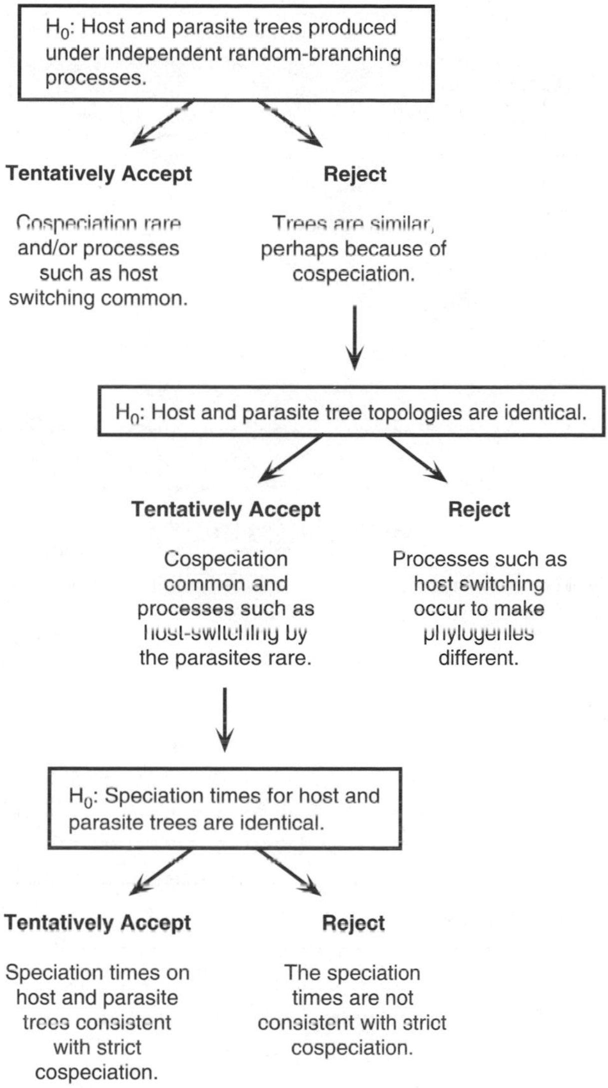

H$_0$: Host and parasite trees produced under independent random-branching processes.

Tentatively Accept

Cospeciation rare and/or processes such as host switching common.

Reject

Trees are similar, perhaps because of cospeciation.

H$_0$: Host and parasite tree topologies are identical.

Tentatively Accept

Cospeciation common and processes such as host-switching by the parasites rare.

Reject

Processes such as host switching occur to make phylogenies different.

H$_0$: Speciation times for host and parasite trees are identical.

Tentatively Accept

Speciation times on host and parasite trees consistent with strict cospeciation.

Reject

The speciation times are not consistent with strict cospeciation.

FIGURE 4.1. Currently, three different null hypotheses of host-parasite association can be tested. Rejecting or tentatively accepting these leads to different biological interpretations.

of figure 4.2 is the host-switching rate (λ) and the y-axis is the topological distance between the host and parasite trees. Here, topological distance is measured by the Robinson and Foulds (1981) metric; a topological distance of $d_T = 0$ means that the host and parasite trees are identical. The Robinson-Foulds metric is the minimum number of branch contractions

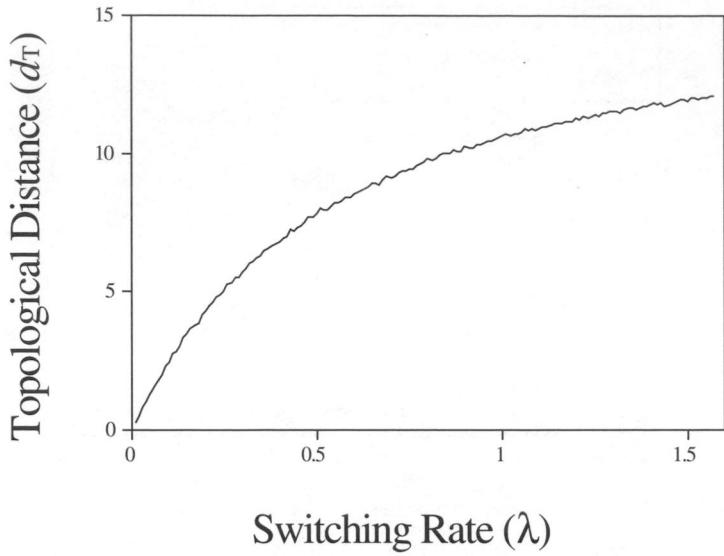

FIGURE 4.2. The topological distance between the host and parasite phylogenies (d_T) increases monotonically with the host-switching rate.

and branch expansions needed to convert one tree into another tree. The maximum topological distance is $(d_T) = 22$ for $s = 13$ species (for rooted trees).

The basic result, summarized in figure 4.2, is that as the host-switching rate increases, the host and parasite trees become less similar. When the host-switching rate (λ) is 0, the host and parasite trees are identical. That is, if the parasite phylogeny were superimposed onto the host phylogeny, the associated species would occupy identical positions. The topological distance between the host and parasite trees monotonically increases with the host-switching rate, λ. When λ becomes very large, the host and parasite trees are effectively independent of each other.

A number of tests have been proposed to examine whether the phylogenies of hosts and parasites appear to be independently derived. The basic steps of these approaches include (1) estimating the phylogenies of the host and parasites, (2) measuring the topological distance between the hosts and parasites, and (3) comparing the observed topological distance with the topological distance expected under a null model of no association between hosts and parasites. One of a large number of measures of the distance between the hosts and parasites, calculated in step 2, can be used. A commonly used distance is the number of components in common between the host and parasite trees (where a component is the sets of taxa formed by removing an internal branch of the phylogeny). The null distribution of the test statistic is calculated under one of several models. For example, one could choose the host and parasite trees from among the sets of all possible labeled rooted trees, all possible unlabeled rooted trees, and all possible labeled histories (fig. 4.3). Of the three alternatives, only the last (choosing host and parasite trees with replacement from the set of all labeled histories) has a clear biological interpretation; under a random branching model of cladogenesis (such as the birth-death process) all labeled histories have equal probability of occurring.

A variation of the component test accommodates several different biological processes that can cause the host and parasite trees to differ; these include a parasite speciation process that is independent of that of the host, host switching, parasite extinction, lineage sorting, or imperfect taxon sampling (collectively referred to as "sorting events" by Page, 1996). The phylogenies of the host and parasites are treated as known without error and the parasite tree is mapped onto the host tree using the TREEMAP procedure (Page, 1994), which maximizes the number of potential cospeciation events. The number of cospeciation events on the original trees is treated as the test statistic and compared with a null distribution that is generated via simulation of many host and parasite trees under a random branching model of cladogenesis (Page, 1996).

As previously noted (Hafner and Nadler, 1990), one weakness of the component test (and related tests) is that they assume that the phylogenies of hosts and parasites are estimated without error. In effect, the phylogenies are treated as the data, not the observed molecular or morphological data. This assumption has not yet been relaxed, and is common to many studies of host-parasite association.

If the null hypothesis of no association is rejected, one biological interpretation is that the parasites have cospeciated with the hosts. Another interpretation is that the hosts and parasites have both been affected by the same biogeographical events (such as vicariance) and that any similarities in phylogeny are simply a result of living in a common geographic area. The hypothesis of cospeciation can be further tested by asking whether the phylogenies of hosts and parasites are identical and by examining the branch lengths of the phylogenetic trees.

The component test was applied to the trees of 13 gopher and louse species shown in figure 4.4. The phylogenies are based on *cytochrome oxidase I* (COI) sequences collected by Hafner et al. (1994). The phylogenies were estimated using maximum likelihood, assuming a model that allows a transition/transversion rate bias and allows the base frequencies to differ (Hasegawa et al., 1984; Hasegawa et al., 1985). Rate variation across sites

A
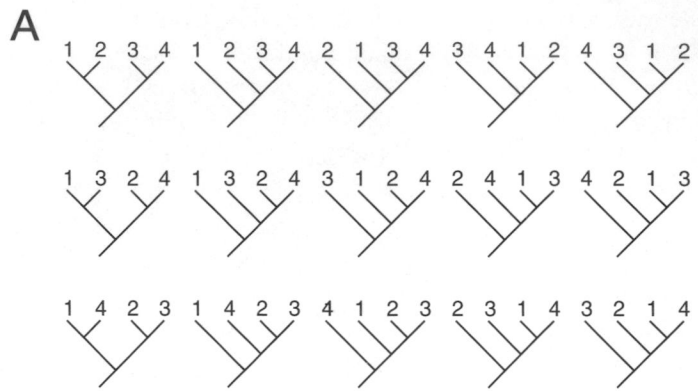

B

C
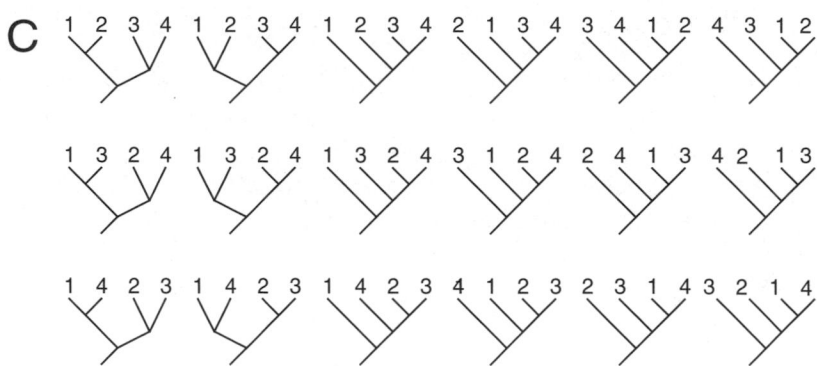

FIGURE 4.3. All possible labeled rooted trees *(A)*, unlabeled rooted trees *(B)*, and labeled histories *(C)* for four species. A labeled history differs from a labeled rooted tree in that the order of the speciation events matters.

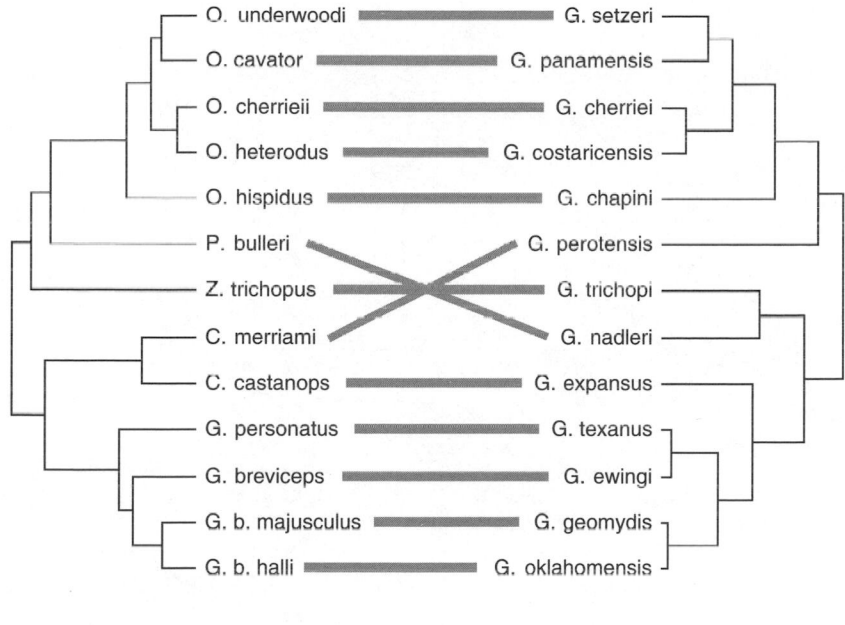

FIGURE 4.4. The maximum likelihood estimates of phylogeny for the gophers (left tree) and lice (right tree). The associations are shown with the gray lines. Phylogenies were estimated under the HKY85Γ model of DNA substitution. Scale bars represent 0.1 substitutions per site.

was accommodated by assuming that the rate at a site is a random variable drawn from a mean–one gamma distribution (Yang, 1993, 1994). The molecular clock hypothesis cannot be rejected for the COI sequences collected from the gophers or lice; the molecular clock hypothesis is used here because it has fewer parameters (is more parsimonious) than models typically used in maximum likelihood analyses (i.e., it allows each branch to have its own rate parameter). The observed topological distance between the gopher and louse trees is $d_T = 10$. Figure 4.5 shows the simulated distribution of d_T when an independent random branching process has produced each phylogeny; the probability of observing $d_T \leq 10$ is less than 0.01. Steel and Penny (1993) provide analytical results for the distribution of the distance between two trees for the Robinson-Foulds metric. The observed topological distance between the gopher and louse trees is much smaller than would be expected under the null hypothesis, leading one to the interpretation that there is some association between speciation events in gophers and lice.

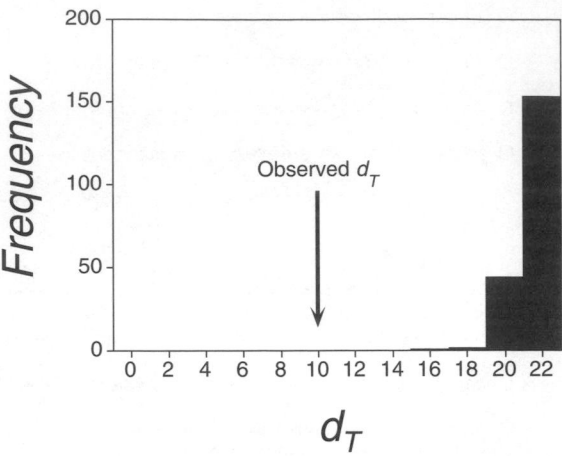

FIGURE 4.5. The observed topological distance between the gopher and louse trees of figure 4.4 (d_T) is much smaller than would be expected if independent random branching processes produced those trees.

Are Host and Parasite Trees Identical?

The previous tests ask whether the phylogenies of hosts and parasites are more similar than would be expected if independent random branching processes produced the host and parasite phylogenies. Rejection of this null hypothesis suggests that some degree of cospeciation has occurred between the host and parasites. A more refined test asks whether the data collected for the hosts and parasites are consistent with the same underlying phylogeny. This is a test of the hypothesis of strict cospeciation between the host and parasite taxa.

Huelsenbeck et al. (1997) proposed two tests to examine the null hypothesis that the host and parasite trees are identical, one based on maximum likelihood and the other on Bayesian inference. The null hypothesis for the likelihood ratio test is that the host and parasite trees are identical. The maximum likelihood under this hypothesis [$\max(\ell_0)$] is calculated under the constraint that the host and parasite trees are identical, but allowing parameters of the substitution model to be different. The alternative hypothesis is that the host and parasite trees are potentially different. The maximum likelihood is also calculated under the alternative hypothesis [$\max(\ell_1)$]. The test statistic is the ratio of the likelihoods

$$\Lambda = \frac{\max(\ell_0)}{\max(\ell_1)}.$$

Because the null hypothesis is a special case of the alternative, $\Lambda \leq 1$; $\Lambda = 1$ when the maximum likelihood trees for the hosts and parasites are identical. The significance of the observed Λ is determined using the parametric bootstrap. Many data sets are simulated under the assumption that the null hypothesis is correct. The parameters of the simulation are the maximum likelihood estimates of phylogeny for host and parasite taxa, as well as the parameters of the substitution model (such as the transition/transversion rate ratio for hosts and parasites and the branch lengths of the trees for hosts and parasites). For each simulated data set, $\max(\ell_0)$ and $\max(\ell_1)$ are calculated and a frequency histogram of the Λ from the simulated data sets constructed. If the observed Λ is greater than 95% of the simulated Λs, the null hypothesis of identical host/parasite trees is rejected. Figure 4.6

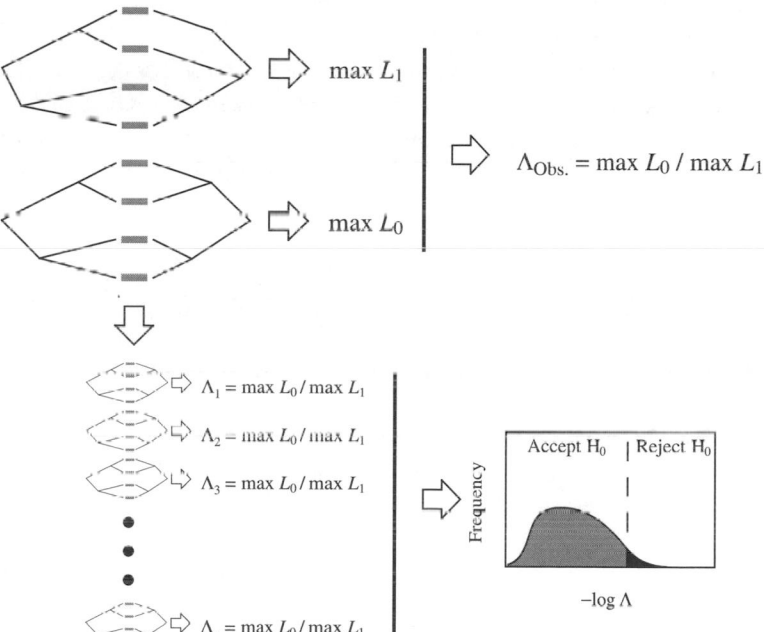

FIGURE 4.6. The likelihood ratio test of the null hypothesis of identical topologies calculates the maximum likelihood under the hypotheses that the trees are identical (L_0) and potentially different (L_1). Many data sets are simulated assuming that the parameters estimated under the null hypothesis are correct and the maximum likelihood under the null and alternative hypotheses calculated for each. If the observed likelihood ratio is larger than 95% of the simulated values, then the null hypothesis is rejected.

illustrates the likelihood ratio test of identical host and parasite topologies. The test does not assume that the host and parasite trees are known without error. However, the method is computationally intensive, involving two full likelihood searches for each of the n simulated data sets.

The Bayesian method directly calculates the probability that the host and parasite trees are identical (Huelsenbeck et al., 1997). Bayesian inference of phylogeny calculates the posterior probability of a phylogeny (τ) given the data (or Pr[τ_i | Sequence Data]) (Rannala and Yang, 1996; Mau and Newton, 1997; Yang and Rannala, 1997; Larget and Simon, 1999; Mau et al., 1999; Newton et al., 1999). Posterior probabilities of trees can be calculated using programs such as BAMBE (Simon and Larget, 1998), PAML (Yang, 1997), and MrBayes (Huelsenbeck and Ronquist, 2001). The probability that the host and parasite trees are identical is simply

$$\sum_{i=1}^{B(s)} \Pr[\tau_i \mid \text{HostData}] \times \Pr[\tau_i \mid \text{ParasiteData}]$$

where $B(s)$ is the number of possible rooted trees for s species. Figure 4.7 illustrates the Bayesian method, which has the advantage of being computationally fast. Moreover, the interpretation of the analysis is straightforward; the result is the probability that the trees are identical.

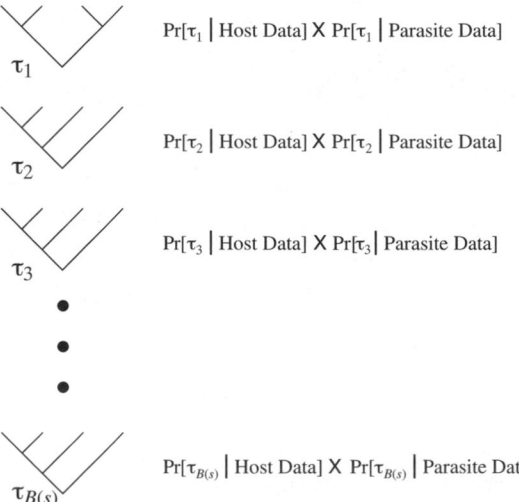

$\Pr[\tau_1 \mid \text{Host Data}] \times \Pr[\tau_1 \mid \text{Parasite Data}]$

$\Pr[\tau_2 \mid \text{Host Data}] \times \Pr[\tau_2 \mid \text{Parasite Data}]$

$\Pr[\tau_3 \mid \text{Host Data}] \times \Pr[\tau_3 \mid \text{Parasite Data}]$

$\Pr[\tau_{B(s)} \mid \text{Host Data}] \times \Pr[\tau_{B(s)} \mid \text{Parasite Data}]$

FIGURE 4.7. The Bayesian method directly calculates the probability that the host and parasite trees are identical. The posterior probability that the host and parasite trees are identical is simply the sum of the joint probabilities of all trees having host and parasite trees that are the same.

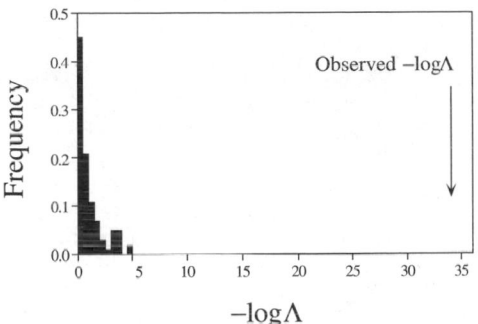

FIGURE 4.8. Application of the likelihood ratio test of the null hypothesis of identical topologies indicates that the gopher and louse trees are not identical. The observed value for the likelihood ratio test statistic is far to the right of all simulated values.

The Bayesian calculation implicitly assumes that the host and parasite trees are independent—that is, the marginal prior probability of observing the host or parasite trees are uniform on the individual tree topologies. This is a potential problem when there are good reasons to expect the host and parasite trees to be interdependent, as is the case when there is a low rate of processes such as host switching. A better solution, implemented below (Huelsenbeck et al., 2000), is to explicitly model the joint prior on the host/parasite trees.

Both the likelihood ratio test and the Bayesian approach have been applied to the gopher and louse data. The likelihood ratio test statistic was $-\Lambda = 34.79$. The null distribution of Λ was obtained using parametric bootstrapping, and is shown in figure 4.8. The observed value falls outside of the 5% tail of the distribution. The observed test statistic is far to the right of all realized values in the parametric bootstrap distribution, indicating overwhelming evidence against the null hypothesis of identical host and parasite trees. The Bayesian approach provides similar results. Figure 4.9 shows the most probable trees for the gophers and lice. The numbers at the interior nodes of the tree represent the posterior probability that the clade is correct. Although the trees have parts in common (e.g., the topology of the top five species), there is virtually no overlap in the topologies of all 13 species for the gopher and louse data. The probability that the two trees are identical is too small to be measured with our computational approach.

Are Speciation Times the Same for Hosts and Parasites?

Phylogenetic trees contain information not only on the relationships of the species, but also on the relative timing of the speciation events on

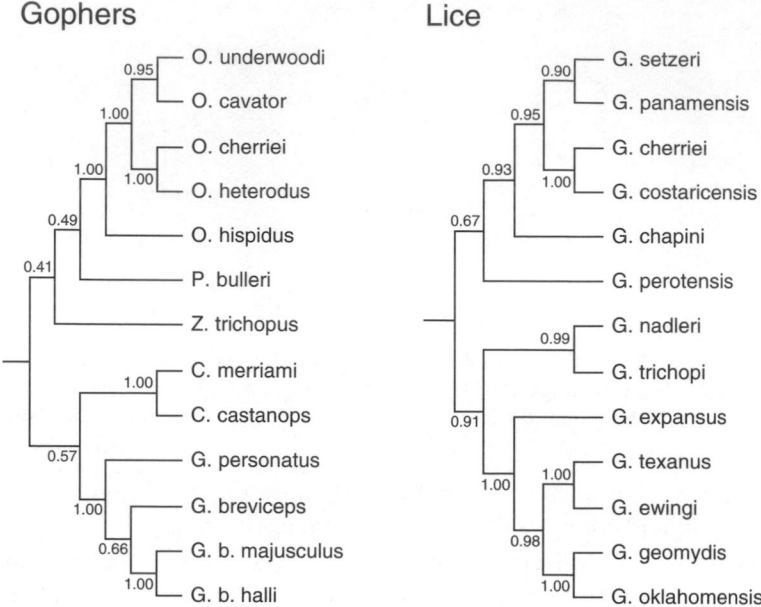

FIGURE 4.9. The best trees for the gophers and lice obtained in a Bayesian analysis under the HKY85+Γ model of DNA substitution. The numbers at the interior branches indicate the posterior probability that the clade is correct.

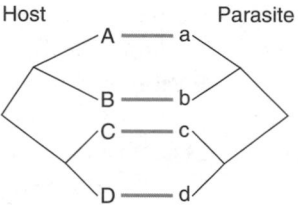

FIGURE 4.10. An example of host and parasite trees consistent with strict cospeciation. Not only are the topologies the same, but the order of the speciation events is consistent with cospeciation.

the tree. Take the example of figure 4.10. Here, the phylogeny of the hosts has the speciation event leading to taxa *A* and *B* occurring before the speciation event leading to taxa *C* and *D*. The same pattern is observed in the phylogeny of the parasites; the speciation event leading to taxa *a* and *b* (which parasitize host species *A* and *B*, respectively) occurred before the speciation event leading to species *c* and *d*. For this hypothetical example, not only are the topologies of the host and parasite trees identical, but the order of the speciation events on the trees is the same. This pattern is suggestive of cospeciation.

Are the branching times on host and parasite phylogenies consistent with a history of strict cospeciation? Several different tests are available to examine this question. The first examines the branch lengths of congruent trees of hosts and parasites (Hafner and Nadler, 1990, Page, 1996). The correlation of speciation times for the host and parasite branch lengths is examined. The significance of the correlation is determined by generating new speciation times under a birth-death process of cladogenesis or by randomizing the observed times (Page, 1996). If the correlation is significant, then the speciation times for the host and parasite phylogenies are considered the same and the biological interpretation is that the hosts and parasites have cospeciated. Figure 4.11 illustrates this test. Here, "speciation depth" refers to the evolutionary distance from the tip of the tree to any internal node of the tree. The speciation times for hosts and parasites for the top panel are consistent with strict cospeciation. The slope of the relationship between host and parasite times is related to the difference in the rates of substitution for hosts and parasites (for the top panel, the hosts evolve more quickly than the parasites). The situation in the bottom panel is not consistent with a history of strict cospeciation between hosts and parasites.

There are several potential problems with the test just described. First of all, it assumes that the topologies and branch lengths of the host and parasite trees are known without error. Again, the topology and branch lengths are essentially treated as the observations, not the DNA sequences. The other problem is that it is not clear that the assumptions of a correlation analysis are met.

Another approach to the problem was formulated by Huelsenbeck et al. (1997), who proposed a likelihood ratio test of the null hypothesis that the speciation times are concurrent. The alternative hypothesis relaxes this assumption, allowing speciation times to be different for hosts and parasites. Because the null hypothesis is a special case of the alternative hypothesis,

$$-2\log \Lambda = -2\log \left[\frac{\max[\ell(H_0)]}{\max[\ell(H_1)]} \right]$$

is asymptotically χ^2 distributed with $s - 2$ degrees of freedom (where s is the number of species in the analysis).

It does not make sense to apply this test if one can reject the hypothesis that the host and parasite topologies are congruent. However, if the host and parasite trees are congruent, then this test examines whether the speciation times are also congruent. If the null hypothesis is tentatively accepted, the biological interpretation is that the hosts and parasites have cospeciated.

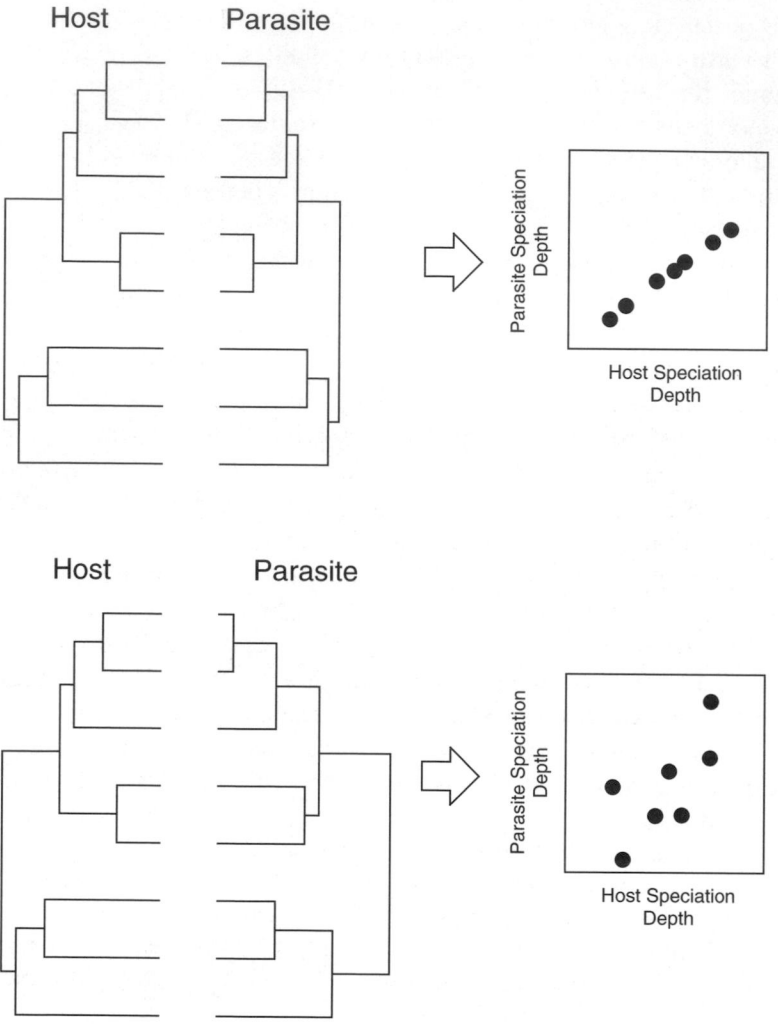

FIGURE 4.11. Some tests of identical speciation times for hosts and parasites examine the correlation in the distance from the tip to a specific speciation event (here called "speciation depth").

Inferring the History of Host-Parasite Association

Although the agreement between host and parasite phylogenies is often remarkable, rarely, if ever, is it complete. Usually, the situation is like that discussed for the gophers and lice. The two trees are more similar than would be expected if independent random branching processes produced the gopher and louse trees, but they are not identical.

Typically, then, both of the null hypotheses discussed above—random agreement and perfect agreement between the host and parasite phylogenies—can be rejected. Hence, although these tests are useful to the evolutionary biologist, they leave him or her in the awkward position of knowing with a high degree of certainty that processes causing the trees to be concordant (such as cospeciation) *and* processes causing the trees to be discordant (such as host switching) have both played a role during the history of the host and parasite association, but still being uncertain about the relative frequency of each kind of process. Several other processes can cause the number of parasite species associated with a host to differ; these include lineage sorting (the segregation of a parasite to only one of the two hosts after a host speciation), parasite speciation, and parasite extinction.

Recent research has focused on reconstructing the history of the association between hosts and parasites. For example, Page (1994) has developed a method, mentioned earlier, called TREEMAP; Charleston (1998) has developed a method called Jungles; and Ronquist (1995, 1997, 1998) has proposed a parsimony-based method. All approach the problem of estimating the history of association between hosts and parasites from a similar angle. First, phylogenies of the host and parasite species are estimated. These phylogenies are then treated as *known without error* and the history of association reconstructed using a maximization (e.g., maximizing the number of cospeciation events: Page, 1994) or minimization (e.g., minimizing the number method of host-switching events) when mapping the parasite tree onto the host tree. Importantly, all three methods effectively treat the phylogenetic trees as observed data rather than unobserved random variables inferred from the data. Moreover, they require user-input weights that specify the relative probabilities of different types of events; these weights are not directly estimated from the data. The jungles and parsimony-based methods are described in chapters by Charleston (chap. 3) and Ronquist (chap. 2).

We have developed an approach for reconstructing the history of host and parasite association that is based upon a stochastic model of host switching (Huelsenbeck et al., 2000). In this section, we describe the method and discuss its strengths and current weaknesses.

A Simple Stochastic Model of Host-Parasite Association

We assume that the host and parasite trees are related by a simple model of host switching. In the absence of host switching by the parasite, the host and parasite trees are identical. Parasites switch hosts at a constant rate, λ. When a host switch occurs, the parasite colonizes a new host, driving the

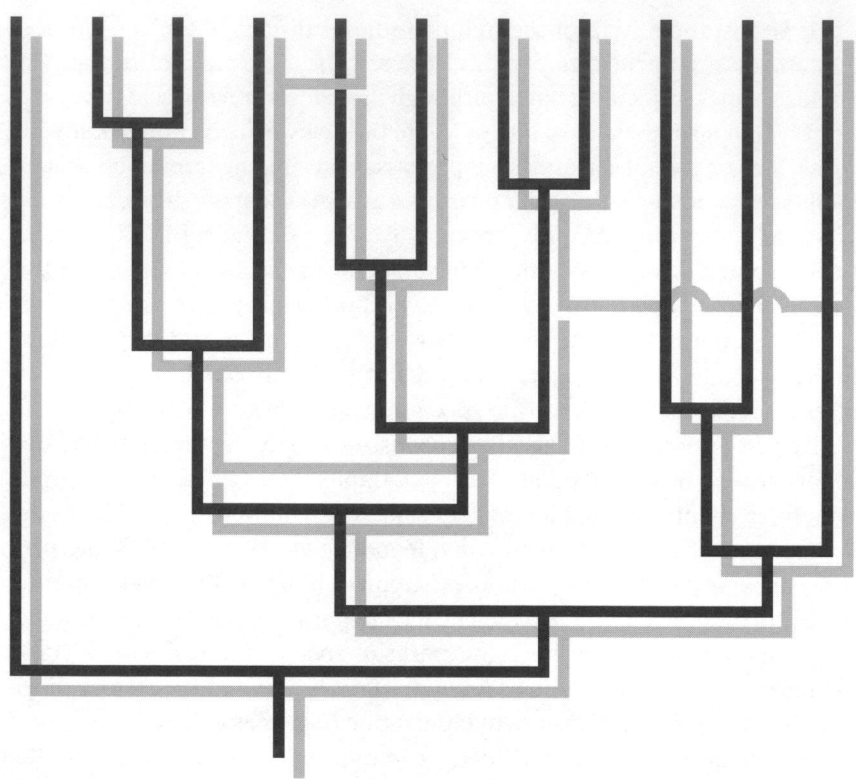

Figure 4.12. An example of the model of host switching assumed by the Bayesian method. The dark tree is the host tree and the lightly shaded tree is the phylogeny of the parasites. There are a total of three host switches on the host phylogeny.

parasite previously associated with that host extinct. Figure 4.12 illustrates the host-switching model; the dark phylogeny represents the host tree and the shaded phylogeny represents the parasite tree. This model of host-parasite association only allows the differences between host and parasite trees to be described by host switches. Moreover, the model predicts a one-to-one relationship between hosts and parasites. Below, we will describe several potential modifications of this model to accommodate processes such as parasite speciation and extinction.

A particular set of host switches is denoted e. The ith event of host switching occurs at a particular time (z_i), and affects a source (γ_i) and target (δ_i) branch (here a source branch is the host lineage on which the host-switching event originates and the target is the host lineage that is colonized). The probability of observing a particular configuration of host switches (e) on the host phylogeny $(\tau_H,\ t)$ is

$$f(e \mid \tau_H, t, \lambda) = \frac{e^{-\lambda T}(\lambda T)^{\xi}}{\xi!} \times \left(\frac{1}{T}\right)^{\xi} \times \prod_{i=1}^{\xi} \frac{1}{b(z_i) - 1}$$

where ξ is the number of host-switching events, T is the sum of the lengths of the branches on the host tree τ_H, and $b(z_i)$ is the number of host lineages in existence at the time of the ith host switch event. The first term in the equation is the Poisson probability of having ξ host switches; the second term is the probability density of the placement of the ξ host switches; and the last term is the probability of the particular host lineages being colonized when a host switch occurs. The probability that there are no host switches $(\xi = 0)$ is $e^{-\lambda T}$.

Stochastic Models of DNA Substitution

Now that we have described a model of host switching that relates the host and parasite trees, we need to model the process of DNA substitution for the DNA sequences sampled from the hosts and parasites. (The aligned DNA sequences are referred to as X for the host sequences and Y for the parasite sequences.) Stochastic models of DNA substitution are widely used in phylogenetic analyses using distances and maximum likelihood. Currently, these models account for many biological processes, such as biases in substitution rates among nucleotide states (Hasegawa et al., 1984, 1985), rate variation across sites (Yang, 1993), and limited correlation among sites (Goldman and Yang, 1994; Muse and Gaut, 1994; Schöniger and von Haeseler, 1994). We use the same types of models implemented by programs such as PAUP* (Swofford, 1999). That is, substitutions are assumed to follow a continuous-time Markov process (see Swofford et al., 1996, for a review of these models). These models allow us to calculate the probability of observing the host DNA sequences

$$f(X \mid \tau_H, t, \theta_H)$$

and the probability of observing the parasite DNA sequences

$$f(Y \mid \tau_H, t, e, \theta_P).$$

The probability of observing the parasite sequences is a function of the host tree and events of host switching, as knowledge of both completely determines the parasite tree. The parameter θ is a vector that holds the parameters of the model of DNA substitution. For example, under the model of DNA substitution first presented by Hasegawa et al. (1985, HKY85), the parameters of the substitution model include the transition/transversion rate ratio (κ) and the stationary base frequencies (π). Huelsenbeck et al. (2000) describe these calculations in more detail.

Estimating the Host-Switching Rate Using Bayesian Inference

The models of host switching and DNA substitution have several unknown parameters. These parameters include the host tree (τ_H) and speciation times *(t)*, host-switching rate (λ), host-parasite history *(e)*, and substitution parameters for the hosts and parasites (θ_H and θ_P). Several of these parameters are of direct interest, such as the host-switching rate and the history of host-parasite association, whereas others are not of direct interest, such as the parameters of the substitution model and the host and parasite phylogenies. Ideally, one would estimate the parameters of interest while accommodating uncertainty in the parameters that are not of direct interest. This can be done in a Bayesian analysis.

Bayesian inference is based upon the posterior probability of a hypothesis. For example, consider the simple case of distinguishing between two hypotheses for the probability of heads for a coin toss. Under the first hypothesis, the coin is fair and the probability of heads equals the probability of tails. Under the second hypothesis the coin has two heads (hence, the probability of tails is zero). Say the coin is tossed twice, resulting in heads both times. The probability that the coin is biased can be obtained using Bayes's formula:

$$\text{Pr[Biased | Two Heads]}$$
$$= \frac{\text{Pr[Two Heads | Biased]} \times \text{Pr[Biased]}}{\text{Pr[Two Heads | Biased]} \times \text{Pr[Biased]} + \text{Pr[Two Heads] | [Fair]} \times \text{Pr[Fair]}}.$$

Here, Pr[Two Heads | Fair] and Pr[Two Heads | Biased] are the likelihood and Pr[Fair] and Pr[Biased] are the prior probabilities of a fair or biased coin, respectively. The likelihood is easy to calculate under a model that has the coin tosses independent:

$$\text{Pr[Two Heads | Biased]} = 1 \times 1 = 1$$

and

$$\text{Pr[Two Heads | Fair]} = \frac{1}{2} \times \frac{1}{2} = \frac{1}{4}.$$

The prior probability of having a fair or biased coin, on the other hand, is more difficult to specify. Several different priors might be tried. For example, one might assume that a coin is equally likely to be fair or biased (e.g., Pr[Fair] = Pr[Biased] = 0.5) or that one is very unlikely to have a biased coin (e.g., Pr[Fair] $\gg$ Pr[Biased]). In this example, we will assume that the coin is drawn at random from a bag with many coins, 99% of which are fair and 1% of which are biased (i.e., Pr[Fair] $= 0.99$ and Pr[Biased] $= 0.01$). The posterior probability that the coin is biased, then, is

$$\text{Pr[Biased | Two Heads]} = \frac{1 \times 0.01}{1 \times 0.01 + 0.25 \times 0.99} = 0.0388.$$

In other words, after observing two heads, the probability that the coin is biased changes from 0.01 (the prior probability that the coin was biased) to 0.0388.

The same reasoning can be used to estimate the parameters of the host-switching model. Our observations are now the aligned DNA sequences from hosts and parasites *(X* and *Y)*. The posterior probability of the host switching rate is

$$f(\lambda \mid X, Y) = \frac{f(X, Y \mid \lambda) f(\lambda)}{f(X, Y)}$$

where $f(X, Y) = \int f(X, Y \mid \lambda) dF(\lambda)$. The likelihood is $f(X, Y \mid \lambda)$ and the prior probability distribution of λ is $f(\lambda)$. The probability of observing the data given a specific value for λ is

$$f(X, Y \mid \lambda) = \int f(X \mid \tau, t, \theta_H) f(Y \mid \tau, t, e, \theta_p) dF(e \mid \tau, t, \lambda) dF(\tau, t) dF(\theta_H) dF(\theta_P)$$

where integration is over all possible host phylogenies and speciation times as well as over all possible numbers of host-switching events. Importantly, the posterior probability of λ is not conditioned on any particular phylogeny, branch lengths, or history of host-parasite association; the posterior probability is instead conditioned on the observed DNA sequences from the hosts and parasites. Individual elements of the above equations can be calculated. For example, the probability of observing the host data given the host tree and branch lengths ($f(X \mid \tau, t, \theta_H)$) is simply the likelihood and is commonly calculated in programs such as PAUP* (Swofford, 1999). However, the integrals and summations cannot be calculated analytically. We used a numerical technique called Markov chain Monte Carlo (MCMC) to approximate the posterior probability of the host-switching rate (Metropolis et al., 1953; Hastings, 1970; Green, 1995).

The results of the analysis of the gopher louse data are summarized in figure 4.13. This figure shows the posterior probability of the host-switching rate (λ) and the number of host switches. There were an average of 9.20 (4, 20) host-switching events on the tree and the rate of host switching was $\lambda = 1.50(0.42, 3.50)$. The interval represents the 95% credibility interval for

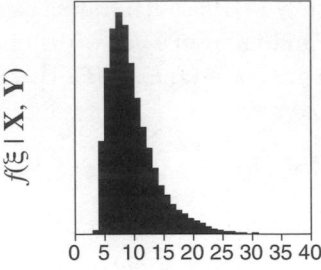

Number of Host Switches

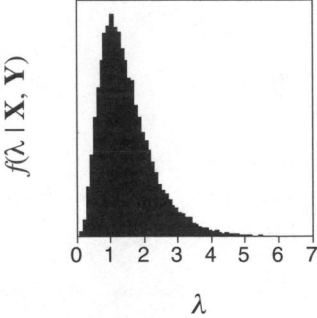

λ

FIGURE 4.13. The posterior probability of the number of host switches and the host-switching rate for the gopher and louse DNA sequences.

the parameters. The biological interpretation of this result is that under the model of host switching, between 4 and 20 host-switching events occurred during the history of the gopher-louse association.

The Bayesian approach also allows one to make inferences about other aspects of the model. For example, the posterior probability of different histories can be calculated using MCMC. Figure 4.14 shows the posterior probability of a host-switching event originating or ending on a lineage. Note that some lineages have a high probability of being associated with a host switch and others have a low probability. The probability that there was a host switch involving lineages 1–5 or lineages 14–16 is particularly small; interestingly, the host and parasite phylogenies are identical for these species. The Bayesian approach also provides information on the specific lineages that are involved in a host-switching event. Figure 4.15 shows the posterior probability of a host switch originating on branch i (the source branch) and colonizing branch j (the target branch) (Huelsenbeck et al.,

2000). The proportion of the square that is shaded represents the posterior probability of a host switch between the specific lineages. Several source-target pairs have a high probability of being involved in a host switch. For example, host-switching events are quite common between branches $7 \leftrightarrow 8$, $10 \leftrightarrow 11$, $12 \leftrightarrow 13$, and $10 \rightarrow 9$.

The Bayesian approach has several advantages over other methods of inferring the history of host and parasite association. First, the inferences are based upon a stochastic model of host switching. This means that probabilities of various host-parasite histories can be calculated and credibility intervals representing the uncertainty of the reconstructed history can be constructed. Second, the inferences do not depend upon any one phylogeny being correct. For example, the estimation of the host-switching rate (λ) was integrated over uncertainty in the phylogenies, branch lengths, parameters of the substitution model, and events of host switching. The inference of the host-switching rate was only conditioned upon the observed DNA sequences. Hence, the Bayesian approach is quite different from approaches such as TREEMAP, jungles, or the parsimony mapping, which all assume that the estimated trees are known without error. Finally, because inferences of host-parasite history are based upon a stochastic model, one can perform model choice using statistical tests such as Bayes factors. This allows the biologist to choose among competing models of host switching to find that model which best explains the data. Huelsenbeck et al. (2000) took this approach to choose between a model that assumes that all lineages have equal probability of being colonized by a new parasite and a model that assumes that a parasite is much more likely to colonize closely related hosts.

Although the Bayesian approach has several advantages, the model of host switching currently used is too simplistic because it only allows host switching by the parasite to cause incongruences between host and parasite trees. A more complete model of host switching would allow (1) parasites to speciate in the absence of a speciation event in the host, (2) parasites to go extinct independently of the host, (3) more than one parasite to persist on a host, and (4) host extinction and incomplete host sampling. At this time, these additional processes have not been incorporated into the Bayesian analysis. However, it should be possible to accommodate them by modifying the model of host switching. Practically speaking, the only aspect of the analysis that would change would be the prior probability of a specific host-parasite history, and the MCMC program would have to integrate over several additional parameters.

Once these additional processes are accommodated in the Bayesian framework, discussed above, several interesting questions can be addressed.

Host-switching event originating from lineage...

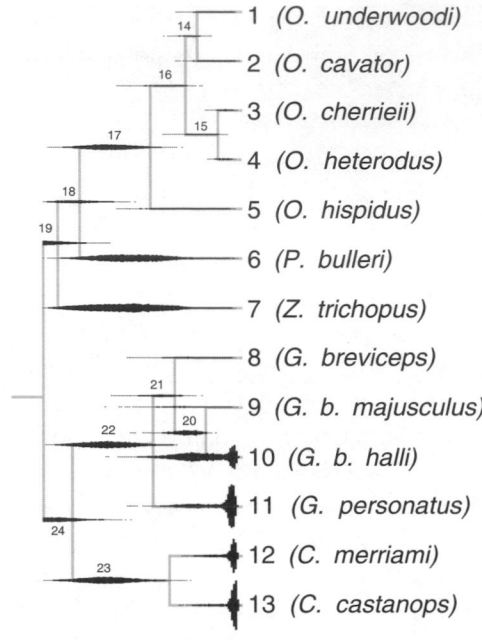

FIGURE 4.14. The width of the branches is proportional to the probability of having a host switch event originate or terminate on the branch. Very few host switches were considered by the Bayesian method on the part of the gopher and louse trees that was concordant (branches 1–5 and 14–16).

First, the relative importance of different processes that make host and parasite trees different can be examined. In the gopher-louse study discussed in this chapter, only one process (host switching) was accommodated. However, other processes, such as parasite speciation, extinction in the absence of speciation, or extinction by the hosts and lineage sorting, may be more frequent than host switches. Second, models of host switching can be compared with the object of finding the model that best explains the data without introducing superfluous parameters. In other words, the importance of different processes can be investigated in a statistical framework. Finally, a more general stochastic model of host switching will help delimit the questions that can be asked from simple comparisons of phylogenies. For example, it may be that one can explain DNA sequences from hosts and parasites equally well with a high rate of process A and a low rate of process B or a low rate of process A and a high rate of process B. If this is the case, then there may be little hope of estimating the rate of host switching separately

...and colonizing lineage...

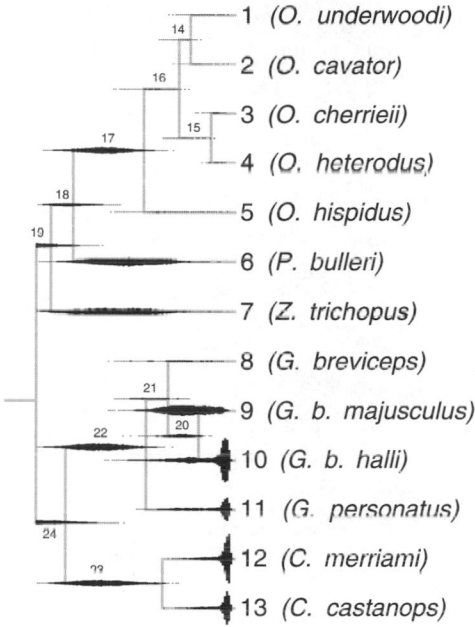

FIGURE 4.14 (continued).

from the rates of other processes that make host and parasite trees differ. To some extent, such a result would be rather depressing, as it would indicate that there is not as much information in host and parasite trees about the various processes that can make the trees different. However, the result can also be looked at in a positive light: if it is true that there is little information in DNA sequences sampled from host and parasites about the history of association, then a better approach would be to measure these rates ecologically and then fix the relative rates in the Bayesian analysis (by specifying informative priors for these parameters). Importantly, the Bayesian approach discussed in this chapter allows the possibility of discovering these phenomena in the first place; other methods for inferring the history of host and parasite association have little hope of informing us about the limits of our inferences.

Summary

Inferring the history of host and parasite association is a difficult statistical problem. In this chapter we reviewed some of the current tests that examine the null hypotheses that the host and parasite trees have independent

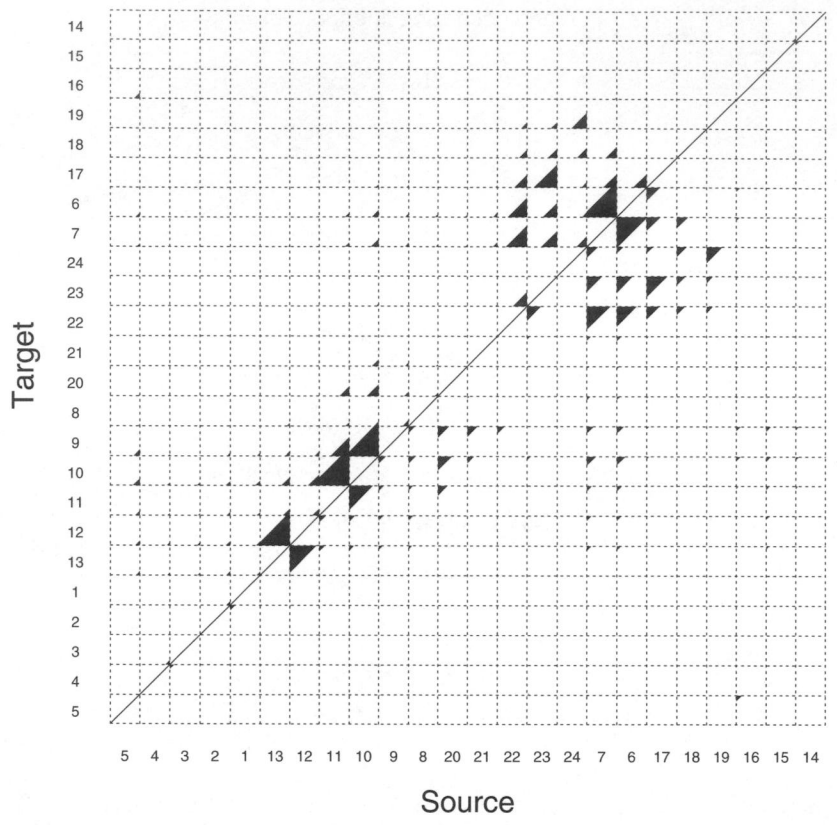

FIGURE 4.15. The posterior probability of having a host switch that originates on branch *i* (source) and colonizes on branch *j* (target).

histories, have identical topologies, and have identical speciation times. One important limitation of many of these tests is that they assume that the host and parasite phylogenies are known without error. However, even a simple inspection of bootstrap values for different nodes on a phylogenetic tree reveals that most phylogenetic trees have some degree of uncertainty; in many cases the uncertainty about the branching order of a tree can be quite large.

The Bayesian analysis of host-parasite cospeciation accommodates uncertainty in the phylogenies of hosts and parasites as well as in other parameters. Moreover, because a stochastic model of host-parasite association is assumed, the parameters of the model can be estimated using standard statistical methods, the uncertainty in the estimates can be evaluated, and different models of host-parasite association can be compared with the object of finding the model that best fits the observations (in this case, DNA sequences from homologous genes sampled from hosts and parasites). The main weakness of the Bayesian approach is not with the method for estimating the parameters of the host-switching model but rather that the model is too simplistic to describe many of the features of host-parasite associations that are known to be important. Future work should accommodate these processes.

References

Brooks, D. R. 1979. Testing the context and extent of host-parasite coevolution. *Systematic Zoology* 28:299–307.

———. 1981. Hennig's parasitological method: A proposed solution. *Systematic Zoology* 30:229–49.

Charleston, M. A. 1998. Jungles: A new solution to the host/parasite phylogeny reconciliation problem. *Mathematical Biosciences* 149:191–223.

Cheng, T. C. 1964. *The biology of animal parasites.* Philadelphia: W. B. Saunders Co.

Esch, G. W., T. C. Hazen, and J. M. Aho. 1977. Parasitism and r- and K-selection. In *Regulation of parasite populations,* edited by G. W. Esch, 9–62. New York: Academic Press.

Goldman, N., and Z. Yang. 1994. A codon-based model of nucleotide substitution for protein-coding DNA sequences. *Molecular Biology and Evolution* 11:725–36.

Green, P. J. 1995. Reversible jump Markov chain Monte Carlo computation and Bayesian model determination. *Biometrika* 82:711–32.

Hafner, M. S., and S. A. Nadler. 1990. Cospeciation in host-parasite assemblages: Comparative analysis of rates of evolution and timing of cospeciation. *Systematic Zoology* 39:192–204.

Hafner, M. S., P. D. Sudman, F. X. Villablanca, T. A. Spradling, J. W. Demastes, and S. A. Nadler. 1994. Disparate rates of molecular evolution in cospeciating hosts and parasites. *Science* 265:1087–90.

Hasegawa, M., H. Kishino, and T. Yano. 1985. Dating of the human-ape splitting by a molecular clock of mitochondrial DNA. *Journal of Molecular Evolution* 22:160–74.

Hasegawa, M., T. Yano, and H. Kishino. 1984. A new molecular clock of mitochondrial DNA and the evolution of Hominoids. *Proceedings of the Japan Academy,* ser. B, 60:95–98.

Hastings, W. K. 1970. Monte Carlo sampling methods using Markov chains and their applications. *Biometrika* 57:97–109.

Huelsenbeck, J. P., B. Rannala, and B. Larget. 2000. A Bayesian framework for the analysis of cospeciation. *Evolution* 54:352–64.

Huelsenbeck, J. P., B. Rannala, and Z. Yang. 1997. Statistical tests of host-parasite cospeciation. *Evolution* 51:410–19.

Huelsenbeck, J. P., and F. Ronquist. 2001. MRBAYES: Bayesian inference of phylogeny. *Bioinformatics* 17:754–55.

Larget, B., and D. Simon. 1999. Markov chain Monte Carlo algorithms for the Bayesian analysis of phylogenetic trees. *Molecular Biology and Evolution* 16:750–59.

Mau, B., and M. A. Newton. 1997. Phylogenetic inference for binary data on dendograms using Markov chain Monte Carlo. *Journal of Computational Graphics and Statistics* 6:122–31.

Mau, B., M. A. Newton, and B. Larget. 1999. Bayesian phylogenetic inference via Markov Chain Monte Carlo methods. *Biometrics* 55:1–12.

Metropolis, N. A., W. Rosenbluth, M. N. Rosenbluth, A. H. Teller, and E. Teller. 1953. Equations of state calculations by fast computing machines. *Journal Chemical Physics* 21:1087–91.

Muse, S., and B. Gaut. 1994. A likelihood approach for comparing synonymous and non-synonymous substitution rates, with application to the chloroplast genome. *Molecular Biology and Evolution* 11:715–24.

Newton, M. A., B. Mau, and B. Larget. 1999. Markov chain Monte Carlo for the Bayesian analysis of evolutionary trees from aligned molecular sequences. In *Statistics in molecular biology and genetics,* edited by F. Seillier-Moseiwitch, 143–62. IMS Lecture Monograph Series, 33.

Page, R. D. M. 1988. Quantitative cladistic biogeography: Constructing and comparing area cladograms. *Systematic Zoology* 37:254–70.

———. 1994. Parallel phylogenies: Reconstructing the history of host-parasite assemblages. *Cladistics* 10:155–73.

———. 1996. Temporal congruence revisited: Comparison of mitochondrial DNA sequence divergence in cospeciating pocket gophers and their chewing lice. *Systematic Biology* 45:151–67.

Price, P. W. 1980. *Evolutionary biology of parasites.* Princeton, N.J.: Princeton University Press.

Rannala, B., and Z. Yang. 1996. Probability distribution of molecular evolutionary trees: A new method of phylogenetic inference. *Journal of Molecular Evolution* 43:304–11.

Robinson, D. F., and L. R. Foulds. 1981. Comparison of phylogenetic trees. *Mathematical Biosciences* 53:131–47.

Ronquist, F. 1995. Reconstructing the history of host-parasite associations using generalized parsimony. *Cladistics* 11:73–89.

———. 1997. Dispersal-vicariance analysis: A new approach to the quantification of historical biogeography. *Systematic Biology* 46:195–203.

———. 1998. Phylogenetic approaches in coevolution and biogeography. *Zoologica Scripta* 26:313–22.

Schöniger, M., and A. von Haeseler. 1994. A stochastic model for the evolution of autocorrelated DNA sequences. *Molecular Phylogenetics and Evolution* 3:240–47.

Simon, D., and B. Larget. 1998. Bayesian analysis in molecular biology and evolution (BAMBE), version 1.01 beta. Department of Mathematics and Computer Science, Duquesne University.

Steel, M. A., and D. Penny. 1993. Distributions of tree comparison metrics—Some new results. *Systematic Biology* 42:126–41.

Swofford, D. L. 1999. *PAUP*. Phylogenetic Analysis Using Parsimony (*and other methods)*. Sunderland, Mass.: Sinauer Associates.

Swofford, D. L., G. J. Olsen, P. J. Waddell, and D. M. Hillis. 1996. Phylogenetic inference. In *Molecular Systematics,* edited by D. M. Hillis, C. Moritz, and B. K. Mable, 407–514. Sunderland, Mass.: Sinauer Associates.

Yang, Z. 1993. Maximum likelihood estimation of phylogeny from DNA sequences when substitution rates differ over sites. *Molecular Biology and Evolution* 10:1396–401.

———. 1994. Maximum likelihood phylogenetic estimation from DNA sequences with variable rates over sites: Approximate methods. *Journal of Molecular Evolution* 39:306–14.

———. 1997. PAML: A program package for phylogenetic analysis by maximum likelihood. *Computer Applications in the Biosciences* 15:555–56.

Yang, Z., and B. Rannala. 1997. Bayesian phylogenetic inference using DNA sequences: A Markov chain Monte Carlo method. *Molecular Biology and Evolution* 14:717–24.

5

POPULATION GENETICS AND
COSPECIATION: FROM PROCESS TO PATTERN

Bruce Rannala and Yannis Michalakis

Introduction

The observation that the phylogeny of a monophyletic group of parasites may mirror the phylogeny of their hosts dates at least to morphological studies of the evolutionary history of host and parasite species conducted in the early 1900s (see review by Klassen, 1992). Numerical algorithms were proposed in the 1970s to assess the level of agreement between host and parasite phylogenies constructed using morphological characters (reviewed in Brooks, 1985). One problem with morphology-based approaches is that a realistic model of morphological evolution is not available, so the statistical significance of the associations cannot be easily assessed. Formal statistical methods were developed in the 1980s for inferring phylogenies using nucleotide sequence data and for comparing gene trees of hosts and parasites. We use the term *gene tree* to highlight the fact that such phylogenies convey the history of a particular region of DNA sequence and not necessarily that of the population or species. The development of these methods has allowed standard statistical techniques such as likelihood ratio tests to be employed in testing the significance of similarities between host and parasite gene trees (Hafner and Page, 1995; Huelsenbeck et al., 1997).

Concordant phylogenetic trees for hosts and parasites have often been interpreted as evidence for causally linked (and possibly synchronous) speciation events in the two groups. Identical inferred host and parasite gene trees, it has been suggested, confirms a hypothesis of host-parasite "cospeciation" (e.g., Hafner et al., 1994). Cospeciation viewed from this perspective is a pattern, rather than a mechanism, of speciation, and identical host and parasite phylogenies can result from one of several evolutionary mechanisms. These involve different combinations of sympatric and allopatric modes of speciation in hosts and parasites. A goal of this chapter is to develop theory relating mechanisms of speciation (and population demographic structure) to expected patterns of cospeciation. Concordant host and parasite gene trees can result when gene flow and migration are simultaneously eliminated between populations of parasites and hosts. Migration is mentioned in addition to gene flow because host switching by parasites (the colonization of a new host species by a parasite species) can be viewed as migration without gene flow (assuming that the newly colonizing parasite does not exchange genetic material with any preexisting parasites). Such events of synchronous genetic and physical isolation of hosts and associated parasites can occur at several levels: among individual hosts (for parasites with vertical transmission); among populations (restricted gene flow between newly founded populations leading to similarities between the gene tree relating populations of hosts and that relating populations of parasites); or among species (speciation events simultaneously eliminating gene flow between incipient host and parasite species and producing identical species phylogenies).

Because the observation of *cospeciation* (concordant gene trees between hosts and parasites) can occur at several levels, we explore the probability that identical host and parasite gene trees are observed for hosts and parasites sampled at these different levels using explicit population genetic models. In situations where branch lengths are of interest, we will also examine the expected length differences between corresponding branches in host and parasite gene trees when their topologies are identical. Our approach will differ from that of most systematists studying host-parasite cospeciation in that we consider a hierarchy of levels, both above and below the species, at which such associations might occur. At the lowest level, one can potentially use rapidly evolving parasites, such as viruses or bacteria, to reconstruct the genealogical relationship among infected hosts (for vertically transmitted parasites) or the pattern of transmission among hosts (for horizontally transmitted parasites). At a higher level, one can consider the correspondence between gene trees of populations of parasites and their hosts (and potentially reconstruct population histories), while at the highest level one can consider the correspondence between gene trees of different species of parasites and hosts (or some higher taxonomic level).

This chapter provides a framework for interpreting the results of empirical studies of host and parasite gene trees, allowing one to predict when a study of phylogenetic association among hosts and parasites can potentially confirm, or refute, models of population structure, of parasite transmission, or of the speciation process. It will be shown that the probability of identical host and parasite gene trees is negligibly small for certain kinds of comparisons, suggesting a priori that pursuing such studies may not be worthwhile. We focus exclusively on comparisons of gene trees for hosts and parasites derived from sequence data rather than phylogenetic trees reconstructed from other kinds of data (e.g., morphological or ecological characters) because gene trees are amenable to mathematical modeling using conventional population genetics theory. We assume that gene trees can be accurately inferred for both hosts and parasites. We do not consider the related statistical question of whether host and parasite gene trees are identical in particular cases; taking account, for example, of the sampling error of inferred phylogenies (Huelsenbeck et al., 1997), or whether the trees are more similar than would be expected by chance if all trees were equally likely a priori (Hafner and Page, 1995). Such questions have already received much study. Moreover, we consider only the effects of genetic drift, assuming that the gene trees are inferred using neutral regions of DNA (e.g., introns, pseudogenes, etc.).

Causes of Phylogenetic Associations

A generic instance of host-parasite cospeciation is illustrated in figure 5.1. In this figure, and in our subsequent analysis, it has been assumed that

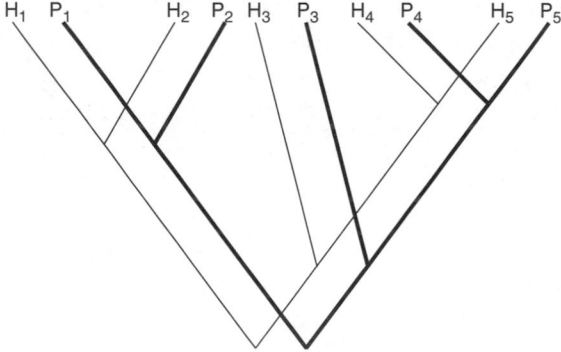

H_1 P_1 H_2 P_2 H_3 P_3 H_4 P_4 H_5 P_5

FIGURE 5.1. An example of host and parasite gene trees that fit a pattern of cospeciation. The symbol H_i denotes the sequence sampled from host i (which may be from the ith population or the ith species, depending on the level of analysis), and P_i denotes the sequence sampled from parasite i. The parasite sequence P_i may be sampled from host individual H_i (for a within-population study), from host population i (for a between-population study), or from host species i (for a between-species study).

a single sequence is examined from each parasite population and from each individual host. In general, this need not be the case; for example, several parasites could be collected from a single host and sequenced for a given locus, multiple parasites could be sequenced from different hosts (of the same species), or multiple individuals of the same host species could be sequenced. Modern techniques of PCR-based DNA amplification and automated sequencing make multisequence comparisons easily achievable, and such data sets may soon become commonplace. It will greatly simplify the analysis, however, if we restrict our study to gene trees having a one-to-one association of host and parasite sequences as shown in figure 5.1. We therefore assume that each host sequence is associated with a single parasite sequence, and if this sequence is sampled from a larger population (a parasite sampled from a population of parasites on a host, for example), it is assumed to have been sampled at random (i.e., all extant sequences are equally likely to be chosen). An important question for future study is how multiple sequences from parasites and/or hosts might be analyzed simultaneously and what kinds of patterns would then be expected.

Theoretical Framework

Assume that a single parasite, denoted P_i, is sampled from host individual H_i, where $1 \leq i \leq s$ and s parasites are sampled in total, each from one of s hosts. Host H_i may be either the ith individual sampled from within a closed population (within-population association studies), a randomly sampled individual from the ith population (between-population association studies), or a randomly sampled individual from the ith species (between-species association studies). In our analysis, we will typically assume that the locus examined for either hosts or parasites is evolving solely under the forces of mutation and neutral genetic drift. For simplicity, it is assumed that the host effective population size is constant (although this assumption may be relaxed), and we will denote this by $N_e^{(h)}$. In the case of between-population association studies, this will represent the effective size of each host population, while for between-species comparisons this will be the effective population size for each host species. The effective population size of the parasite population within each host will be denoted $N_e^{(p)}$ and may vary over time due to population bottlenecks induced by the process of transmission of parasites from one host to another. It is assumed that hosts are diploid and host gene trees are constructed using nuclear genes (gene trees for haploid genomes can be used by setting $2N_e^{(h)} = N_e^{(h)}$ in the formulae). Parasites are assumed to be haploid (a nuclear gene tree for a diploid parasite species could instead be used by setting $N_e^{(p)} = 2N_e^{(p)}$ in the formulae). A discrete generation Fisher-Wright model is used to model the population structure of both hosts and parasites, and we use the usual continuous-time approximation. Time is measured in units of generations; it is assumed that hosts and parasites have identical generation times. This assumption is not needed and is adopted purely to simplify the presentation of the theory.

Within-Population Associations

The first situation that we consider is a random sample of infected hosts from a panmictic population with no immigration. A total of s infected hosts are sampled and a single parasite is sampled from each host. The first question we address is whether the gene tree of a neutral locus sequenced for the parasites may contain information about either the host genealogy or the temporal-spatial pattern of host infections. Throughout, we will assume that the mutation rates at the locus in parasites are high enough that the gene tree can be accurately inferred. This is typically the case for antigenic proteins of RNA viruses, for example. The first factor to consider is the mode of transmission of parasites; this determines the potential information contained in the parasite gene tree. We will focus on two cases: (1) parasites with vertical transmission; and (2) parasites with horizontal transmission. The life cycle of many parasites includes a mixture of horizontal and vertical transmission. Such complications can be dealt with by considering a model with migration of parasites among hosts; a discussion of this model is deferred to the section on among-population gene trees. The gene genealogy of a parasite with vertical transmission carries potential information about the genealogical relationships of infected hosts. That of a parasite with horizontal transmission carries potential epidemiological information about the patterns of parasite transmission among hosts.

The pattern of transmission of a parasite (with strict vertical transmission) among a sample of 5 hosts from a large outbreeding population is illustrated in figure 5.2. In this case, the parasite gene tree conveys potential information about the genealogical relationships of infected host individuals (i.e., their path of descent through a pedigree). If we knew the times in the past (indicated as v_1, v_4, $v_1 + v_6$, $v_5 + v_8$ in the figure) at which the 5 hosts first share exactly 4, 3, 2, and 1 ancestors that transmitted the parasite (as well as which hosts shared infected ancestors), then we could compare this "ancestral infection graph" with a gene tree of sequences from parasites sampled from these hosts. The ancestral infection graph (AIG) is normally unknown, and an alternative approach is then to infer a nuclear gene tree for the infected hosts (assuming that the host mutation

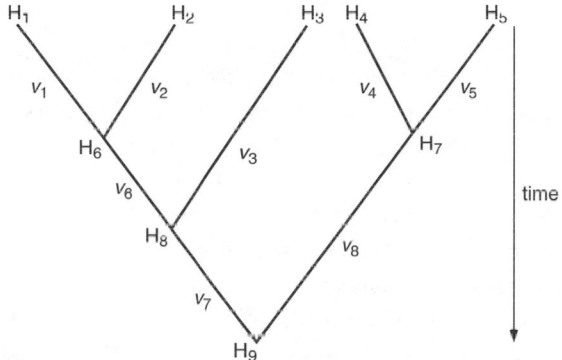

FIGURE 5.2. Tree representing the times at which hosts (and parasites) diverged from one another. The interpretation of the tree depends on the level of analysis. If the parasites are sampled from individual hosts within a population, this will be the ancestral infection graph; if they are sampled from different host populations, it will be the tree of population subdivision events; and if they are sampled from different host species, it will be the species tree. The branch lengths v_i are proportional to time (in units of generations) and the ancestral nodes (H_6, H_7, H_8, H_9) represent either infected ancestors, ancestral populations, or ancestral species, depending on whether a within-population, between-population, or between-species study is carried out.

rate is high enough for this to be possible) and to compare the gene trees of infected hosts and their parasites for agreement.

The host and parasite gene trees can be expected to be similar (ignoring spurious similarities) only in the case that the sampled host sequences are inherited from a single most recent infected common ancestor. In that case, we will say that they are identical by descent (IBD: Malécot, 1948). This requirement is necessary to generate similar gene trees because a host always transmits its infection to an offspring but transmits a particular gene only half the time. In figure 5.2 the most recent infected common ancestor is labeled H_9. The probability that a randomly chosen pair of sequences at a nuclear locus, each from one of a pair of infected hosts, are IBD with a particular sequence found in their most recent infected common ancestor (existing t generations ago) is

$$1/2^{(2t+1)}.$$

The probability that a sample of s sequences from infected hosts are IBD through a most recent common infected ancestor in the AIG therefore becomes negligibly small even if the infected common ancestor existed as recently as 10 generations in the past. This suggests that it is pointless to compare nuclear gene trees of infected hosts with parasite gene trees to identify similarities due to their shared AIG. If sequences of genes from haploid genomes are examined for the hosts instead, and the mode of transmission of the haploid genome is identical to the mode of transmission of the parasite, then the AIG and the host gene tree will be identical. An example would be the comparison of a gene tree from a parasite that is always transmitted in an infected oocyte from a mother to her offspring with a host gene tree for a sequence from a maternally transmitted organelle (e.g., a mitochondrial sequence). Even in this ideal case, however, there still remains at least one additional process that may cause differences between host and parasite gene trees, so-called lineage sorting due to the within-host coalescent process for parasites (see Avise, 1994, for a nontechnical description of this process).

Lineage sorting occurs because of a phenomenon known as the population coalescent process (Kingman, 1982). The basic idea is that multiple genetic lineages can persist in an ancestral population and can have genealogical relationships that differ from the AIG (or equivalently the population or species histories). Reviews of the coalescent theory of population genetic structure can be found in Hudson (1990) and Griffiths and Tavaré (1994). The coalescent theory describes the probability distribution of gene genealogies underlying a sample of neutral sequences. It was developed for use in analyzing sequences sampled at random from a large population; in situations where we apply the theory to describe host (or parasite) gene genealogies we will implicitly assume that these conditions apply. A coalescent process operates within populations of parasites from individual hosts. This concept is illustrated in appendix 5.1.

Let us now suppose that either (1) the host AIG is known (based on direct observations); (2) the host gene tree underlying sequences of a gene from a uniparentally transmitted organelle is compared with the gene tree of a uniparentally transmitted parasite; or (3) we are interested in using a parasite gene tree to predict the AIG of the hosts. Should the parasite gene tree then be identical to the AIG of the hosts? The answer is a qualified yes, but only in the special case that a single parasite lineage is passed from each parent to its infected offspring. In that case, nodes in the parasite gene tree are identical to points of host transmission. If instead multiple genetic lineages of the parasite are passed from an infected host to its offspring, lineage sorting can lead to gene trees that differ from the AIG.

If the AIG is known, we can calculate the probability that it is identical to the parasite gene tree by mathematically taking account of the within-host coalescent process. The calculation is very similar to that used in predicting the probability that a gene tree is identical to a species tree

(Tajima, 1983; Pamilo and Nei, 1988). The parasite gene tree is identical to the AIG if either the gene sequences of parasites sampled from each host always coalesce in a first shared ancestral host lineage (i.e., there are never more than two ancestral parasite lineages in an ancestral host), or the lineages that coalesce in an ancestral host having 3 or more ancestral parasite lineages happen to be identical to the AIG of the hosts. Since all possible coalescences of parasite lineages are equally likely in an ancestral host having more than two ancestral parasite lineages, the probability that a parasite gene tree is identical to the AIG in such cases is generally small (see appendix 5.1). We can then obtain a good approximation for the probability that the parasite gene tree has a branching structure that is identical to the AIG by calculating the probability that all pairs of parasite lineages coalesce in their first shared ancestral host.

If N_e is the effective population size of the parasites infecting each host, then the probability that sequences from a pair of randomly chosen parasites (inhabiting a common host) coalesce to a single ancestral sequence before generation t in the past is

$$\int_0^t \frac{1}{N_e} e^{-y/N_e} dy = 1 - e^{-t/N_e},$$

and the probability that a parasite gene tree, denoted as τ_p, for a neutral locus is identical to the AIG, denoted as T, is approximately

$$\Pr(\tau_p = T) = \prod_{j=s+1}^{2s-2} \left(1 - e^{-v_j/N_e}\right), \tag{5.1}$$

where the v_j ($j \geq s + 1$) are the lengths of the internal branches in the AIG. For the AIG depicted in figure 5.2, if we set $v_6 = 0.1$, $v_7 = 0.1$, $v_8 = 0.1$, and $v_9 = 0.1$ (in units of N_e generations), the probability that the parasite gene genealogy is identical to the AIG is only 8.2×10^{-5}. If all the internal branch lengths are increased by a factor of 100 (so that $v_6 = 10.0$, etc.), then this probability instead becomes 0.999. In general, if N_e is very small, or sampled infected hosts are distantly related, the parasite gene genealogy should provide a good estimate of the AIG. A useful rule of thumb is that the AIG should be accurately reflected by the parasite gene tree if the number of generations (or transmission events) separating the infected hosts is more than 10 times the parasite effective population size. For closely related infected hosts, or a large parasite population in each infected host, the correspondence between the parasite gene tree and the AIG will be poor.

The simple formulation presented above assumes that no population bottleneck occurs during the parasite transmission process and all hosts are infected by the same (effective) number of parasites. The analysis quickly becomes intractable if biologically realistic models of parasite transmission and within-host reproduction are considered. The most promising approach for making probability calculations for particular species will undoubtedly be numerical analysis and Monte Carlo simulation. It is worthwhile, however, to consider a slightly more realistic model of the process of parasite transmission and growth that, although simple, still roughly describes the situation for many parasite species. The model assumes that N_0 parasites are transmitted by each host (whether to an offspring of an infected parent, in the case of vertical transmission, or to a newly-infected host, in the case of horizontal transmission), and the parasite then replicates for T generations in the newly infected host until it has reached a population of size N_T. The parasite can then be transmitted to other (uninfected) hosts. The expected growth rate, r, under this model, given T, N_T, and N_0, is

$$r = \frac{1}{T} \log \left(\frac{N_T}{N_0} \right).$$

We now consider the coalescent process for a pair of parasite sequences sampled from different hosts at the point, in the past, at which their ancestral lineages first infect a common ancestral host (i.e., immediately prior to the transmission event infecting the pair of offspring that leave infected descendants ancestral to the pair of sampled sequences). The probability that the pair of ancestral lineages coalesce in this first common ancestral host (see Slatkin and Hudson, 1991) is

$$\phi(N_0, T, r) = \int_0^T \frac{1}{N_0 e^{-ry}} \exp \left\{ -\int_0^y \frac{1}{N_0 e^{-rz}} dz \right\} dy$$

$$\phi(N_0, T, r) = \int_0^T \frac{1}{N_0 e^{-ry}} \exp \left\{ -\int_0^y \frac{1}{N_0 e^{-rz}} dz \right\} dy,$$

$$= 1 - \exp \left\{ \frac{1 - e^{rT}}{N_0 r} \right\}, \tag{5.2}$$

where $N_0 \geq 1$. The probability that the parasite lineages coalesce in a shared ancestral host exactly t generations (for a vertically transmitted parasite), or equivalently t host transmission events (for a horizontally transmitted parasite), after they first share an ancestral host is

$$(1 - \phi(N_T, T, r))^{t-1}\phi(N_T, T, r),\tag{5.3}$$

and the average time until they coalesce is

$$E(t) = \left(1 - \exp\left\{\frac{1 - e^{rT}}{N_0 r}\right\}\right)^{-1}.$$

If the sequences do coalesce, then both descend from a single lineage infecting that host. If this is the case for all pairs of ancestral parasite lineages coalescing in the history of the sample, then the parasite gene tree will mirror the AIG.

The results for a parasite with horizontal transmission are similar to those for the vertically transmitted parasite. The main difference is that the AIG now represents the pattern of transmission events among the ancestors of the infected hosts, assuming each host is infected by exposure to only one other infected host (a host may infect multiple individuals but is infected only once). The AIG can then be informative about the sources of parasite infections in different groups of hosts. Once again, the gene tree of the parasites will correspond to the AIG only in the case that lineage sorting in hosts infected with more than two ancestral parasite lineages does not occur. The probability of identical topology between the AIG and the parasite gene tree is calculated for horizontally transmitted parasites using the same formulas as derived above.

Between-Population Associations

We now consider the probability that the gene trees of hosts and parasites from different partially isolated populations have an identical topology. A single host and a single parasite are sampled at random from each population and both hosts and parasites are sequenced for a neutral locus. We will be specifically interested in a model of populations that have been successively founded by colonists each from a single (possibly different) source population. For simplicity, we assume that each founding event produces one additional population. The history of the population founding events can then be represented as a binary tree. The binary tree of populations will have the same basic form as the AIG in figure 5.2, except that the Hs now represent host populations rather than individuals and the branch lengths represent the times of population founding events rather than parasite transmission events. Such a model is appropriate for studying populations in chains of emerging volcanic islands, for example, or habitat patches emerging following a period of glaciation. Each patch is inhabited by $N_e^{(h)}$ diploid hosts (for simplicity, we assume that host population sizes are equal, but this assumption can be easily relaxed). It is assumed that a fraction m_h of hosts in each patch are replaced by migrants from other patches; the source of the migrants does not matter for our calculations.

Let $N_e^{(p)}$ be the effective size of the population of parasites inhabiting each infected host and assume that either parasites are transmitted vertically or each host is infected by exposure to exactly one other infected host. Let m_p be the probability that a host is infected by parasites from an immigrant host in each generation (or at each transmission event). With strict vertical transmission this occurs when an ancestral host (in a particular generation) is an immigrant and therefore $m_p = m_h$. Assume that a single host sequence is sampled from each population. Our interest will focus on calculating the probability that the sampled sequences from each pair of populations linked by a founding event coalesce in their first shared ancestral population; that is, that they coalesce before the time in the past at which sequences sampled from other populations enter the ancestral population through a second population founding event (this is analogous to the calculation we performed earlier to determine the probability that pairs of sampled parasite sequences coalesced in their first ancestral host lineage). The host tree will be identical to the population history if the sampled lineages are not migrants and each pair of sequences (i.e., the sequence from a newly founded population and its source population) coalesce in their first shared ancestral population. This is not the only way in which the population history could match the topology of the gene tree, but it will account for most of the probability when the migration rate is low and the number of populations exchanging migrants is large.

Consider a pair of sequences whose ancestral lineages are first found in a shared ancestral population at time 0 (i.e., the time of a population founding event at which one of the populations from which a host sequence was sampled gave rise to the other). Let v_i be the time until the next population founding event (going back in time). The probability density that neither lineage is a migrant in any generation during the interval $(0, v_i)$ and the lineages coalesce precisely at time $0 \leq t \leq v_i$ is

$$f(t) = e^{-2m_h t} e^{-t/2N_e^{(h)}} \frac{1}{2N_e^{(h)}} e^{-m_h(v_i - t)}. \tag{5.4}$$

This equation can be understood as follows: the probability that neither lineage is a migrant during the interval $(0, t)$ is $e^{-2m_h t}$, the probability that no coalescence event occurs during $(0, t)$ is $e^{-t/N_e^{(h)}}$, the probability that a coalescence event occurs at generation t is $1/2N_e^{(h)}$, and the probability that the single remaining lineage (following the coalescent event) is not a migrant during (t, v_i) is $e^{-m_h(v_i - t)}$. The probability that in the ancestry of the pair of sequences exactly one coalescence event and no migration events occur during the interval $(0, v_i)$ is then

$$\int_0^{v_i} f(t)dt = \frac{e^{-m_h v_i}\left(1 - e^{-v_i\left(m_h + 1/2N_e^{(h)}\right)}\right)}{2N_e^{(h)}m_h + 1}.$$ (5.5)

The approximate probability that the host gene tree (denoted τ_h) is identical to the population history (denoted T) is then

$$\Pr(\tau_h = T) = \prod_{i=1}^{s} e^{-m_h v_i} \times \left(2N_e^{(h)}m_h + 1\right)^{2-s} \prod_{j=s\,|\,1}^{2s-2} e^{-m_h v_{ji}}\left(1 - e^{-v_j\left(m_h + 1/2N_e^{(h)}\right)}\right)$$

$$= \frac{\exp\left[m_R \sum_{i=1}^{2s-2} v_i\right]}{\left(2N_e^{(h)}m_h + 1\right)^{s-2}} \prod_{j=s+1}^{2s-2}\left(1 - e^{-v_j\left(m_h + 1/2N_e^{(h)}\right)}\right).$$ (5.6)

Calculating the probability that a parasite gene tree matches the tree of population founding events is more involved. Once the lineages ancestral to a pair of parasite sequences enter a common population, two events must occur in order for the lineages to coalesce: (1) the lineages must enter a common host, and (2) once in a common host, the lineages must coalesce. This is illustrated in figure 5.3. Let t_p be the waiting time for two parasite sequences, each from a different randomly chosen host, to coalesce once they are found in the same population. We can represent t_p as a convolution of two waiting times: (1) the waiting time, t_B, until the two sequences enter a common host; and (2) the waiting time, t_W, for the two sequences to coalesce once they share a common host. For parasites with strict vertical transmission, each waiting time follows an independent exponential distribution

$$f_B(t_B) = \frac{1}{N_e^{(h)}}e^{-t_B/N_e^{(h)}},$$

$$f_W(t_W) = \frac{1}{N_e^{(p)}}e^{-t_W/N_e^p}.$$ (5.7)

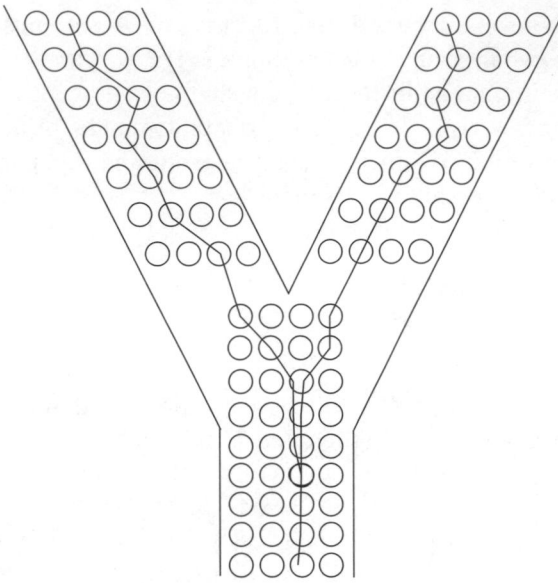

FIGURE 5.3. An illustration of the coalescent process for two parasite sequences sampled from different host populations, where the host population is composed of only four individuals. The lines represent the parasites lineages through time, the Y-shaped outer tree represents the host populations and the host population founding event, and the circles represent individual hosts. The parasite sequences must first enter a common host before they can coalesce.

The moment generating function (see Casella and Berger, 1990) of t_p is then

$$M(x) = E[e^{xt_p}] = \frac{1}{\left(1 - N_e^{(h)}\right)\left(1 - N_e^{(p)}\right)},$$

and the mean and variance are

$$E\lfloor t_p \rfloor = M'(0),$$
$$= N_e^{(h)} + N_e^{(p)},$$
$$\mathrm{Var}\,[t_p] = M''(0) - E\,[t_p]^2,$$
$$= N_e^{(h)^2} + N_e^{(p)^2}. \tag{5.8}$$

The mean and variance of the coalescence time for a pair of sequences from hosts found in the same population are $E[t_h] = 2N_e^{(h)}$ and $\mathrm{Var}[t_h] = 4N_e^{(h)^2}$. The expected coalescence time for sampled parasite sequences is of the same order of magnitude as that of the hosts if $N_e^{(p)} = N_e^{(h)}$ (ignoring differences in generation times between hosts and parasites). The parasite gene tree (denoted τ_p) is identical to the population history with probability

$$\Pr(\tau_p = T) = \prod_{i=1}^{s} e^{-v_i m_p} \prod_{j=s+1}^{2s-2} \int_0^{v_j} f_p(t) e^{-2m_p t} e^{-m_p(v_j - t)} dt \qquad (5.9)$$

For general values of $N_e^{(h)}$ and $N_e^{(p)}$ the integral in the above equation will need to be evaluated numerically as the probability density function $f_p(.)$ of t_p has no simple form. In the special case that $N_e^{(h)} = N_e^{(p)} = N_e$, the density function reduces to a gamma density with parameters $\alpha - 2$ and $\beta = N_e$. In that case, the integral has the explicit solution

$$\int_0^{v_j} f_p(t) e^{-2m_p t} e^{-m_p(v_j - t)} dt = \frac{(1 - e^{-(1/N_e + m_p)v_j}) - (1/N_e + m_p) v_j e^{-(1/N_e + m_p)v_j}}{e^{m_p v_j} (N_e m_p + 1)^2}.$$

$$(5.10)$$

The probability that host and parasite gene trees are identical is approximated by the probability that both match the population history, which is obtained as the product

$$\Pr(\tau_p = \tau_h) \approx \Pr(\tau_p = T) \times \Pr(\tau_h = T)$$

The effect of migration rate and population size on the probability that host and parasite gene trees match the population history is illustrated for representative values of $N_e^{(h)}$, $N_e^{(p)}$, and m in figure 5.4. In general, a decrease in the migration rate results in an increase in the probability that host and parasite gene trees are identical to the population history when $N_e^{(\cdot)} > 1$. Increasing $N_e^{(\cdot)}$ decreases the probability that either of the host or parasite trees matches the population history when the internal branch lengths are less than about $5N_e^{(\cdot)}$ generations. The effect of changing the branch lengths is more complicated. Increasing the internal branch lengths increases the probability of a match when branch lengths are short and lineage sorting is a problem; however, it also increases the probability that a migration event occurs, therefore decreasing the probability of a match. In general, there is a high probability of a match between host and parasite trees (and the population history) when $v_j > 5_e^{(\cdot)}$ and $N_e^{(\cdot)} m < 1$. Host and parasite gene trees have about the same probability of matching the population history in this case.

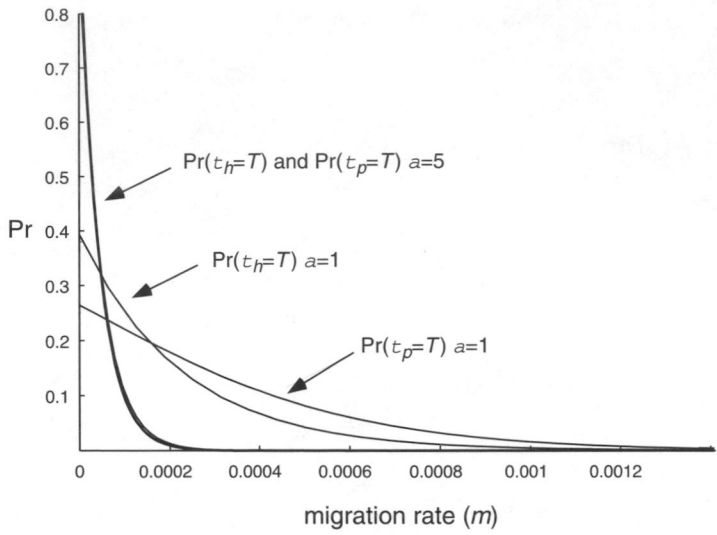

FIGURE 5.4. Effect of migration and population size on the probability that host and parasite gene trees have topologies identical to the tree of population founding events. $N_e = 1000\alpha$, τ_p denotes the parasite gene tree, τ_h denotes the host gene tree, and T denotes the tree of population founding events (assumed to be the same for both hosts and parasites).

Between-Species Associations

As mentioned earlier, the definition of cospeciation most often used in the literature has been an empirical one: a host phylogeny mirroring that of associated parasites. There are at least four possible causes of such an association. The classical cospeciation scenario is one in which the phylogenetic correlation arises because both hosts and parasites speciate allopatrically at about the same time (see, e.g., Hafner and Nadler, 1988; Hafner et al., 1994). For example, a geographical barrier arises that isolates two populations of hosts and obligate parasites, creating incipient host and parasite species. A second possibility is that hosts speciate allopatrically, yet the barrier that has caused speciation in the hosts is not a barrier to migration of parasites. In this case, parasites will not speciate unless they do so by a sympatric speciation event; possibly this is driven by coadaptive responses to the evolution of the host (incipient host species may be diverging due to either neutral genetic drift or selection in the different habitats). A third possibility is that the hosts are sympatric but the parasites are in allopatry and undergo speciation. In this case, hosts might speciate as a result of coadaptive changes (defense strategies, for example) in response to the local parasite species (the possibility of speciation by adaptation

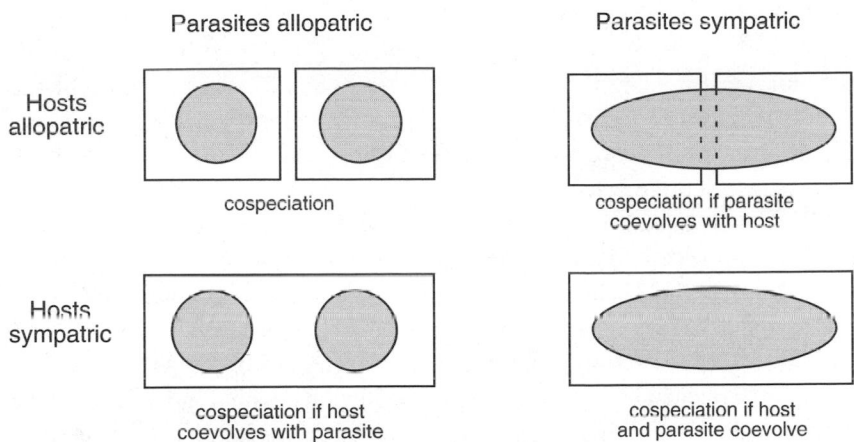

FIGURE 5.5. Modes of speciation in hosts and parasites that can lead to the observation of cospeciation (identical gene trees of hosts and parasites). Host populations are represented as square boxes and parasite populations as ellipses.

of parasites to their hosts and vice versa is discussed extensively in Price, 1980). A fourth possibility is that both hosts and parasites speciate in sympatry. One possible scenario is a process of coevolution (Erlich and Raven, 1964) whereby a subpopulation of hosts or parasites evolves a new strategy and the resulting evolutionary arms race creates new species of both hosts and parasites (see the two-locus model proposed by Bush, 1975). Figure 5.5 illustrates these different models of host-parasite cospeciation.

The degree of similarity between gene trees of related species of hosts and parasites under the different speciation scenarios described above will depend mainly on how rapidly genetic isolation arises in the host and parasite species and the duration of the intervals between speciation events. If the events giving rise to the species under study follow one another closely in time and/or there are extended periods of gene flow during the formation of the species, then the lineage sorting effects that are the main subject of this chapter can lead to incongruences between host and parasite gene trees.

It is clear from the discussion above that identical gene trees for host and parasite species do not necessarily provide evidence of host-parasite coevolution as envisioned by Erlich and Raven (1964). It is possible, however, that the patterns of molecular evolution expected under an allopatric isolation model may be different from those expected under a coevolutionary model (with some form of sympatric speciation). Once more is known, at the level of functional genomics, for example, about processes of adaptation in hosts and parasites, it is conceivable that patterns of evolutionary change in particular genomic regions (synonymous versus nonsynonymous substitution rates, for example) might allow one to distinguish the cause of the observed patterns of cospeciation in particular instances. Such adaptive associations should prove a rich source of questions for future studies, both empirical and theoretical.

We now focus on the probability that gene trees of associated host and parasite species are identical. We again assume the sampled sequences are neutral and their genealogies arise due to the forces of genetic drift alone. For simplicity, we assume a binary tree of host speciation events and no gene flow (hybridization or lateral gene transfer) between species (either hosts or parasites). The tree of speciation events is analogous to the tree of population founding events, and the AIG, considered in earlier sections and illustrated in figure 5.2 except that the Hs now represent species and the branch lengths the times between speciation events. We make no explicit assumptions regarding the mechanism of speciation in hosts and parasites; we assume only that the mechanism acts such that incipient species of hosts and parasites become genetically isolated at similar times in the past. Let T be the tree of speciation events and let τ_h and τ_p be the host and parasite gene trees, respectively. A good approximation for the probability that the host gene tree is identical to the tree of speciation events is

$$\Pr(\tau_h = T) = \prod_{i=s+1}^{2s-2} \left(1 - e^{-v_i/2N_e^{(h)}}\right).$$

The approximate probability that the parasite gene tree is identical to the tree of speciation events is

$$\Pr(\tau_p = T) = \prod_{j=s+1}^{2s-2} \int_0^{v_j} f_p(t)dt.$$

The exact form of the pdf of the time until a pair of parasite lineages coalesce in the jth ancestral host species depends on the model of parasite transmission. For vertically transmitted parasites, the formula is similar to equation 5.10 given above (assuming for simplicity that $N_e^{(h)} = N_e^{(p)} = N_e$) but with $m_p = 0$. In that case, the solution to the integral is

$$\int_0^{v_i} f_p(t)dt = 1 - (1 + 1/N_e)e^{v_j/N_e}.$$

For horizontal transmission, one can model the process by thinking of the parasite population within the population of hosts as a whole as analogous to a subdivided population of free-living organisms but with transmission among hosts replacing immigration. This approach is particularly simple if $N_e^{(p)} \approx 1$. In that case, once two parasite lineages occupy the same host they coalesce to a common ancestral lineage (in that host) with probability 1. The time until a pair of parasite sequences coalesce once they inhabit in a common host species is then completely determined by the rate of transmission among hosts.

Let γ be the probability a parasite is transmitted to another host in each host generation (or equivalently, the expected fraction of parasites exchanged among infected or susceptible hosts per host generation), and let N_i be the number of infected and susceptible hosts (for simplicity, we assume that N_i is constant). We define the effective prevalence of the parasite ε to be

$$\varepsilon = \frac{N_i}{N_e^{(h)}}.$$

Using the same continuous time scale that we have used for studying the coalescent process, one can show that for small γ and large N_i the probability that the lineages coalesce at time t is

$$f_p(t) = \frac{-2\gamma}{\varepsilon N_e^{(h)}} \exp\left\{\frac{-2t\gamma}{\varepsilon N_e^{(h)}}\right\}, \tag{5.11}$$

and the probability that the parasite lineages coalesce in the first ancestral host species is

$$\int_0^{v_i} f_p(t)dt = 1 - \exp\left\{\frac{-2v_i\gamma}{\varepsilon N_e^{(h)}}\right\}.$$

Exploring a range of values for the parameters of the models derived above allows some general conclusions. For vertically transmitted parasites, the gene trees of hosts and parasites are most likely to agree if both host and parasite effective population sizes are small and the internal branch lengths in the species tree are large. For the simple model of horizontally transmitted parasites, the gene trees of hosts and parasites are most likely to agree when the rate of transmission among hosts is high, the fraction of hosts infected (or susceptible) is low, and the internal branch lengths in the species tree are large.

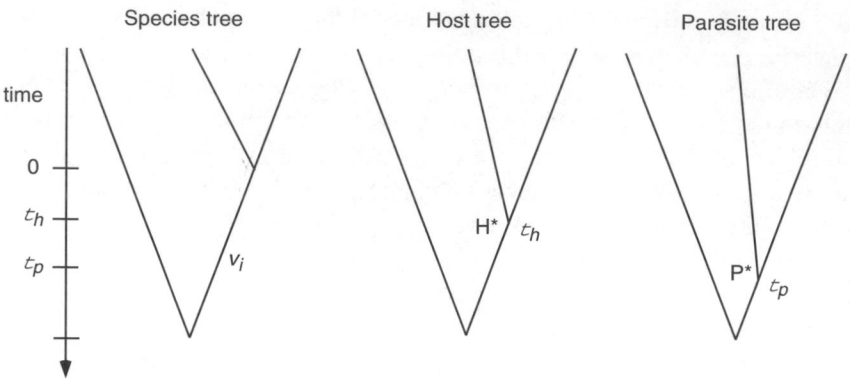

FIGURE 5.6. Relationship between the time of a speciation event in a species tree ($t = 0$ on the scale at left of figure) and the times of a coalescence event in the gene tree of sequences sampled from individual hosts and parasites. The time of the coalescence event in the parasite tree is denoted t_p and the time of the coalescence event in the host tree is denoted t_h.

We now consider the expected similarities of branch lengths of host and parasite gene trees when the species have undergone strict cospeciation (i.e., there is a common species tree for hosts and parasites) and when the topologies of host and parasite gene trees are identical to each other and to the species tree. We consider a particular internal branch i on the species tree of length v_i and we assume that $t_p \leq v_i$ and $t_h \leq v_i$. Here t_h and t_p are the amounts by which the host and parasite coalescence times exceed the speciation time (see fig. 5.6). The above inequalities imply that both the parasite and host gene tree lineages coalesce in this first set of common ancestral species. If $t_p = t_h$, the gene trees of the hosts and parasites have identical ages for ancestors H^* and P^* of the sequences examined (assuming equal generation times for hosts and parasites). To quantify the expected difference between the coalescence times observed for host and parasite sequences, we consider the systematic difference between t_p and t_h as well as the variance of the difference. The systematic difference is quantified by considering the difference in the expected value of t_p versus t_h. These expectations must be obtained for the conditional distributions (i.e., assuming that $t. \leq v_i$ where $t.$ denotes either t_p or t_h). For the hosts this conditional expectation is

$$E[t_h \mid t_h \leq v_i] = 2N_e^{(h)} - \frac{v_i}{e^{v_i/2N_e^{(h)}} - 1}. \tag{5.12}$$

For simplicity, we will only consider the model of horizontally transmitted parasites with $N_e^{(p)} \approx 1$ described above. In that case, the conditional expectation of t_p is

$$E[t_p \mid t_p \leq v_i] = \alpha^{-1} - \frac{v_i}{e^{\alpha v_i - 1}}, \tag{5.13}$$

where $\alpha = 2\gamma/(\varepsilon N_e^{(h)})$. For large $v_i/N_e^{(h)}$ the effects of conditioning on $t. \leq v_i$ are negligible and the expected difference between t_p and t_h is approximately

$$E[t_p - t_h] = N_e^{(h)} \left\{ \frac{\varepsilon}{2\gamma} - 2 \right\}. \tag{5.14}$$

The variance of the difference is approximately

$$\mathrm{Var}[t_p - t_h] = N_e^{(h)^2} \left\{ \frac{\varepsilon^2}{4\gamma^2} + 4 \right\}. \tag{5.15}$$

The above analysis implies that coalescence times in host and parasite gene trees may differ from each other considerably, even if associated hosts and parasites have speciated in synchrony. The magnitude of the expected difference is proportional to the host effective population size. The degree of systematic difference obtained for the simple model with horizontal transmission depends on the incidence of infected hosts and the transmission rate among hosts. If the prevalence is four times the transmission probability, then, on average, the coalescence times are identical. If the prevalence is more than four times the transition probability, then host sequences will coalesce sooner than those of parasites (on average); otherwise parasite sequences will tend to coalesce sooner than those of hosts. The variance of the differences in coalescence times between branches of host and parasite gene trees is proportional to the square of the effective population size of the hosts. For large host effective population sizes, one can then expect to see much greater variation of coalescence times in individual gene trees of hosts and parasites.

Conclusions

In this chapter, we have discussed the effects of population genetic processes (migration and genetic drift) within species and populations of hosts and parasites on the probability that host and parasite gene trees have an identical topology and also potentially similar branch lengths (coalescence times). Our approach to this problem differs from that of many systematists studying host-parasite cospeciation in that we consider potential associations between gene trees of hosts and parasites within populations, among populations, and among species. We have shown that, in some cases, the characteristic footprint of cospeciation (identical host and parasite gene trees) may potentially be observed between individual hosts, or host populations, as well as between species of hosts and parasites.

In our analysis, we have considered a "best-case" scenario in which there is a common tree of speciation, population founding, or transmission events for hosts and associated parasites and the gene trees of hosts and parasites can be inferred without error. Even in this ideal situation, the probability that host and parasite gene trees have identical topologies may be small; this is the case when effective population sizes of hosts and/or parasites are large, for example, and the times between speciation events, for example, are short. Thus, population demographic factors should also be considered (in addition to factors such as host switching by parasite species or multiple ancestral species of parasites) in formulating explanations for observed differences between host and parasite gene trees. Demographic factors alone can also account for differences (both random and systematic) between branching times in host and parasite gene trees. Ad hoc explanations for dissimilar branch lengths in host and parasite gene trees such as "delayed cospeciation" (Hafner and Nadler, 1990) should be considered only after direct demographic explanations have been excluded. Because all the coalescent processes considered in this paper operate with rates principally determined by either $v_j / N_e^{(p)}$ or $v_j / N_e^{(h)}$, some generalizations are possible about situations in which lineage sorting will be important. In general, if there is no gene flow between populations (or species), then if $v_j > 10 N_e$ (where N_e denotes both host and parasite effective population size), lineage sorting effects will not obscure the pattern of cospeciation.

In carrying out our analyses of the effects of population-level processes on patterns of host-parasite cospeciation, we have been forced to use very simple models of parasite transmission and host (and parasite) demographic structure. Nonetheless, we expect that the general pattern of effects of population size, transmission rate, branch lengths, and other parameters predicted by our theory should often apply to more complex models. Much remains to be done to develop realistic models and to explore the range of effects of different parameters on the probabilities that gene tree topologies and branch lengths are identical (or correlated) among associated hosts and parasites. In this chapter, we have presented a unified framework for exploring such questions using the coalescent theory of population genetics. We view these theoretical developments as only a first step.

Acknowledgments

Much of the analysis in this chapter was carried out while B.R. was a visiting professor at the Universite Pierre et Marie Curie, Laboratoire D'Ecologie, Paris, during December of 1997. The authors are grateful to the École Normale Supérieure, which provided the financial support for this visit.

REFERENCES

Avise, J. C. 1994. *Molecular markers, natural history and evolution.* New York: Chapman and Hall.

Brooks, D. R. 1985. Historical ecology: A new approach to studying the evolution of ecological variation. *Annals of the Missouri Botanical Garden* 72:660–80.

Bush, G. L. 1975. Modes of animal speciation. *Annual Review of Ecology and Systematics* 6:339–61.

Casella, G., and R. L. Berger. 1990. *Statistical inference.* Belmont, Calif.: Duxbury Press.

Erlich, P. R., and P. H. Raven. 1964. Butterflies and plants: A study in coevolution. *Evolution* 18:586–608.

Griffiths, R. C., and S. Tavaré. 1994. Ancestral inference in population genetics. *Oxford Surveys in Evolutionary Biology* 9:307–19.

Hafner, M. S., and S. A. Nadler. 1988. Phylogenetic trees support the coevolution of parasites and their hosts. *Nature* 332:258–59.

———. 1990. Cospeciation in host-parasite assemblages: Comparative analysis of rates of evolution and timing of cospeciation. *Systematic Zoology* 39:192–204.

Hafner, M. S., and R. D. M. Page. 1995. Molecular phylogenies and host-parasite cospeciation: Gophers and lice as a model system. *Philosophical Transactions of the Royal Society of London,* ser. B, 349:77–83.

Hafner, M. S., P. D. Sudman, F. X. Villablanca, T. A. Spradling, J. W. Demastes, and S. A. Nadler. 1994. Disparate rates of molecular evolution in cospeciating hosts and parasites. *Science* 265:1087–90.

Hudson, R. R. 1990. Gene genealogies and the coalescent process. *Oxford Surveys in Evolutionary Biology* 7:1–44.

Huelsenbeck, J. P., B. Rannala, and Z. Yang. 1997. Statistical tests of host-parasite cospeciation. *Evolution* 51:410–19.

Kingman, J. F. C. 1982. On the genealogy of large populations. *Journal of Applied Probability* 19A:27–43.

Klassen, G. J. 1992. Coevolution: A history of the macroevolutionary approach to studying host-parasite associations. *Journal of Parasitology* 78:573–87.

Malécot, G. 1948. *The mathematics of heredity.* San Francisco: Freeman.

Pamilo, P., and M. Nei. 1988. Relationships between gene trees and species trees. *Molecular Biology and Evolution* 5:568–83.

Price, P. W. 1980. *Evolutionary biology of parasites.* Princeton, N.J.: Princeton University Press.

Slatkin, M., and R. R. Hudson. 1991. Pairwise comparisons of mitochondrial DNA
    sequences in stable and exponentially growing populations. *Genetics* 129:555–
    62.
Tajima, F. 1983. Evolutionary relationships of DNA sequences in finite populations.
    Genetics 105:437–60.

APPENDIX 5.1: PROBABILITY GENE AND SPECIES TREES ARE IDENTICAL

To illustrate how the coalescent process can cause gene trees of associated hosts
and parasites to differ from their common species tree, we consider a simple case
of three species of associated hosts and parasites. A single sequence is sampled
from each species of host and parasite. The species tree and three possible host
gene trees (one that matches the species tree and two that do not) are shown in
figure 5.A.1 (*a, b, c,* and *d,* respectively). It is assumed that the associated hosts

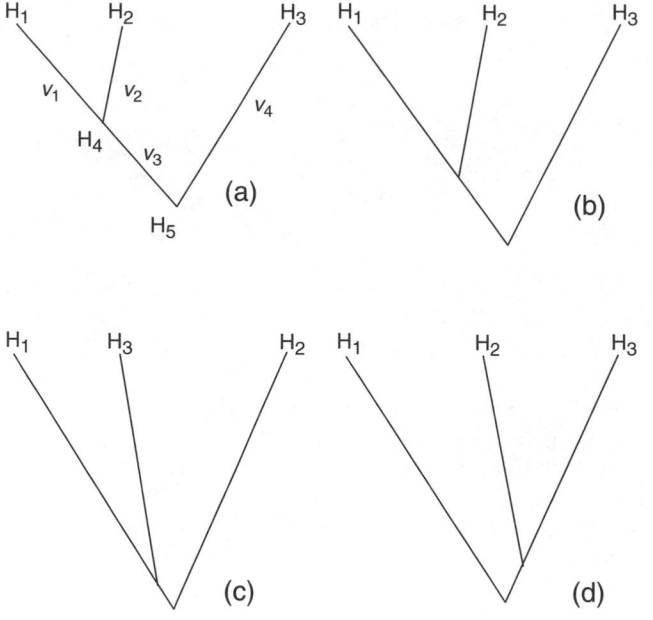

FIGURE 5.A.1. A tree of the speciation events giving rise to three host species is shown
at upper left in *(a)*. The branch lengths v_1, v_2, v_3, and v_4 are in units of generations.
A possible gene tree for the species is shown at upper right in *(b)*. In this tree,
sequences from host species H_1 and H_2 coalesce during the interval v_3 and the gene
tree is therefore identical to the species tree. At the lower left *(c)* and right *(d)* of the
figure two gene trees are shown for which the first coalescence event occurs after the
interval v_3 in ancestral species H_5, allowing coalescence events that result in gene trees
having a topology that is different from that of the species tree.

and parasites speciated at precisely the same times in the past. Host species H_1 and H_2 and associated parasite species P_1 and P_2 arose at time v_1 in the past and host species H_1 and H_4 and parasite species P_1 and P_4 arose at time $v_4 = v_1 + v_3$. There are two ways in which a host gene tree can have an identical topology to the species tree: (1) sequences from hosts H_1 and H_2 coalesce in a first common ancestral species (H_4) during v_3. The probability that this occurs is

$$1 - e^{-v_3/N_e^{(h)}},$$

where $N_e^{(h)}$ is the effective population size of ancestral host species H_4 (see fig. 5.A1b); or (2) they do not coalesce in ancestral host species H_4 ($t \geq v_1$) but instead coalesce in ancestral host species H_5. In that case, the gene tree is identical to the species tree only if lineages ancestral to sequences from H_1 and H_2 coalesce first (see fig. 5.A1b) rather than those ancestral to H_1 and H_3 (see fig. 5.A1c) or H_2 and H_3 (see fig. 5.A1d).

All three possible patterns of coalescence are equally probable, so the probability of this is

$$\frac{1}{3}e^{-v_3/N_e^{(h)}}.$$

The probability the host gene tree is identical to the species tree is then

$$\Pr[\tau_h = T] = 1 - e^{-v_3/N_e^{(h)}} + \frac{1}{3}e^{-v_3/N_e^{(h)}},$$

$$= 1 - \frac{2}{3}e^{-v_3/N_e^{(h)}}.$$

For $v_3/N_e^{(h)} \gg 1$ this is well approximated as $1 - e^{-v_3/N_e^{(h)}}$. Similarly, the parasite gene tree will agree with the species tree if the lineages either coalesce in the first common ancestral species P_4, or instead coalesce in ancestral species P_5, but then P_1 and P_2 happen to coalesce in that ancestor, rather than P_1 and P_3 or P_2 and P_3. With more than three species, there is the possibility that no coalescence events occur until four or more lineages co-occur in a single ancestral species. Pamilo and Nei (1988) suggest an approximation for the probability that a gene tree has a topology identical to a species tree based on the assumption that coalescence events occur in the first shared ancestral species for each pair of lineages. They show that this approximation is quite accurate for about five or more species. We use essentially the same approximation in this paper to derive the probability that host and parasite gene trees are identical.

6

NEW EVIDENCE FOR HYSTRICOGNATH
RODENT MONOPHYLY FROM
THE PHYLOGENY OF THEIR PINWORMS

J.-P. Hugot

Introduction

Rodents are currently divided into two suborders, the Sciurognathi and the Hystricognathi (Anderson and Knox Jones Jr., 1984; Luckett and Hartenberger, 1985). The Hystricognathi can be further subdivided into the Old World Phiomorpha and the New World Caviomopha. The evolutionary relationships among hystricognath rodents and the geographical origin of the Caviomorpha (which has been endemic to the Neotropics for most of the Tertiary) have generated considerable debate. Some of the arguments used concerned the distribution of the parasite pinworms of these rodents. Quentin (1973a) first noticed the intriguing distribution of the pinworm nematode genus *Wellcomia* Sambon, 1907. This genus includes all the species of pinworms known from both the Old World and New World porcupines; if *Wellcomia* is monophyletic, its present-day geographic distribution calls into question the different hypotheses concerning the phylogeny and the geographical origin of these groups. Recent collections have made it possible to redescribe and check the host specificity and the precise collection locality of the species in this genus, and the related genera (also parasite of hystricognath rodents) *Helminthoxys* Freitas Texeira, Lent and Almeida, 1937; *Petronema* Hugot, 1983a; and *Protozoophaga* Diesing, 1851. Using this new information, a data set of morphological characters was assembled to test the monophyly of each genus and to propose a phylogeny of the whole group. Finally, the cladogram of the parasites was used as an tool for testing different hypotheses previously proposed for the classification of the Hystricognathi. The following questions were examined: does the phylogeny of their specific pinworms support (1) the monophyly of the Hystricognathi; (2) the monophyly of the Phiomorpha; (3) the monophyly of the Caviomorpha; and/or (4) any new (or previously proposed) grouping within these groups; and (5) which scenarios can be proposed to explain the current distribution of the Hystricognathi?

Material and Methods

Species and Specimens

The cladistic analysis includes 19 species of pinworm parasite in hystricognath rodents. The outgroup includes two species: *Hilgertia hilgerti* from a sciurognath rodent of the family Ctenodactylidae, and *Acanthoxyurus beecrofti* from a sciurognath rodent of the family Anomaluridae. Appendix 6.1 lists the 21 species analyzed and their hosts and localities. The identification of the hosts at species and family levels follows Wilson and Reeder (1993).

Characters

The cladistic analysis is based on 22 morphological characters from various organ systems (see appendices 6.2 and 6.3). All characters are unordered. When necessary we have used the method proposed by Maddison (1993) for coding multistate characters. Inapplicable characters (sensu Maddison, 1993) were coded "-". For cladistic analysis, character weights were scaled so that each character receives the same total weight regardless of the number of states observed, with the base weight for each character being equal to 100. Character states used for this analysis are explained and figured in a previous work (Hugot, 1988).

Cladistic Analysis

Parsimony analyses were conducted using PAUP 4.0b3a (Swofford, 1999). MacClade 3.08a (Maddison and Maddison, 1992) was used for data and tree handling and for computation of tree statistics. Autodecay 2.9.5 (Eriksson, 1996) and TreeView 1.6.2 (Page, 1996a) were used for computation and printing of decay indices on trees. PAUP runs were performed using heuristic search with tree bisection-reconnection (TBR) branch swapping, random addition sequence with 1,000 replicates, MULPARS option in effect, steepest descent option not in effect, branches collapsed (creating polytomies) if maximum branch length $= 0$, and multistate taxa interpreted as polymorphism.

Comparison of Host and Parasite Cladograms

Comparing host and parasite cladograms was performed using TreeMap 1.0b (Page, 1994). Following Page (1996b), speciation events are divided into three categories: cospeciation, duplication, and host switch. Sorting events (or losses) cover any case in which a parasite species can be expected on one host species and has not been observed.

Results

Cladistic Analysis

The analysis gave a single tree, which is represented on figure 6.1 and has the following statistics: tree length = 26.73, consistency index (CI) = 0.835, homoplasy index (HI) = 0.177, retention index (RI) = 0.918, and rescaled consistency index (RC) = 0.768. The decay indices are figured at the nodes of the cladogram of figure 6.1. The parasites of Hystricognathi appear as a monophyletic group in which two subgroups can be distinguished: (1) the monophyletic *Wellcomia* with *Protozoophaga* as a sister group, and (2) the monophyletic *Helminthoxys* with *Petronema* as a sister group. Within *Wellcomia* the parasites of the New World porcupines *Erethizon, Coendou,* and *Sphiggurus* (Erethizontidae) are associated with the parasite of *Dinomys* (Dinomyidae), and the parasites of the Old World porcupines *Atherurus* and *Hystrix* (Hystricidae) are associated as sister species; these two groups are associated and opposed to the parasite of *Dolichotis* (Dolichotidae). Within *Helminthoxys* the parasites of *Microcavia* spp. (Caviidae) and *Lagostomus* spp. (Chinchillidae) are associated as sister groups and opposed successively to (1) the parasite of *Abrocoma* (Abrocomidae), (2) the parasites of *Trichomys* (Echimyidae) and *Octodon* (Octodontidae), (3) the parasites of *Capromys* spp. (Capromyidae), and finally, (4) the parasite of *Dasyprocta* (Dasyproctidae).

Host Phylogenies

Until now no consensus has been established for the phylogenetic classification of the Hystricognathi, and the most recent attempts have given highly divergent results. An additional difficulty is that different studies have frequently sampled different taxa, and have focused on family-level relationships. Thus, it is impossible to compare the phylogeny of the parasites with the different host trees at the species level. In addition, comparison at the family level is not easy because of the different taxonomic sampling, and because some of the families that host pinworms have never been included in a cladistic analysis. In this study I have used three different topologies (fig. 6.2). Two trees are based on molecular data: *Host-tree-1* is adapted from Huchon et al.'s (1999, 2000) analysis of von Willebrand factor gene sequences, and *Host-tree-2* is from Nedbal et al.'s (1994) analysis of

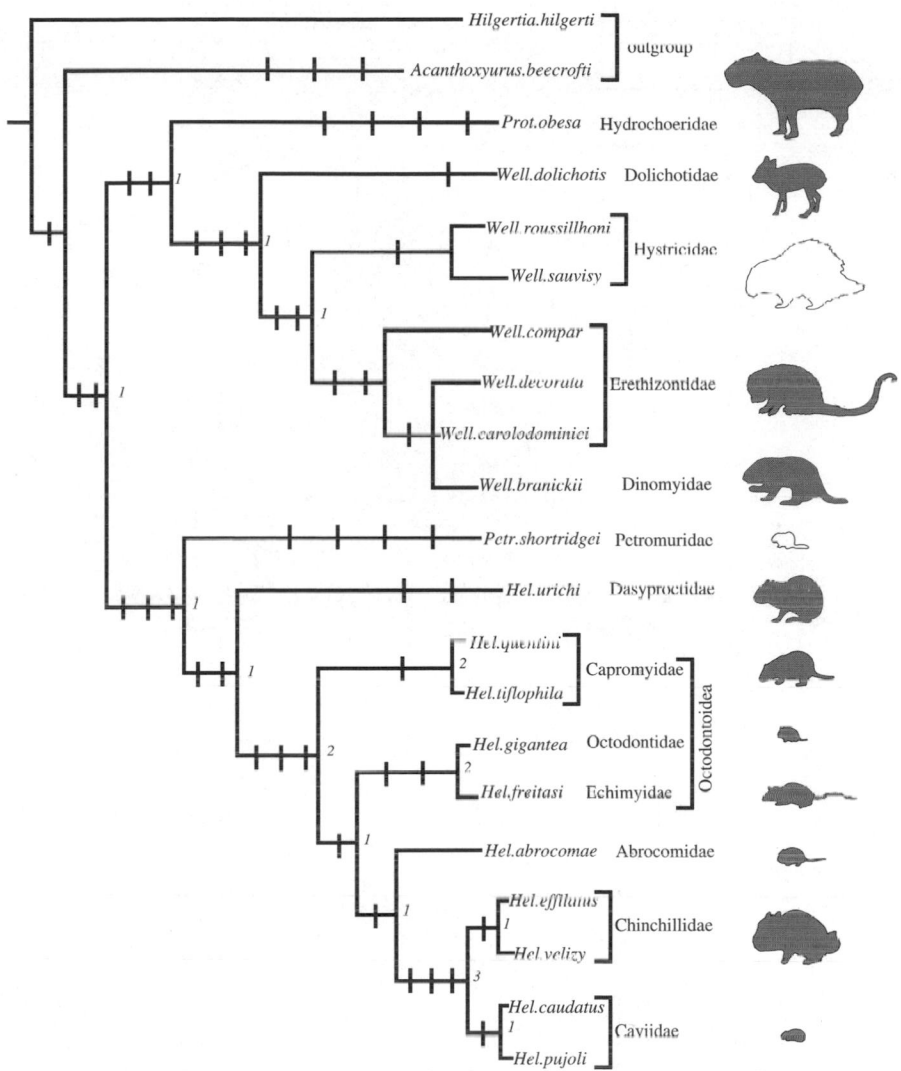

FIGURE 6.1. Results of cladistic analysis of hystricognath pinworms (21 taxa and 22 characters; see appendix 6.3). Tick marks with numbers indicate synapomorphies (character:character state). Numbers in italics are decay indices (when greater than zero). *Prot.* = *Protozoophaga; Well.* = *Wellcomia; Petr.* = *Petronema; Hel.* = *Helminthoxys.* The names of the host families are after Wilson and Reeder (1993). Silhouettes of host family representatives are shaded either gray or white to indicate New World or Old World distributions, respectively.

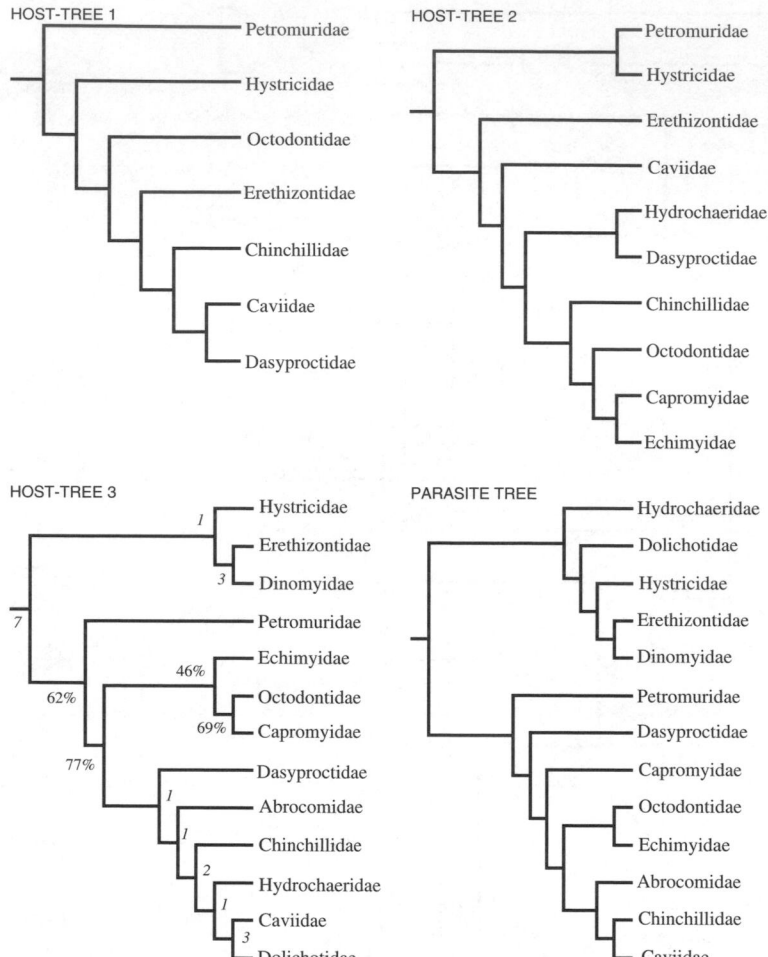

FIGURE 6.2. Parasite and host trees used for tree comparisons. *Host-tree-1* after Huchon et al. (1999, 2000); *Host-tree-2* after Nedbal et al. (1994); *Host-tree-3* results of the cladistic analysis of the matrix represented in appendix 6.4. On *Host-tree-3*, numbers in italics are decay indices (when greater than zero); plain numbers are the percentage occurrence of clades in the 13 equally parsimonious trees (when different from 100%). In the parasite tree the names of the parasites have been switched with the names of the host families, as shown on figure 6.1.

mitochondrial 12S rRNA sequences. The third tree, *Host-tree-3,* is based on morphological data compiled from Hartenberger (1985, p. 27), Lavocat and Parent (1985, p. 351), Bugge (1985, p. 373), George (1985, p. 458), Woods and Hermanson (1985, p. 525), and Bryant and McKenna (1995, p. 23). This yielded 102 characters that, after exclusion of uninformative characters,

resulted in the 55-character matrix in appendix 6.4. This matrix was sub-
mitted to parsimony analysis with the Anomaluridae and Ctenodactylidae
defined as outgroups. The analysis resulted in 13 equally parsimonious trees
(tree length = 102, CI = 0.598, HI = 0.402, RI = 0.692, RC = 0.414) from
which the majority rule consensus tree has been computed. The terminal
taxa in the host trees (fig. 6.2) are restricted to the families known to host
the parasites. Figure 6.2 also shows a cladogram for the parasites in which
the names of the parasite taxa have been replaced with the name of the
family to which their respective host(s) belongs.

Tree Comparison

The different analyses were performed using the exhaustive search rou-
tine in TREEMAP. The reliability of the different scenarios was investigated
using the following criteria: (1) the highest number of cospeciation events;
(2) the number of other evolutionary events (duplications, switches, and
sorting events) needed by a particular scenario (column total of tables 6.2,
6.3, and 6.4); and (3) the minimum number of parasite lineages isolated
before the complete separation of the Phiomorpha and Caviomorpha.
Criterion 1 reflects how many nodes are congruent; criterion 2 specifies
the most parsimonious scenario; and criterion 3 gives the minimal number
of trans-Atlantic migration events that have to be postulated for explanation of the current distribution of the Caviomorpha parasites.

Comparison of the Different Phylogenetic Hypotheses
for the Hystricognathi

In this part, the parasite tree of figure 6.2 was considered an additional
phylogenetic hypothesis for the hystricognath rodents. Then, the trees represented on figure 6.2 were reduced to these terminal taxa, which are shared
by the four trees, as represented in figure 6.3. Finally, TREEMAP was used to
determine how many congruent nodes could be found when considering
successively each pair of trees. The results are given in table 6.1. The lowest values are found when comparing together *Host-tree-1* and *Host-tree-2*,
with only three congruent nodes out of six (50%). The best fit is observed
between *Host-tree-3* and the *Parasite-tree* with five congruent nodes out
of six (83%), and a very low probability ($p = .001$) for getting as much
consistency by chance.

Comparison of the Parasite Tree with the Host Trees in Figure 6.3

Table 6.2 gives the statistics for the different scenarios obtained when
comparing the *Parasite-tree* of figure 6.3 with *Host-trees 1, 2,* and *3* of the

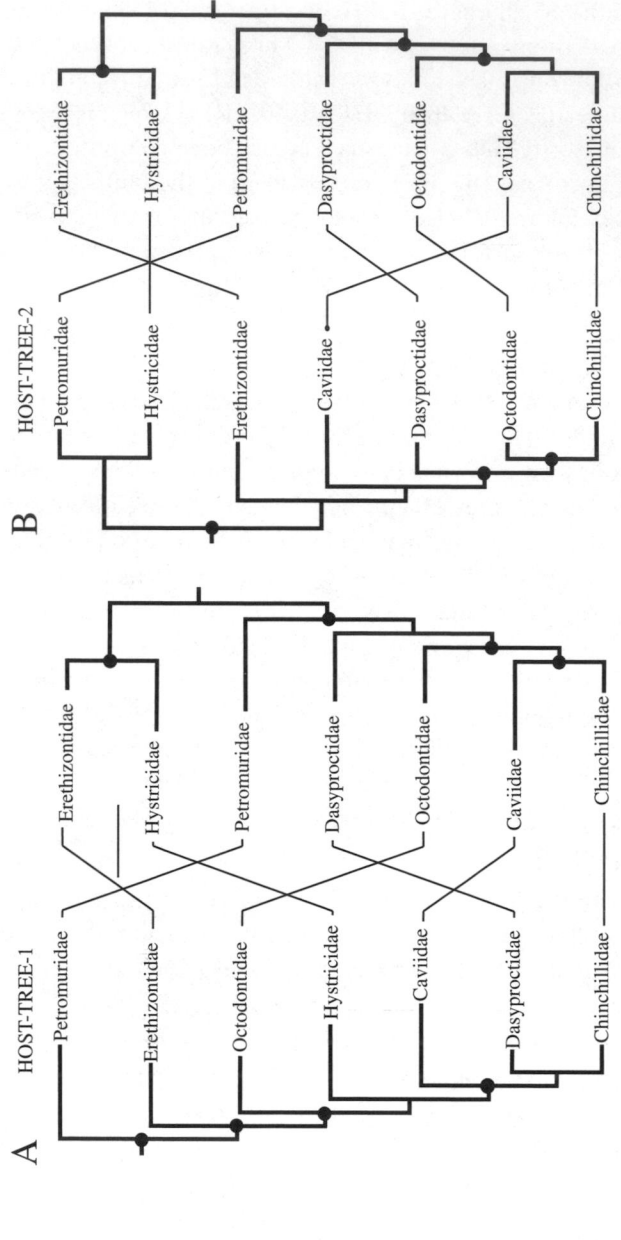

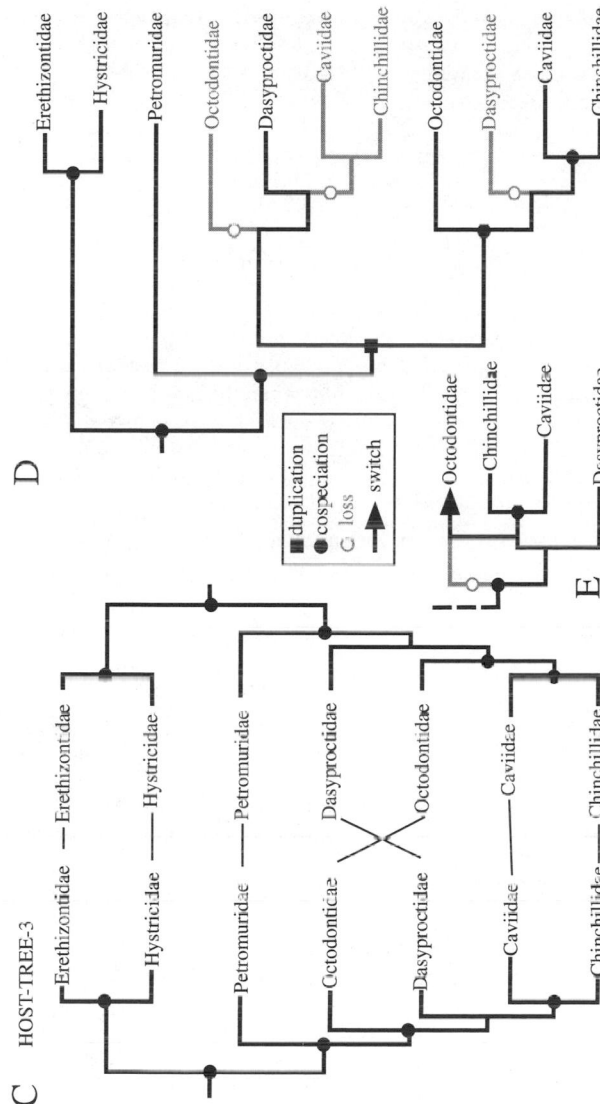

FIGURE 6.3. The trees represented in figure 6.2 when restricted to these terminal taxa that are shared by the four trees. *A, B,* and *C* show the tanglegrams obtained by comparison of the *Parasite-tree* with *Host-trees* 1, 2, and 3, respectively. *D* and *E* show reconstruction scenarios 3-1 and 3-2 of table 6.2, obtained using the trees in tanglegram *C*

TABLE 6.1 Upper diagonal: number of congruent nodes (out of a maximum number of 6) between the different phylogenies of the hystricognath rodents depicted in figure 6.3. Lower diagonal: probability of getting as many congruent nodes between each pair of trees by chance based on a Markovian test generating 1,000 random trees.

	host-tree-1	host-tree-2	host-tree-3	parasite-tree
host-tree-1		3	4	4
host-tree-2	.424		3	4
host-tree-3	.056	.361		5
parasite-tree	.050	.052	.001	

TABLE 6.2 Statistics for reconstructions of the parasite tree when using the different hystricognath rodent phyloogenies depicted in figure 6.3. p is the probability of independence between host and parasite phylogenies when the host tree was compared with 1,000 random parasite cladograms generated using a Markovian model. The last column gives the minimal number of trans-Atlantic migration events needed to explain the current distribution of the parasites of the Caviomorpha (see text for explanation). Scenarios in bold are sketched on figures 6.3D and E, respectively.

Host Tree	Scenario	Cospeciations	Other Events				p	Migrations
			Duplications	Switches	Sorting	Total		
1	1	4	2	0	10	12	.059	2
1	2	4	1	1	8	10	.044	2
1	3	4	1	1	8	10	.057	3
1	4	4	0	2	4	6	.047	3
1	5	4	0	2	6	8	.042	2
2	1	4	1	1	5	7	.052	2
3	**1**	**5**	**1**	**0**	**3**	**4**	**.005**	**2**
3	**2**	**5**	**0**	**1**	**1**	**2**	**.003**	**2**
3	3	5	0	1	1	2	.001	2

same figure. *Host-tree-1* gave 5 different scenarios with a maximum number of 4 cospeciation events, 6 to 12 additional evolutionary events (= additional hypothesis), and 2 to 3 migration events. *Host-tree-2* gave 1 scenario with 4 cospeciation events, 7 additional evolutionary events, and 2 migration events. *Host-tree-3* gave 3 different scenarios with a maximum number of 5 cospeciation events, 2 to 4 additional evolutionary events, and 2 migration events. Figure 6.3A–C show the tanglegrams created by comparison of the *Parasite-tree* (on the right side) with each of the *Host-trees*.

The most parsimonious scenarios were obtained using *Host-tree-3* (scenarios 3-1, 3-2, and 3-3 in table 6.2). Figure 6.3D–E represent scenarios 3-1 and 3-2, respectively. Scenario 3-1 postulates cospeciation events at the root of the trees and at the base of the clade Erethizontidae + Hystricidae, followed by a duplication of the parasites giving birth to two sister lineages; three subsequent losses (sorting events) are required to reconcile the trees. Scenario 3-2 also assumes two cospeciation events on the basal nodes of the trees, but in this scenario, the duplication is changed into the loss of the parasite of the Octodontidae, followed by a switch of the parasites from the common ancestor of the clade Chinchillidae + Caviidae to the Octodontidae. In Scenario 3-3, the duplication is changed into the loss of the parasite of the common ancestor of the pair Chinchillidae + Caviidae, followed by a switch from the Octodontidae to the common ancestor of the pair Chinchillidae + Caviidae. All three scenarios imply that two different parasite lineages, at least, existed before the complete separation of the Phiomorpha and Caviomorpha (i.e., the parasites of the Erethizontidae and the common ancestor of the parasites of all the other Caviomorpha).

Comparison of the Parasite Tree with Host-tree-2 of Figure 6.2

For this analysis the parasite tree was restricted to the host families, which are represented in the *Host-tree-2*. Hence, the Dinomyidae, Dolichotidae, and Hydrochaeridae have been pruned from the *Parasite-tree*. Table 6.3 gives the statistics for the 8 different scenarios obtained in this analysis. The scenarios have 5 cospeciation events (5/9 = 56% of congruent nodes) and 12 to 19 additional evolutionary events, and all of them imply 3 migration events. Figure 6.4A represents the tanglegram for the *Parasite-tree* (on the right side) with *Host-tree-2*. Figure 6.4B represents scenario 2-8 of table 6.3. This scenario, with only 12 additional evolutionary events, is the most parsimonious. It assumes an initial cospeciation event on the basal node of the host tree (the common ancestor of

TABLE 6.3 Statistics for reconstructions of the *Parasite-tree* and *Host-tree-2* of figure 6.2. Scenario in bold is shown in figure 6.4B. For explanation of column headings, see table 6.2.

| Host Tree | Scenario | Cospeciations | Other Events | | | Total | p | Migrations |
			Duplications	Switches	Sorting			
2	**8**	**5**	**0**	**4**	**8**	**12**	**0.141**	**3**
2	5	5	1	3	9	13	0.155	3
2	6	5	1	3	11	15	0.147	3
2	7	5	1	3	12	16	0.166	3
2	2	5	2	2	12	16	0.171	3
2	3	5	2	2	13	17	0.169	3
2	4	5	2	2	14	18	0.153	3
2	1	5	3	1	15	19	0.149	3

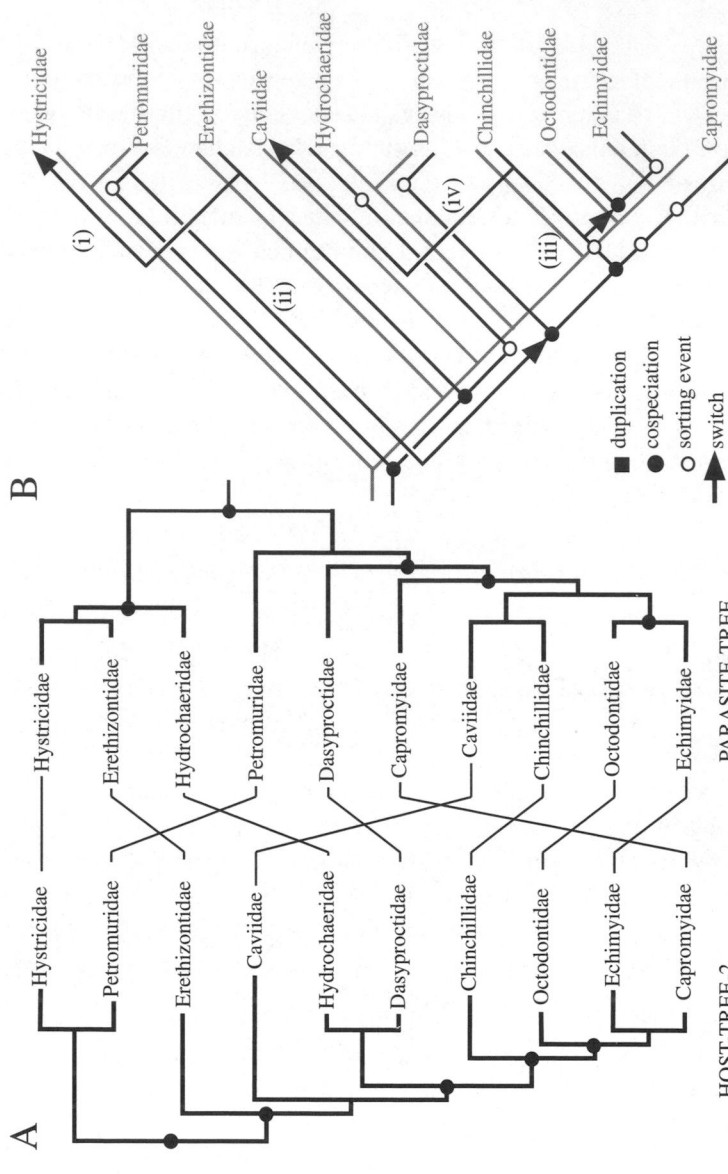

FIGURE 6.4. *A*, Tanglegram obtained by comparison of *Host-tree-2* and the *Parasite-tree* shown in figure 6.2. *B*, Scenario 2-8 of table 6.3. Host switches (i)–(iv) are discussed in the text.

the Hystricognathi), and a second one on the common ancestor of the Caviomorpha. This part gives birth to the specific parasites of the Petromuridae, Erethizontidae, and Hydrochaeridae, respectively, accompanied by losses on the other branches (3 sorting events). Four host switches are also observed: (1) from the Erethizontidae to the Hystricidae, and (2) from the Petromuridae to the common ancestor of the group including the Dasyproctidae, Chinchillidae, and Octodontoidea (Octodontidae, Echimyidae, Capromyidae); and two additional switches, from the Chinchillidae toward (3) the common ancestor of the Octodontidae/ Echimyidae, and (4) the Caviidae. This last part also allows 3 cospeciations and needs 5 sorting events. Host switch (1) implies that the parasites of the Erethizontidae, or at least their common ancestor, existed before the separation of the Phiomorpha and Caviomorpha; this in turn assumes that the lineage leading to the parasites of the Hydrochaeridae also existed before this separation. Host switch (2) assumes that the common ancestor of the parasites of the Dasyproctidae, Chinchillidae, and Octodontoide existed before this separation. This requires that three distinct parasite lineages, at least, accompanied the migration of the Caviomorpha to the Neotropics.

Comparison of the Parasite Tree with Host-tree-3 of Figure 6.2

Table 6.4 gives the statistics for the 33 different scenarios obtained in this analysis. All the scenarios have 8 cospeciation events (8/12 = 67% of congruent nodes) and 8 to 28 additional evolutionary events, and imply 4 to 9 migration events. The scenarios are distributed in two different groups: in *Group-1* (upper part of table 6.4), the root of the *Parasite-tree* corresponds to the common ancestor of the Hystricognathi; in *Group-2* (lower part of table 6.4), the root of the *Parasite-tree* fits with the common ancestor of the Petromuridae and the Caviomorpha minus the Hystricidae, Erethizontidae, and Dinomyidae. The scenarios in table 6.4 have been sorted within each group by increasing number of duplications, decreasing number of host switches, and increasing number of additional evolutionary events, successively. Figure 6.5A represents the tanglegram for the *Parasite-tree* (on the right side) with the *Host-tree*. Figures 6.5B, C, and D represent scenarios 3-26, 3-31, and 3-12 of table 6.4, respectively. With 8 additional evolutionary events, scenario 3-26 is one of the three most parsimonious scenarios within *Group-1*. With 12 additional evolutionary events, scenario 3-31 is the most parsimonious scenario within *Group-2*. All the scenarios of *Group 2* imply 8 or 9 migration events. Within *Group-1*, most of the scenarios also imply 8 or 9 migration events, but 6 of them imply only 4 migration events. Scenario 3-12, with 23 additional evolutionary events, is one of the most parsimonious of these last 6.

TABLE 6.4 Statistics for reconstructions of the *Parasite-tree* and *Host-tree-3* of figure 6.2. Scenario in bold is shown in figure 6.4B. For explanation of column headings, see table 6.2. Scenarios in bold are sketched on figures 6.5B, C, and E, respectively.

| Host Tree | Scenario | Cospeciations | Other Events | | | | p | Migrations |
			Duplications	Switches	Sorting	Total		
3	**26**	**8**	**0**	**4**	**4**	**8**	**0.004**	**9**
3	27	8	0	4	4	8	0.005	9
3	28	8	0	4	4	8	0.006	9
3	16	8	1	3	6	10	0.007	9
3	15	8	1	3	8	12	0.004	9
3	17	8	1	3	11	15	0.003	9
3	18	8	1	3	11	15	0.008	8
3	19	8	1	3	11	15	0.007	8
3	6	8	2	2	9	13	0.002	9
3	8	8	2	2	13	17	0.006	9
3	7	8	2	2	15	19	0.005	9
3	**12**	**8**	**2**	**2**	**19**	**23**	**0.003**	**4**
3	13	8	2	2	19	23	0.009	4
3	14	8	2	2	19	23	0.002	4
3	2	8	3	1	16	20	0.003	9
3	4	8	3	1	21	25	0.001	4
3	5	8	3	1	23	27	0.006	4
3	1	8	4	0	24	28	0.009	4
3	**31**	**8**	**0**	**4**	**8**	**12**	**0.004**	**9**
3	33	8	0	4	9	13	0.006	9
3	32	8	0	4	9	13	0.004	9
3	30	8	0	4	9	13	0.005	9
3	29	8	0	4	11	15	0.003	9
3	25	8	1	3	10	14	0.006	9
3	24	8	1	3	10	14	0.004	9
3	23	8	1	3	10	14	0.007	9
3	22	8	1	3	11	15	0.002	9
3	20	8	1	3	12	16	0.009	9
3	21	8	1	3	13	17	0.007	8
3	11	8	2	2	10	14	0.001	8
3	10	8	2	2	14	18	0.001	8
3	9	8	2	2	14	18	0.007	9
3	3	8	3	1	15	19	0.008	8

Discussion

Validity of the Parasitological Data

Patterson and Wood (1982) challenged the interpretation of the parasitological data proposed by Quentin (1973a), because the morphological data used to argue for the monophyly of *Wellcomia* partly originated from old and incomplete descriptions, and also because some of the material was collected from hosts in zoos, which made questionable the specificity and origin of the parasites. All the specimens used in the present work were collected in the natural range of their respective host. This also includes several recently described new species (Hugot, 1982, 1983b, 1985, 1986; Sutton and Hugot, 1987; Hugot and Sutton, 1989; Hugot and Gardener, 2000). Thus, our data set relies on parasite specimens for which host specificity and precise collection locality cannot be disputed. In addition, three new species (not included in the present analysis because their description is still in progress) have been recently discovered. The first two were collected in an *Acanthion brachyura* and an *Atherurus macrourus* (Hystricidae) from Thailand and are identified as new species of genus *Wellcomia*. The third was collected in a *Proechimys guairae* (Echimyidae) from Venezuela and is a new member of genus *Helminthoxys*. The results presented here verify the identity and specificity of all the species previously described in the four genera involved in the present analysis; show that the host range of the genera *Wellcomia*, *Helminthoxys*, *Petronema*, and *Protozoophaga* seems to be strictly restricted to the Hystricognath; and confirm that all pinworm parasites from porcupines—both New and Old World—can be classified in genus *Wellcomia*.

Hystricognathi Monophyly

For a long time, two different hypotheses have been in competition concerning the origins and classification of the Hystricognathi. According to one hypothesis (Wood, 1950, 1972; Patterson and Wood, 1982; Wood, 1985) the similarity between the Phiomorpha (Old World Hystricognathi) and the Caviomorpha (New World Hystricognathi) resulted from convergent characters. The origins of the Phiomorpha ought to be found within the old endemic stock of African rodents, the origins of the Caviomorpha being within an extinct group of North American rodents, the Franimorpha. In the second hypothesis (Lavocat, 1969; Hoffstetter, 1972), the Phiomorpha and Caviomorpha were sister taxa, together classified as a monophyletic group (the Hystricognathi); and the origins of the Caviomorpha had to be sought in Africa, from where they migrated to South America. During a congress organized in Paris by Luckett and Hartenberger (1985), the arguments supporting each hypothesis were debated, leading to a severe challenge to Wood's hypothesis. In addition, several authors—using different character sets—gave strong support to the monophyly of the Hystricognathi (Beintema, 1985; Bugge, 1985; George, 1985; Lavocat and Parent, 1985; Woods and Hermanson, 1985). The monophyly of the Hystricognathi also

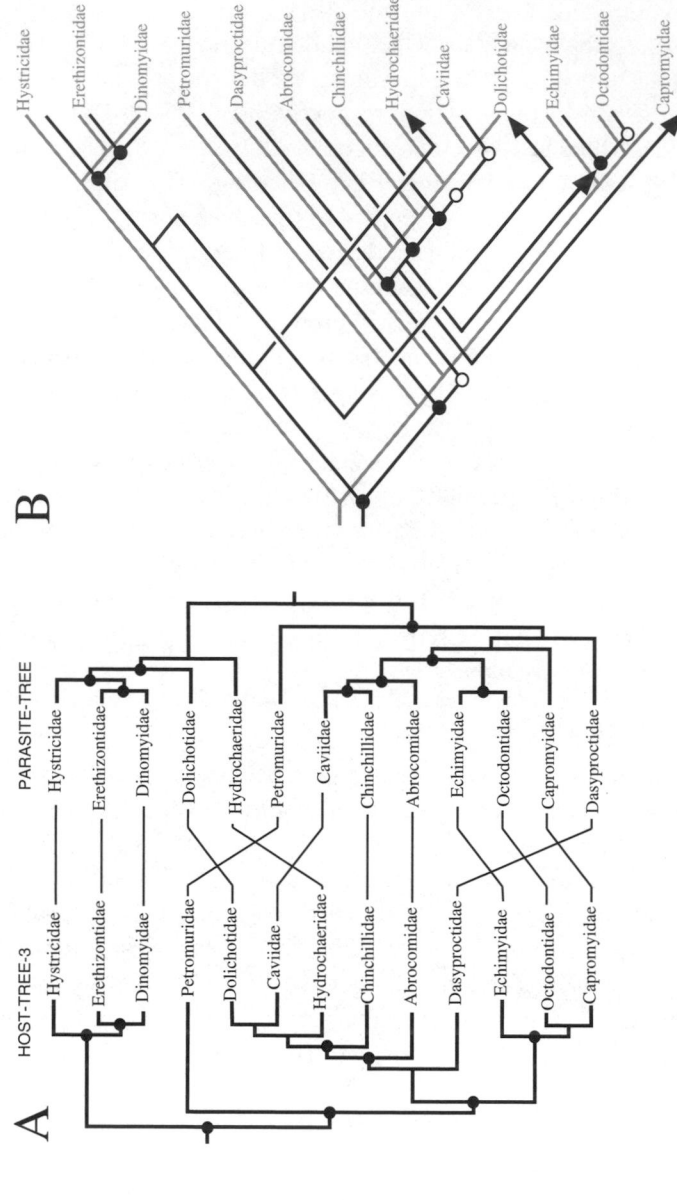

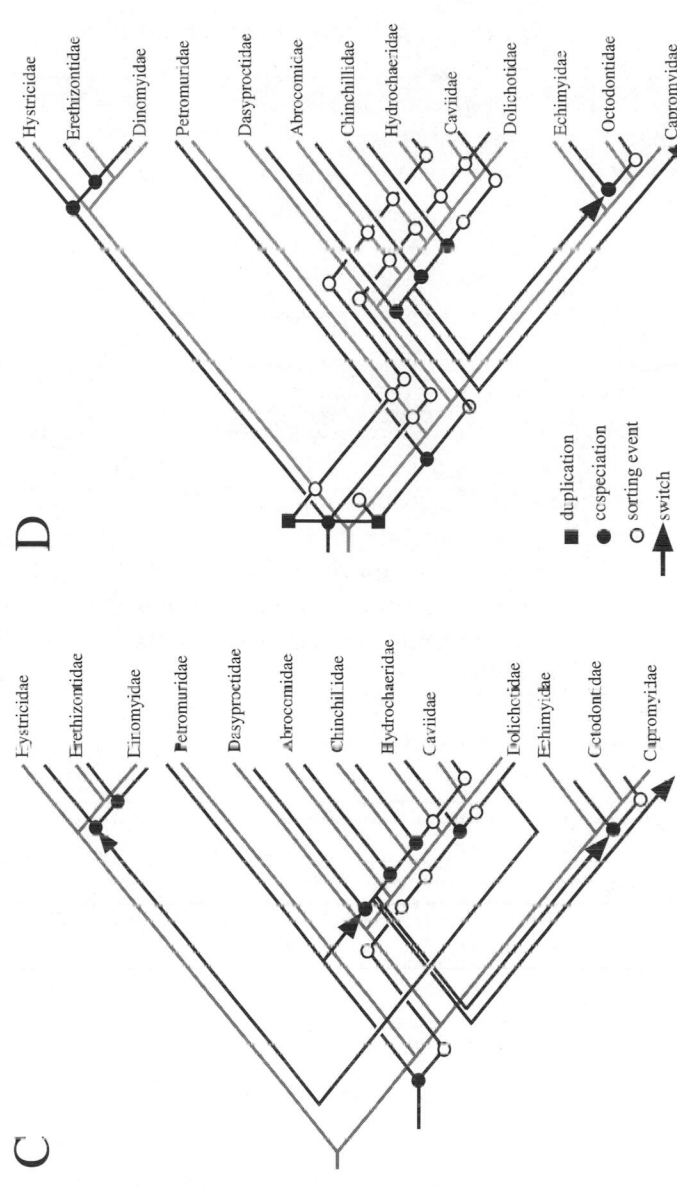

FIGURE 6.5. *A*, Tanglegram obtained by comparison of *Host-tree-3* and the *Parasite-tree* shown in figure 6.2, and scenarios 3-26 (*B*), 3-31 (*C*), and 3-12 (*D*) of table 6.4.

is accepted by most recent workers, either based on molecular data (Nedbal et al., 1994; Huchon et al., 1999, 2000) or morpho-anatomical characters (Martin, 1994; Bryant and McKenna, 1995; Landry Jr., 1999). Finally, the cladistic analysis of the different genera of pinworm parasites of Hystricognathi implies that these parasites together form a monophyletic group, and this characteristic is consistent with the monophyly of their hosts.

Relationships within the Hystricognathi

Comparison of the different cladograms represented on figure 6.2 (table 6.1) shows that in general, they strongly disagree with one another. Only *Host-tree-2* supports the monophyly of the two main subgroups generally distinguished within the Hystricognathi: the Phiomorpha and Caviomorpha. *Host-tree-1* supports the monophyly of the Caviomorpha. *Host-tree-3* and the *Parasite-tree* refute the monophyly of both subgroups. *Host-trees 1, 2,* and *3* also disagree on the arrangement of the different families, with the exception of the association Octodontidae, Echimyidae, and Capromyidae, which is common to *Host-trees 2* and *3*. This grouping corresponds with the superfamily Octodontoidea supported by Woods and Hermanson (1985), Nedbal et al. (1994), and Bryant and McKenna (1995), but recently refuted by Landry (1999).

The best fit is observed between the *Parasite-tree* and *Host-tree-3,* which divide the Hystricognathi into two subgroups. The first one includes the Hystricidae as a sister-group of the pair Dinomyidae + Erethizontidae. The second group includes the Petromuridae together with most of the remaining Caviomorpha. Both trees also agree on the association of the Caviidae and Chinchillidae with the Abrocomidae. However, the *Parasite-tree* refutes the Octodontoidea and associates the Hydrochaeridae and Dolichotidae with the Hystricidae, Erethizontidae, and Dinomyidae.

Are the Old World and New World Porcupines Closely Related?

The cladogram of the parasites supports close relationships between the pinworms parasitic on porcupines (Hystricidae and Erethizontidae) and also groups the parasites of Erethizontidae and Dinomyidae. This poses two different questions: (1) what support can be found for these relationships in the different studies concerning the hosts? (2) how reliable is the support for the classification of these parasites in the same genus *(Wellcomia)*?

The grouping of the Dinomyidae and Erethizontidae had previously been proposed by Grant and Eisenberg (1982) and Woods and Hermanson (1985), who also suggested a close association of these two families with the Hystricidae. Recently, Landry (1999) proposed grouping all three families in a superfamily Erethizontoidea. All these works were based on morpho-anatomical data. To date, molecular studies have ignored the Dinomyidae and consequently have not provided any evidence to support or refute their affinities with the porcupines. Correspondingly, none of the recent studies based on molecular data have given any support to the grouping of the Hystricidae and Erethizontidae (Nedbal et al., 1994; Huchon et al., 1999, 2000). However, they generally classified the Hystricidae or Erethizontidae as the most divergent branch in their respective groups (Phiomorpha or Caviomorpha). Repeatedly, the support for the branching of these two groups on the cladogram of the Hystricognathi (decay indices or bootstrap values) appears to be very weak. In addition, molecular studies generally suffer from insufficient taxonomic sampling. The different species analyzed are not a comprehensive representation of the different families recognized within the Hystricognathi—each family included in a particular study was represented by one, rarely two, different species. Finally, if we also consider that the topologies proposed by different molecular analyses are highly divergent, the proximity of the Hystricidae and Erethizontidae, which has been repeatedly supported by morphological data, cannot be considered seriously challenged by current molecular analyses.

The second question relies on the morphological characters, which can be considered synapomorphic for genus *Wellcomia* as a whole and, within this genus, for the parasites of Hystricidae, Erethizontidae, and Dinomyidae (cladogram in fig. 6.1). The monophyly of genus *Wellcomia* is supported by characters 4:2 and 11:3. Character 4:2 corresponds to a differentiation of the impaired part of the female genital tract that secretes a sticky substance for holding the eggs together as they are hatched. Character 11:3 is a differentiation of the ventral cuticle of the males, the area rugosa, which in *Wellcomia* comprises 15 to 20 ranks of large cuticular crests. During mating the male pinworms arch the posterior part of their body, forming several loops around the body of the female. The area rugosa helps the male to steadily embrace the female. The grouping of the parasites of Hystricidae, Erethizontidae, and Dinomyidae relies on characters 6:2 and 9:1. Character 6:2 describes the particular disposition of the cephalic papillae (cephalic nervous endings) in this group. Character 9:1 deals with the presence in this group of a neoformation: the buccal triangular blades. This neoformation accompanies an important alteration of the symmetry

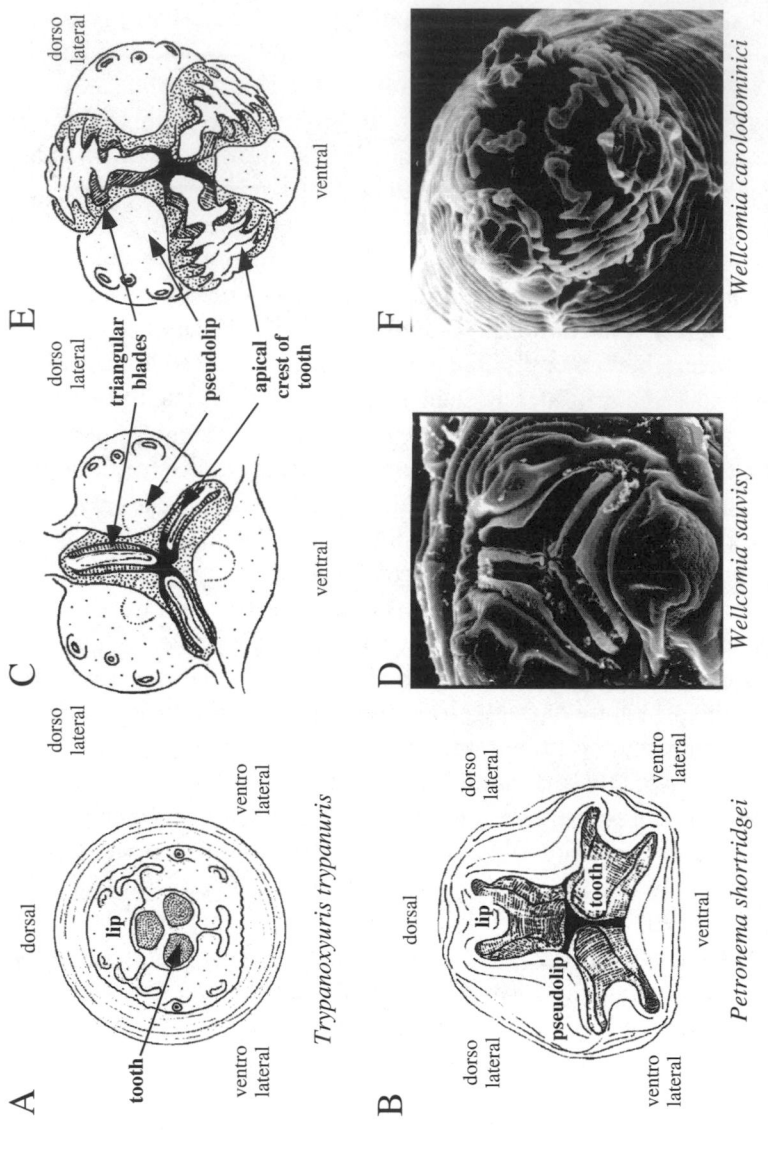

FIGURE 6.B.1. *A*, Head in apical view of *Trypanoxyuris trypanuris* Vever, 1923, (Oxyurida). *B*, Head in apical view of *Petronema shortridgei*. *C* and *D*, Head in apical view of *Wellcomia sauvisy*. *E* and *F*, Head in apical view of *Wellcomia carolodominici*. (Figures not at same scale; *D* and *F* are original scanning electron micrographs).

Box 6.1. Pinworm characters

A gives the general disposition of the head in apical view, which can be observed in most of the Oxyurida. Three esophageal teeth are visible in the buccal cavity. They are partly covered by three lips; the teeth and lips are disposed following one dorsal and two ventro-lateral axes of symmetry. In *B* the esophageal teeth are partly covered by three very reduced lips disposed following the same three axes. As three well-developed neoformations, the pseudolips are also visible: they are disposed following one ventral and two dorso-lateral axes of symmetry. In the genera *Protozoophaga, Wellcomia,* and *Helminthoxys,* the lips are absent and solely the pseudolips are present. Thus, in the parasites of Hystricognathi the symmetry of the teeth follows the general pattern observed in the Oxyurida, but the disposition of the pseudolips follows the opposed three axes of symmetry. Genus *Petronema,* in which three very reduced lips can be observed, represents a transitional arrangement between the generalized pattern and the derived pattern observed in the other three genera. Within genus *Wellcomia* this modification of the buccal symmetry is accompanied, in a group restricted to the parasites of Hystricidae, Erethizontidae, and Dinomyidae, by an additional neoformation: the triangular blades. These neoformations are three cuticular extensions, which develop under each pseudolip and are visible along the perimeter of the buccal cavity. In addition, in this group, an apical crest also develops at the top of each tooth such as the respective edges of the triangular blades and crests are corresponding. In different species the edges of these structures either can be linear, as observed in *Wellcomia sauvisy* (*C* and *D*), or deeply indented, as in *W. carolodominici* (*E* and *F*). In a previous paper (Hugot, 1982) I have pointed out that different species in genus *Wellcomia* can be arranged into a transformation series in which all the intermediates can be observed between the simple pattern of *C* and the complicated pattern of *E*.

and general disposition of the organs of the buccal cavity in *Wellcomia*, which is common to all the parasites of Hystricognathi, and is described in box 6.1. None of these characters has been observed in any other group within the Oxyurida. The inversion of the buccal symmetry, which is common to all the parasites of Hystricognathi (box 6.1), has no equivalent in any other group of Nematoda.

The numerous similarities observed within the members of genus *Wellcomia* can be alternatively interpreted as either reflecting close relationships among their hosts, or resulting from convergent evolution. The characters that can be considered synapomorphic for this genus (and particularly for the parasites of porcupines) are highly derived and profoundly affect the organization of the genital tract or sexual characters in both sexes, or the arrangement of the buccal structures. This phenomenon probably refutes the notion that these modifications would be the result of adaptive mechanisms dealing with host-parasite interactions. Thus, interpreting the evolution of these characters as homoplasic means that convergent modifications independently occurred in both the hosts (porcupines) and their parasites. This looks to me very improbable; I believe that taken together, these characters provide important arguments for close relationships among the hosts.

Evolutionary Scenarios and Migration Events

The Caviomorpha appeared simultaneously with the Platyrrhini (New World monkeys) in the Deseadan (late Oligocene) of Bolivia (Lavocat, 1969; Hoffstetter and Lavocat, 1970; Lavocat, 1971; Hoffstetter, 1972; Huchon et al., 1999). However, new discoveries suggest that the presence of these groups in South America could be of earlier origin. A caviomorph mandible referred to the Dasyproctidae has recently been reported from the late Eocene from Chile (Wyss et al., 1993). Because South America was an island continent during most of the Tertiary, and because pre-Deseadan localities have provided only mammal fossils referring to the old endemic fauna (xenarthrans, marsupials, and notoungulates), the origin of the Caviomorpha has to be questioned. Given that Wood's (1985) hypothesis of a North American origin of the ancestors of the Caviomorpha, crossing the Caribbean sea by island hopping or rafting, has been refuted (see above), the sole hypothesis is an Old World origin of the Caviomorpha as postulated by Hofstetter and Lavocat (Lavocat, 1969; Hoffstetter and Lavocat, 1970; Lavocat, 1971; Hoffstetter, 1972). This hypothesis requires that the ancestors of the Caviomorpha crossed the South Atlantic via rafting. Several authors also have postulated that the Erethizontidae (the New World porcupines) could have derived independently from African immigrants (Lavocat, 1969; Hoffstetter, 1972; Beintema, 1985; Lavocat and Parent, 1985; Woods and Hermanson, 1985). This supposes at least two independent trans-Atlantic migration events for the Caviomorpha.

The current distribution of the pinworms requires at least 2 (table 6.2) and up to 9 (table 6.4) independent migrations to America for the Caviomorpha. The scenarios needing 8 or 9 trans-Atlantic migration events (table 6.4, fig. 6.5B–C) assume that the Phiomorpha exchanged their parasites with the Caviomorpha after most of the current families recognized within the Caviomorpha had differentiated. This supposes that most of the radiation of the New World Hystricognathi took place in Africa, and these scenarios can be considered highly implausible. Thus, within the scenarios resulting from the comparison of the best documented and best fitting host and parasite trees, the most parsimonious and most reliable one is the scenario represented on figure 6.5D. This scenario supposes at least four migration events to explain the current distribution of the parasites of Hystricognathi.

Long-distance oceanic rafting strains credulity, and this increases if multiple trans-Atlantic colonizations have to be postulated. However, Landry (1999) remarked that

> the acceptance of anthropologists . . . that the higher primates, too, must have reached their present distribution by trans-Atlantic rafting [and] since it is unlikely that monkeys and porcupines crossed on the same log . . . there have been at least two examples of trans-Atlantic rafting of vertebrates during the Tertiary [thus] if two are conceded, further crossings are more defendable.

In connection with long-distance oceanic rafting, Landry (1999) also mentioned Censky et al. (1998) documenting the colonization of an island 300 km away by iguanas landing from a storm-driven log raft, and cites Lavocat (in Rowlands and Weir, 1974; discussion, p. 56) reporting the finding of "a raft with trees and five rodents 1000 kilometers from any coast."

Conclusion

The phylogeny of the pinworm parasites of hystricognath rodents (1) supports the monophyly of the Hystricognathi; (2) does not support the monophyly of the Phiomorpha or Caviomorpha; (3) supports close relationships among the Hystricidae, Dinomyidae, and the Erethizontidae, thus their classification in a superfamily Erethizontoidea; and (4) challenges the hypothesis of a single immigration event for the arrival of the Caviomorpha in America. These points also are supported by most of the works based on morpho-anatomical studies and dealing with the phylogeny and evolutionary history of the Hystricognathi. Conversely, the hypotheses based on the distribution and phylogeny of the parasites generally disagree with the conclusions of work based on molecular data.

Parasitological data were initially challenged (Patterson and Wood, 1982) because the specificity and origin of the parasites were questionable. Our data set relies on parasite specimens whose host specificity and precise collection locality cannot be disputed. The very specialized characters of the parasites used in the cladistic analysis are not closely linked with host-parasite interactions, thus the probability that the appearance of similar morphological characters in different parasite groups is the result of convergence looks to be low. Therefore, the different scenarios based on the cladogram of the parasites provide important arguments for discussion of the relationships among and within the host groups. Additional studies of the phylogenetic relationships within the hosts or the parasites are now needed for a better resolution of the points under controversy. Especially, the monophyly of the Caviomorpha needs to be tested using several different molecular markers. On the other hand, morpho-anatomical data have not been collected from a large sample of species, and have yet to be analyzed extensively. The exact relationships between the Old World and New World porcupines also have to be made clearer. To complete the description of the parasites, the search for new specimens is also necessary.

Acknowledgments

This work was partly funded by a NATO collaborative research grant no. CRG 920612 to S. L. Gardner, J. P. H., and S. Morand and by the Groupe de Travail Morphométrie et Analyse de Forme du Musém National d'Histoire Naturelle. This is a publication of EA 2586.

REFERENCES

Anderson, S., and J. Knox Jones Jr. 1984. *Orders and families of recent mammals of the world.* New York: Wiley.

Barus, V. 1972. Remarks on the Cuban species of the genus *Helminthoxys* (Nematoda, Syphaciidae). *Folia Parasitologica (Praha)* 19:105–11.

Baylis, H. A. 1922. Notes on some parasitic nematodes. I.—On the genus *Wellcomia* Sambon, and a new species of that genus. II.—A new species of *Ascaris* from Armadillo. III.—Note on two species of *Porrocaecum* from birds. *Annals and Magazine of Natural History,* ser. 9, 9:494–504.

Beintema, J. J. 1985. Amino acid sequence data and evolutionary relationships among Hystricognaths and other rodents. In *Evolutionary relationships among rodents. A multidisciplinary analysis,* edited by W. P. Luckett and J.-L. Hartenberger, 549–65. New York: Plenum Press.

Bryant, J. D., and M. C. McKenna. 1995. Cranial anatomy and phylogenetic position of *Tsaganomys altaicus* (Mammalia: Rodentia) from the Hsanda Gol formation (Oligocene), Mongolia. *American Museum Novitates* 3156:1–42.

Bugge, J. 1985. Systematic value of the carotid arterial pattern in rodents. In *Evolutionary relationships among rodents. A multidisciplinary analysis,* edited by W. P. Luckett and J.-L. Hartenberger, 355–80. New York: Plenum Press.

Cameron, T. W. M., and M. R. Reesal. 1951. Studies of endoparasitic fauna of Trinidad Mammals. VII. Parasites of hystricomorph rodents. *Canadian Journal of Zoology* 29:276–89.

Censky, E. J., K. Hodge, and J. Dudley. 1998. Over-water dispersal of lizards due to hurricanes. *Nature* 395:556.

Diesing, K. M. 1851. *Systema Helminthium. II.* Berlin: N.p.

Eriksson, T. 1996. Autodecay, version 2.9.5 (Hypercard stack distributed by the author). Stockholm: Botaniska institutionen, Stockholm University.

Freitas Texeira, J. F., L. Lent, and L. L. Almeida. 1937. Pequena contribuiçao ao estudo da fauna helmintologica de Argentina (Nematoda). *Memorio del Instituto Oswaldo Cruz, Rio de Janeiro* 32:195–209.

George, W. 1985. Reproductive and chromosomal characters of ctenodactylids as a key to their evolutionary relationships. In *Evolutionary relationships among rodents. A multidisciplinary analysis,* edited by W. P. Luckett and J.-L. Hartenberger, 453–74. New York: Plenum Press.

Grant, T. I., and J. F. Eisenberg. 1982. On the affinities of the Dinomyidae. *Säugetierkunde Mitteilungen* 30:151–57.

Hartenberger, J.-L. 1985. The order Rodentia: Major questions on their evolutionary origin, relationships and suprafamilial systematics. In *Evolutionary relationships among rodents. A multidisciplinary analysis,* edited by W. P. Luckett and J.-L. Hartenberger, 1–34. New York: Plenum Press.

Hoffstetter, R. 1972. Origine et dispersion des Rongeurs hystricognathes. *Comptes Rendus de l'Académie des Sciences Série III—Sciences De La Vie-Life, Paris* 274:2867–70.

Hoffstetter, R., and R. Lavocat. 1970. Découverte dans le Déséadien de Bolivie des genres pentalophodontes appuyant les affinités africaines des rongeurs caviomorphes. *Comptes Rendus de l'Académie des Sciences Série III—Sciences De La Vie-Life, Paris* 271:172–75.

Huchon, D., F. M. Catzeflis, and E. J. P. Douzery. 1999. Molecular evolution of the nuclear von Willebrand factor gene in mammals and the phylogeny of rodents. *Molecular Biology and Evolution* 16:577–89.

———. 2000. Variance of molecular datings, evolution of rodents, and the phylogenetic affinities between Ctenodactylidae and Hystricognathi. *Proceedings of the Royal Society of London,* ser. B, 267:393–402.

Hugot, J. P. 1982. Sur le genre *Wellcomia* (Oxyuridae, Nematoda), parasite de Rongeurs archaïques. *Bulletin du Muséum National d'Histoire naturelle, Paris 4° série* 4:25–48.

———. 1983a. Deux Oxyures parasites de *Petromus typicus,* un Rongeur sud-africain archaïque. *Bulletin du Muséum National d'Histoire naturelle, Paris 4° série* 5:187–99.

———. 1983b. Redescription d'*Helminthoxys tiflophila* et considération sur la systématique des Oxyuridae parasites de Rongeurs. *Annales de Parasitologie Humaine et Comparée* 58:255–65.

———. 1985. Sur le genre *Acanthoxyurus* (Oxyuridae, Nematoda). Etude morphologique. *Bulletin du Muséum National d'Histoire naturelle, Paris 4° série* 7:157–79.

———. 1986. Etude morphologique d'*Helminthoxys urichi* (Oxyurata, Nematoda), parasite de *Dasyprocta aguti* (Caviomorpha, Rodentia). *Bulletin du Muséum National d'Histoire naturelle, Paris 4° série* 8:133–38.

————. 1988. Les nématodes Syphaciinae parasites de Rongeurs et de Lagomorphes. Taxonomie. Zoogéographie. Evolution. *Mémoires du Muséum National d'Histoire naturelle, Paris,* Sèrie A, Zoologie 141:1–153.

Hugot, J.-P., and S. L. Gardner. 2000. Description of *Helminthoxys abrocomae* n. sp. (Nematoda: Oxyurida) from *Abrocoma cinerea* in Bolivia. *Systematic Parasitology* 47:223–30.

Hugot, J.-P., and C. A. Sutton. 1989. Etude morphologique de deux oxyures appartenant au genre *Helminthoxys. Bulletin du Muséum National d'Histoire naturelle, Paris 4° série* 11:387–95.

Landry Jr., S. O. 1999. A proposal for a new classification and nomenclature for the Glires (Lagomorpha and Rodentia). *Mitteilungen des Museums für Naturkunde, Zoologische Reihe* 75:283–316.

Lavocat, R. 1969. La systématique des Rongeurs hystricomorphes et la dérive des continents. *Comptes Rendus de l'Académie des Sciences, Paris* 269:1496–97.

————. 1971. Affinités systématiques des caviomorphes et des phiomorphes et origine africaine des caviomorphes. *Anais da Academia Brasileira de Ciències (suplemento)* 43:515–22.

Lavocat, R., and J. P. Parent. 1985. Phylogenetic analysis of middle ear features in fossil and living rodents. In *Evolutionary relationships among rodents. A multidisciplinary analysis,* edited by W. P. Luckett and J.-L. Hartenberger, 333–54. New York: Plenum Press.

Leidy, J. 1856. A synopsis of Entozoa and some of their ectocongeners observed by the author. *Proceedings of the Academy of Natural Sciences of Philadelphia* 8:42–58.

Luckett, W. P., and J.-L. Hartenberger. 1985. Comments and conclusions. In *Evolutionary relationships among rodents. A multidisciplinary analysis,* edited by W. P. Luckett and J.-L. Hartenberger, 685–712. New York: Plenum Press.

MacLure, G. W. 1932. Nematode parasites of mammals with a description of a new species *Wellcomia branickii* from specimens collected in the New York Zoological Park, 1930. *Zoologica (New York)* 15:1–28.

Maddison, W. P. 1993. Missing data versus missing characters in phylogenetic analysis. *Systematic Biology* 42:576–81.

Maddison, W. P., and D. R. Maddison. 1992. MacClade: Analysis of phylogeny and character evolution, version 3. Sunderland, Mass.: Sinauer Associates.

Martin, T. 1994. African origin of caviomorph rodents is indicated by incisor enamel microstructure. *Paleobiology* 20:5–13.

Mönnig, H. O. 1931. A second species of the nematode genus *Acanthoxyurus. Report of the Veterinary Research Union of South Africa* 17:269–72.

Nedbal, M. A., M. W. Allard, and R. L. Honeycutt. 1994. Molecular systematics of hystricognath rodents: Evidence from the mitochondrial 12S rRNA gene. *Molecular Phylogenetics and Evolution* 3:206–20.

Page, R. D. M. 1994. Parallel phylogenies: Reconstructing the history of host-parasite assemblages. *Cladistics* 10:155–73.

————. 1996a. Temporal congruence revisited: Comparison of mitochondrial DNA sequence divergence in cospeciating pocket gophers and their chewing lice. *Systematic Biology* 45:151–67.

———. 1996b. TreeView: An application to display phylogenetic trees on personal computers. *Computer Applications in the Biological Sciences* 12:357–58.

Parra Ormeño, M. S. B. E. 1953. Estudio de dos nuevos Helminthos intestinales de *Lagidium peruanum*. *Publicaciones del Museo de Historia Natural, Seria A, Zoologia* 11:1–26.

Patterson, B., and A. E. Wood. 1982. Rodents from the Deseadan Oligocene of Bolivia and the relationships of Caviomorpha. *Bulletin of the Museum of Comparative Zoology* 149:371–543.

Quentin, J. C. 1969. *Helminthoxys freitasi* n. sp., Oxyure parasite d'un Rongeur Echimyidae du Brésil. *Bulletin du Muséum National d'Histoire naturelle, Paris* 2° série 41:579–83.

———. 1973a. Affinités entre les Oxyures parasites de Rongeurs Hystricidés, Erethizontidés et Dinomyidés. Intérêt paléobiogéographique. *Comptes Rendus de l'Académie des Sciences, Paris* 276:2015–17.

———. 1973b. Les Oxyurinae de Rongeurs. *Bulletin du Muséum National d'Histoire naturelle, Paris* 3° série 167:1045–96.

Quentin, J. C., S. S. Courtin, and J. A. Fontecilla. 1975. *Octodontoxys gigantea* n. gen., n. sp. Nuevo nemàtodo parásito de un roedor caviomorfo de Chile. *Boletín Chileno de Parasitología* 30:21–25.

Rowlands, I. W., and B. J. Weir. 1974. The biology of the hystricomorph rodents. *Boletín Chileno de Parasitología* 34:1–492.

Sambon, L. 1907. Descriptions of some new species of animal parasites. *Proceedings of the Zoological Society, London* 1:282–83.

Sandground, J. H. 1928. Some new cestode and nematode parasites from Tanganyka Territory. *Proceedings of the Boston Society of Natural History* 39:131–50.

Schurmans-Stekhoven, J. H. 1951. Nematodos parásitos de anfibios, pájaros y mamíferos de la República Argentina. *Acta Zoológica Lilloana* 10:315–400.

Seurat, L. G. 1915. Nématodes parasites. Expédition de M.M. Walter Rothschild, E. Hartet et C. Hilgert dans le sud Algérien. *Novitates zoologicae* 22:1–25.

Sutton, C. A., and J. P. Hugot. 1987. Contribution à la connaissance de la faune parasitaire d'Argentine, XVIII. Etude morphologique de *Wellcomia dolichotis* n. sp. (Oxyuridae, Nematoda), parasite de *Dolichotis patagonica*. *Systematic Parasitology* 10:85–93.

Swofford, D. L. 1999. *PAUP*. Phylogenetic Analysis Using Parsimony (*and Other Methods)*. Sunderland, Mass.: Sinauer Associates.

Travassos, L. 1923. Informacoes sobre a fauna helmintologica de Matto Grosso. *Folha medical, Rio de Janeiro* 4:58–60.

Vigueras, I. P. 1943. Un genero y cinco especias nuevas de helminthos cubanos. *Universidad Habana* 8:315–56.

Wilson, D. E., and D. M. Reeder. 1993. *Mammal species of the world. A taxonomic and geographic reference.* 2d ed. Washington, D.C.: Smithsonian Institution Press.

Wood, A. E. 1950. Porcupines, paleogeography and parallelism. *Evolution* 4:87–98.

———. 1972. An Eocene hystricognathous rodent from Texas: Its significance in interpretation of continental drift. *Science* 175:1250–51.

———. 1985. The relationships, origin and dispersal of the hystricognathous rodents. In *Evolutionary relationships among rodents. A multidisciplinary analysis,* edited by W. P. Luckett and J.-L. Hartenberger, 475–513. New York: Plenum Press.

Woods, C. A., and J. W. Hermanson. 1985. Myology of hystricognath rodents: An analysis of form, function and phylogeny. In *Evolutionary relationships among rodents. A multidisciplinary analysis,* edited by W. P. Luckett and J.-L. Hartenberger, 515–48. New York: Plenum Press.

Wyss, A. R., J. J. Flynn, M. A. Norell, C. C. Swisher, R. Charrier, M. J. Novacek, and M. C. McKenna. 1993. South America's earliest rodent and recognition of a new interval of mamalian evolution. *Nature* 365:433–37.

APPENDIX 6.1

Taxa analyzed in this study.

Family Oxyuridae Cobbold, 1864
Subfamily Syphaciinae Railliet, 1916

Genus *Hilgertia* Quentin (1973b): *H. hilgerti* (Seurat, 1915); host taxa: *Ctenodactylus gundi* (Rothmann), **Algeria.**

Genus *Acanthoxyurus* Sandground (1928): *A. beecrofti* Hugot (1985); host taxa: *Anomalurus beecrofti* Peters, **Gabon.**

Genus *Petronema* Hugot (1983): *P. shortridgei* (Mönnig, 1931); host taxa: *Petromus typicus* A. Smith, **South Africa.**

Genus *Wellcomia* Sambon (1907): *W. branickii* MacLure (1932); host taxa: *Dinomys branickii* Peters, **Bolivia.** *W. carolodominici* Hugot (1982); host taxa: *Coendou prehensilis* (L.), **French Guyana.** *W. compar* (Leidy, 1856); host taxa: *Erethizon dorsatum* L., **Nearctic.** *W. samboni* Baylis (1922); host taxa: *Sphiggurus spinosus* Cuvier, **Paraguay.** *W. dolichotis* Sutton and Hugot (1987); host taxa: *Dolichotis patagonica* (Zimmermann), **Argentina.** *W. roussilhoni* Hugot (1982); host taxa: *Hystrix cristata* L., **Senegal.** *W. sauvisy* Hugot, 1982; host taxa: *Atherurus africanus* Gray, **Gabon.**

Genus *Protozoophaga* Travassos (1923): *P. obesa* (Diesing, 1851); host taxa: *Hydrochaeris hydrochaeris* (L.), [Venezuela].

Genus *Helminthoxys* Freitas Texeira, Lent and Almeida (1937): *H. abrocomae* n. sp. Hugot and Gardner (2000); host taxa: *Abrocoma cinerea* Thomas, **Bolivia.** *H. caudatus* Freitas Texeira, Lent and Almeida (1937); host taxa: *Microcavia australis* (Geoffroy & d'Orbigny), **Argentina.** *H. effilatus* Schurmans-Stekhoven (Schurmans-Stekhoven, 1951); host taxa: *Lagidium viscacia* (Molina), **Argentina.** *H. freitasi* Quentin (1969); host taxa: *Trichomys aperoides* Lund, **Brazil.** *H. gigantea* Quentin (1975); host taxa: *Octodon degus* (Molina), **Chile,** and *Octodon bridgeri Waterhouse,* **Argentina.** *H. pujoli* Quentin (1973b); host taxa: *Microcavia niata* (Thomas), **Bolivia.** *H. quentini* Barus (1972); host taxa: *Capromys pilorides*

(Say), **Cuba**. *H. tiflophila* (Vigueras, 1943); host taxa: *Capromys prehensilis* Poepping, **Cuba**. *H. urichi* Cameron and Reesal (1951); host taxa: *Dasyprocta agouti* (L.), **Trinidad, French Guyana**. *H. velizy* Parra Ormeño (1953); host taxa: *Lagidium peruanum* Meyen, **Bolivia, Peru**.

Appendix 6.2

List of characters in data matrix (appendix 6.3).

1. *Oviduct walls.*—0 = generalized type, 1 = *Wellcomia* type with toothbrush epithelium.
2. *Spermatheca*—0 = dilatation of oviduct, 1 = S-shaped, 2 = *Wellcomia* type.
3. *Vagina.*—0 = nonevaginated, 1 = evaginated *Hilgertia* type, 2 = evaginated *Wellcomia* type.
4. *Impair part of uterus.*—0 = with a U-shaped strong muscular part, 1 = muscular part atrophied, cuticular part very long, vulva in the posterior part of the body, 2 = secretory, *Wellcomia* type, 3 = Petronema type.
5. *Egg.*—0 = one pole rounded, the other one sharp, 1 = spherical, 2 = oval, 3 = bean-shaped, 4 = lemon-shaped, 5 = globular with an apical operculum.
6. *Cephalic papillae.*—0 = square disposition, 1 = rectangular disposition, 2 = trapezoidal disposition, 3 = grouped laterally, 4 = grouped backward.
7. *Lips.* 0 = developed, 1 = reduced, 2 = absent.
8. *Pseudo-lips.*—0 = absent, 1 = present, not covering the mouth opening, 2 = present, covering the mouth opening, 3 = present, with tonguelike extension, 4 = *Protozoophaga* type.
9. *Interlabia.*—0 = absent, 1 = triangular blade.
10. *Esophageal bulb.*—0 = spherical, 1 = ovale, longer than wide, 2 = piriform.
11. *Area rugosa.*—0 = including slots *Syphacia* type, 1 = including ranks of small-sized cuticular crests fitting with cuticular striation, 2 = including ranks of medium sized cuticular crests, 3 = including ranks of large cuticular crests.
12. *If 11.*— = 2.—0 = first 3 ranks grouped, 1 = first 6 ranks grouped.
13. *Pseudo-mamelon.*—0 = absent, 1 = present.
14. *Secretory-mamelon.*—0 = absent, 1 = present.
15. *Male cuticular striation.*—0 = thin, 1 = broad.
16. *Tip of male tail.*—0 = long and sharp, 1 = short and conical.
17. *Cuticular ornamentation of female tail.*—0 = absent, 1 = present.
18. *Arrangement of the first two pairs of male genital papillae.*—0 = square, pedunculated, 1 = square nonpedunculated, 2 = lined, pedunculated, 3 = lined, nonpedunculated.
19. *Corpus of gubernaculum.*—0 = *Syphacia* type, 1 = globular, 2 = lengthened and narrow, 3 = lengthened, narrow, and very short, 4 = atrophied.
20. *Accessory hook of male gubernaculum.*—0 = *Syphacia* type, 1 = rough-surfaced, 2 = V-shaped, symetric, 3 = V-shaped, asymmetric, 4 = trifid, 5 = atrophied.

21. *Lateral wings.*—0 = simple-crested, rounded on a transversal cut,
1 = simple-crested, broad, triangular on a transversal cut, 2 = simple-crested,
narrow, triangular on a transversal cut, 3 = atrophied, 4 = double-crested.
22. *Cervical wings.*—0 = absent, 1 = present, *Petronema* type, 2 = present,
Acanthoxyurus type, 3 = present, *Helminthoxys* type, 4 = present, *Wellcomia*
type.

APPENDIX 6.3

Data matrix for the parasites (deposited as TreeBASE
study number S540, matrix number M794)

Taxon	Character
	1111111111222
	1234567890123456789012
Hilgertia hilgerti	00100000000-0000000000
Acanthoxyurus beecrofti	11003110000-0000000042
	1
Petronema shortridgei	12034411001-0100012011
Wellcomia branickii	1222323110311001121100
Wellcomia carolodominici	1222323110311000121100
Wellcomia compar	1222323110310000121110
Wellcomia decorata	1222323110311000121100
Wellcomia dolichotis	12222131003-0000021114
Wellcomia roussillhoni	1202223110320001021100
Wellcomia sauvisy	1202223110320001021100
Protozoophaga obesa	12021034003-00010045?0
Helminthoxys abrocomae	12004332012-0100012223
Helminthoxys gigantea	12004332022-0100012313
Helminthoxys freitasi	1200?332022-0100012313
Helminthoxys caudatus	12005332012-0100032403
Helminthoxys pujoli	12005332012-0100032403
Helminthoxys effilatus	12005332012-0110032423
Helminthoxys velizy	12005332012-0110032423
Helminthoxys quentini	12002332012-0100013230
Helminthoxys tiflophila	12002332012-0100013230
Helminthoxys urichi	1201?133002-0100012210

APPENDIX 6.4

Data matrix for the rodents. This matrix combines several data matrices previously published by different authors (Bugge, 1985; George, 1985; Hartenberger, 1985; Lavocat and Parent, 1985; Woods and Hermanson, 1985; Bryant and McKenna, 1995). *Host-tree-3* in figure 6.2 is the consensus majority rule of the 13 equally parsimonious trees for this matrix.

Taxon / Source study	(Hartenberger, 1985)	(Bugge, 1985)	(Lavocat and Parent, 1985)	(Woods and Hermanson, 1985)	(George, 1985)	(Bryant and McKenna, 1995)
Character	11	111	111112222	2222223333	3333334	4444444444555555
	12345678901	234	567890123	4567890123	4567890	1234567890123456
Number in source study	1111122222					
	9037891234	236	249012570	2368280145	1234670	189012345802456
Anomaluridae	?1010100000	010	210001110	??????????	0000000	1101101010000000
Ctenodactylidae	00000000111	100	221121211	??????????	0000111	2111122011000030
Erethizontidae	01000101000	011	011110210	1100000001	1111010	2211011111111111
Dinomyidae	?100101110	111	011110210	??????????	1111010	2211011111111111
Hystricidae	01010101000	111	011110210	1100000001	1111111	1111012011111111
Petromuridae	1?100101101	111	221110120	1010101100	?111111	2111011021111121
Octodontidae	10100101000	111	222220121	0011101110	1111111	2122021020111131
Abrocomidae	00101101111	111	222220121	??????????	1111111	????????????????
Echimyidae	00010101100	111	222220121	??????????	1111111	2122021020111131
Capromyidae	???????????	111	222220121	0011111110	1111111	2122021020111131
Chinchillidae	?1101111111	111	222220121	??????????	1111111	2122021020111111
Caviidae	?1100011111	111	222220121	??????????	1011101	2122021020111111
Dolichotidae	?1100011111	101	222220121	??????????	1011101	2122021020111111
Hydrochaeridae	?1101111111	111	222220121	??????????	1011101	2122021020111111
Dasyproctidae	00001101100	111	222220121	1000010000	1111001	2122021020111111

7

COSPECIATION AND HORIZONTAL TRANSMISSION RATES IN THE MURINE LEUKEMIA–RELATED RETROVIRUSES

Joanne Martin, Peter Kabat, and Michael Tristem

Introduction

For the past several years we have been investigating a variety of questions regarding the evolution of retroviruses (see box 7.1); especially those concerned with retroviral diversity and distribution (Martin et al., 1997; Herniou et al., 1998). Another theme of our work is the study of retroviral horizontal transmission dynamics. There have been several reported cases of the infectious transmission of retroviruses between species, including HIV1 and HIV2 from monkeys to humans (Martin et al., 1997; Herniou et al., 1998), but the overall frequency of retroviral transmission is unclear. Furthermore, the factors influencing the interspecies transmission rate are also poorly understood at present. These factors include the phylogenetic distance between potential hosts and the life history of the host, as well as the viral genotypes themselves.

We have started to address these questions by constructing host versus virus phylogenies. The viruses we have chosen for these studies are termed the murine leukemia-related viruses (MLVs) (Coffin, 1992). This genus of viruses was chosen because it was known that they were harbored by several classes of vertebrates, although sequence information had only been obtained from those within mammalian and avian hosts. The MLVs comprise a large group of exogenous and endogenous viruses, which were first described in association with the etiology of "spontaneous" white-blood-cell cancers in certain strains of mice (Gross, 1951; Levy, 1973), and have since been linked to a variety of diseases. Sequence diversity of the MLV genus has been studied primarily in the context of its member's role in disease, particularly focusing upon those affecting domesticated species and nonhuman primates. While the pathogenicity of MLVs has been well studied in domesticated mammals and birds, the comparatively small amount of

Box 7.1

Retroviruses are enveloped viruses that encode a (+) sense nonsegmented RNA genome. The best known retrovirus is undoubtedly HIV, the causative agent of AIDS in humans, but there are many other types that cause a variety of diseases in a wide variety of vertebrates. Retroviruses have the ability to undergo a process called reverse transcription, in which they convert their RNA genomes into DNA. The viral DNA can then be inserted into the genome of the host organism. Because this insertion event can sometimes occur in germ cells, retroviruses occasionally become fixed in the genome of their host species and can be passed vertically for long periods of time: tens of millions of years in the case of some human viruses. Vertically transmitted viruses, which are termed endogenous, often become defective because they are subject to random mutation while the host is undergoing DNA replication. In contrast, active, nondefective viruses such as HIV are termed exogenous. Exogenous viruses (as well as some nondefective endogenous viruses) are capable of horizontal transmission between individuals of the same species and occasionally between species.

Although there is some variation in genomic organization all retroviruses encode the same basic set of gene products. The *gag* (group-specific antigen) open reading frame (ORF) encodes structural proteins, which make up the retroviral core, whereas the *pol* (polymerase) ORF contains several proteins involved in viral replication, including a Reverse Transcriptase (RT), an Integrase (Int), and a Protease (Pro). Most retroviruses also contain a third ORF, termed *env*, which encodes envelope glycoproteins. Differences in genomic organization have been used to classify retroviruses at the genus level: there are currently eight recognized genera in the family Retroviridae.

data collected from other vertebrates has hindered efforts to piece together the ancestry of these viruses. Until recently, the accepted retroviral classification (partly based on morphology) divided the MLV-related viruses into 3 subgenera (Coffin, 1992): (1) the mammalian C types; (2) the avian reticuloendotheliosis viruses (REVs); and (3) the reptilian C types (for which no sequence information was available). The evolutionary relationships within and between these subgenera were poorly understood due to the lack of sequence information for phylogenetic analyses.

The mammalian MLV subgenus consists of two subgroups, here termed type 1 and type 2. The type 1 subgroup contains the majority of the prototypical MLVs, including three fully sequenced exogenous mammalian viruses termed FeLV (Feline leukemia virus), GaLV (Gibbon ape leukemia virus), MuLV (Murine leukemia virus), and numerous endogenous isolates (Shinnick et al., 1981; Overbaugh et al., 1988; Delassus et al., 1989). The smaller type 2 subgroup contains only endogenous elements at present, and only one of them (the human endogenous retrovirus HERV.E) has been completely sequenced (Repaske et al., 1985; Tristem et al., 1996). HERV-E differs from many of the type 1 mammalian viruses by having a primer binding site (the initiation site for viral (-) strand replication) that complements the 3′ end of the rat glutamic acid tRNA, whereas other mammalian MLVs often have primer binding sites that complement a proline tRNA (Shinnick et al., 1981; Repaske et al., 1985). Several sequence motifs conserved between the type 1 MLVs are conspicuously absent in HERV-E, and the *env* gene region of this element shows little homology with known infectious retroviruses (Repaske et al., 1985; Tristem et al., 1996). A number of HERV-E-related sequences have been detected in the genomes of apes and Old World monkeys, and have been determined to occupy the same genomic locus in each host (Shih et al., 1991; Wilkinson et al., 1994). This indicates that an infectious HERV-E element inserted into the genome of a common ancestor of the apes and Old World monkeys. The point in the primate lineage when apes and Old World monkeys are thought to have shared a common ancestor is an estimated 27–36 million years ago (Purvis, 1995), indicating that the mammalian MLVs are at least this old.

The REVs are a group of MLV-related viruses isolated from domesticated poultry, and all the currently recognized members of this group are exogenous (Payne, 1992). With no evidence for these avian viruses having integrated into the germ line of their hosts, it appears that birds may have acquired REVs relatively recently (Payne, 1992), and that the mammalian MLV subgenus (with numerous endogenous counterparts described) is older than the REVs. While sequence data from the *env* gene are available from a number of REVs, and the *v-rel* oncogene carried by REV-T (of turkeys) is well studied (Wilhelmsen et al., 1984), *pol* gene information is available from only one member of this subgenus, spleen necrosis virus (SNV), a highly pathogenic virus of chickens and ducks (Weaver et al., 1990).

The reptilian C type subgenus comprises just two snake viruses. Until recently they were the only MLV-related viruses to have been identified in cold-blooded vertebrates. Viruses from the Corn snake and Russell's viper were classified with the MLVs on the basis of morphological observations (Zeigel and Clark, 1971; Lunger et al., 1974), but their sequence similarities to mammalian and avian MLVs remained unexplored.

Methods

We investigated the distribution and phylogenetic relationships of the MLVs by amplifying endogenous viruses from a variety of vertebrate hosts using PCR. Our endogenous retroviral sequences are obtained by PCR (Tristem, 1996). Retroviruses have a high degree of sequence variation, but there are some regions of the genome, largely with the polymerase gene, that are reasonably well conserved. We typically amplify a 1Kb region containing the 3' end of the Pro protein and the 5' end of RT. Our PCR primers target one of several well-conserved motifs situated at or near the active sites of the two proteins. In the case of the sequences presented in this chapter, one Pro primer (PRO; 5' GTT/G TTI G/TTI GAT/C ACI GGI G/TC 3', where I represents inosine) was used in combination with one of two RT primers (MLV; 5' AGI GTI GGI GAA/G TTC/T TTA/G AA 3' or CT; 5' AGI AGG TCA/G TCI ACA/G TAG/C TG 3'). Reverse transcriptases contain a series of conserved domains (Xiong and Eickbush, 1990), and the MLV and CT primers target domains 4 and 5, respectively. Amplified products are electrophoresed through 1.3% agarose gels and bands of the appropriate size excised. Because the amplified product is often derived from several different elements, the fragments are cloned before sequencing. All retroviral fragments are back-hybridized to genomic DNA from the host species to confirm their origin.

Amino acid sequences derived from viruses within the same genus are easily aligned across most of amplified region. For example, excluding postinsertion indels and a small linker region between protease and reverse transcriptase, only 4 gaps (two of 1 residue and two of 2 residues) need to be inserted to align all the viruses included in the phylogenies presented here. The phylogenies shown below were constructed in PAUP* (Swofford, 1999) utilizing an alignment 260 amino acid residues in length. Tanglegrams of host versus virus phylogenies were produced using TREEMAP (Page, 1994) and MacClade (Maddison and Maddison, 1992). Statistical significance was calculated by hand, and by using the program STATVIEW.

Results

Over one hundred vertebrate species (14 mammals, 21 birds, 28 reptiles, 24 amphibians, and 21 fish) were screened for MLVs, and a total of 48 of these viruses were isolated from 45 species (table 7.1). MLV-related fragments were identified in all four terrestrial vertebrate classes, but not from fish species. More than one MLV-related element was amplified from 14 taxa, but in most cases (11 of 14), multiple elements from the same taxon were found to be more than 95% similar at the nucleotide level, and are therefore not included in the total above. In the three remaining cases, one taxon was found to harbor two elements with less than 75% similarity at the nucleotide level (Partridge: ~63% similarity; Regent's bower bird: ~60%

TABLE 7.1 Vertebrate host taxa from which retroviral fragments were obtained

Host Taxon*	Retroviral Designation
Mammals	
Phascolarctos cinereus (w)	RV-Koala
Monodelphis spp. (w)	RV-Opossum
Tachyglossus aculeatus (w)	RV-Short-Beaked Echidna
Rattus exulans (w)	RV-Polynesian Rat
Rattus rattus flavipectus (w)	RV-Roof Rat (I + II)
Mus caroli (w)	RV-SE Asian Feral Mouse(*M.car*)
Mus cervicolor (w)	RV-SE Asian Feral Mouse(*M.cer*)
Birds	
Perdix perdix (w)	RV-Partridge (I + II)
Turdus merula (w)	RV-Blackbird
Turdus iliacus (w)	RV-Redwing
Troglodytes troglodytes (w)	RV-Wren
Columba palumbus (w)	RV-Wood Pigeon
Sericulus bakeri (w)	RV-Regent's Bower Bird (I + II)
Corvus frugilegus (w)	RV-Rook
Phasianus colchicus (w)	RV-Pheasant
Corvus monedula (w)	RV-Jackdaw
Erithacus rubecula (w)	RV-Robin
Pica pica (w)	RV-Magpie
Gallus gallus domesticus (c)	RV-Chicken
Sturnus vulgaris	RV-Starling
Reptiles—Lizards	
Varanus komodoensis (w)	RV-Komodo Dragon
Anolis carolinensis (c)	RV-Green Anole
Reptiles—Crocodilians	
Osteolaemus tetraspis (w)	RV-African Dwarf Crocodile
Tomistoma schlegelii (w)	RV-False Gharial
Reptiles—Snakes	
Boa constrictor (?)	RV-Boa Constrictor
Bitis arietans (w)	RV-Puff Adder
Bothrops jararaca (w)	RV-Pit Viper
Thamnophis sirtalis (c)	RV-Garter Snake
Viper berus (w)	RV-European Adder
Nerodia harteri (w)	RV-Concho Water Snake
Atheris nitscheri (w)	RV-Bush Viper
Spilote pullatus (w)	RV-Chicken Snake
Lamprophis fuliginosus (w)	RV-Brown House Snake
Philothamnus heterodermus (w)	RV-Colubrid Snake spp.
Amphibians—Gymnophiona	
Epicrionops spp. (w)	RV-Caecilian (Epi)
Ichthyophis bannanicus (w)	RV-Caecilian (Ban)
Ichthyophis kohtaoensis (w)	RV-Caecilian (IchI+II)
Ichthyophis cf *tricolor* (w)	RV-Caecilian (Tri)

TABLE 7.1 (*continued*).

Host Taxon*	Retroviral Designation
Gegeneophis raraswanii (w)	RV-Caecilian (Geg)
Gymnophus syntrema (w)	RV-Caecilian (Gym)
Microcaecilia spp. (w)	RV-Caecilian (Mic)
Schistometopum thor (w)	RV-Caecilian (Sch)
Dermophis mexicanus (w)	RV-Caecilian (Der)
Amphibians—Anura	
Rana esculenta (c)	RV-Edible Frog
Bufo calamita (?)	RV-Natterjack Toad

* DNA was isolated from a wild-caught animal (w), a captive animal (c), or an animal of unknown status (?).

similarity; Caecilian (Ich): ~70% similarity). In these cases both elements were included in subsequent phylogenetic reconstruction.

The 48 fully characterized endogenous retroviral fragments amplified using the PRO/CT primer pair ranged in length between 912 and 984 base pairs (average 949 base pairs), with the exception of one of the partridge clones, which had two large deletions and was only 810 base pairs in length. Fragments amplified using the PRO/MLV primer pair (extending only into domain 4 of the RT gene) ranged between 849 and 888 base pairs in length. Eleven elements contained the entire region of the *pol* gene sequenced in one uninterrupted open reading frame (as is the case with infectious exogenous viruses), while the remaining 30 fragments contained one or more in-frame stop codon and/or probable frame shift mutations.

An amino acid alignment was prepared based on the entire region of the genome amplified (excluding a small region between Pro and RT), which included 44 novel MLV-related fragments and 20 previously described members of the MLV genus, all but one of which were derived from mammalian hosts (see table 7.2). The previously described exogenous avian reticuloendotheliosis virus, SNV, was also included to investigate its relationship to the novel endogenous elements derived from 13 species of (mainly nondomesticated) birds.

Phylogenetic analyses were performed on an amino acid alignment of 260 residues consisting of 78 residues derived from Pro and 182 derived from RT. Numerous trees were generated using the neighbor-joining and maximum parsimony approaches. Trees were rooted either on the HERV.I-related viruses (Martin et al., 1997) or on more distantly related group III retroviruses (Herniou et al., 1998). Trees based on alignments of the protease gene alone, and each individual domain of the reverse transcriptase gene, were also constructed. Because there were no clear

TABLE 7.2 Other retroviral isolates and corresponding host taxa discussed in this chapter

Host Taxon	Retroviral Designation
Mammals	
Hylobates lar (lar gibbon)	GaLV* (Gibbon ape leukemia virus)
Felis domesticus (domestic cat)	FeLV* (Feline leukemia virus)
Mus musculus (house mouse)	MuLV* (Murine leukemia virus)
Mus musculus (house mouse)	MuRRS (Murine retrovirus–related sequence)
Papio sp. (baboon)	BaEV (Baboon endogenous virus)
Vulpes vulpes (red fox)	VuEV (*Vulpes* endogenous virus)
Meles meles (Eurasian badger)	MeEV (*Meles* endogenous virus)
Mustela vison (American mink)	MiEV (Mink endogenous virus)
Tadarida brasiliensis (free-tailed bat)	TaEV (*Tadarida* endogenous virus)
Halichoerus grypus (gray seal)	HaEV (*Halichoerus* endogenous virus)
Bos taurus (domestic cow)	BoEV (Bovine endogenous virus)
Ovis aries (domestic sheep)	OvEV (Ovine endogenous virus)
Homo sapiens (human)	HC2 (human clone 2)
Oryctolagus cuniculus (rabbit)	OrEV (*Oryctolagus* endogenous virus)
Sus scrofa (domestic pig)	PERV (Porcine endogenous retrovirus)
Birds	
#Duck	SNV (Spleen necrosis virus)

* Exogenous isolate.

Natural host is the duck, but the sequence here was actually derived from an isolate integrated into the vaccine strain of the Fowlpox virus.

differences in these phylogenies compared with those constructed using the complete alignment, we think it unlikely that there has been much recombination in our data set.

Murine Leukemia–Related Virus Phylogeny

Figure 7.1 shows a typical maximum-parsimony-derived tree recovered during our analyses. All trees were consistent in clustering the snake, crocodile, type 1 mammalian isolates, and most of the avian viruses into distinct, well-supported clades. Neighbor-joining trees placed the two crocodile fragments as the most ancestral elements, whereas maximum parsimony trees placed the crocodilian and the avian elements into a basal clade. Although not well supported by bootstrap analyses, most trees also placed the type 2 mammalian MLVs into a single lineage. We had previously observed that the type 1 and type 2 mammalian groups clustered together in both neighbor-joining and maximum parsimony analyses. However, this was not supported with high bootstrap values (Martin et al., 1999), and the inclusion of extra caecilian sequences has now had the effect of splitting

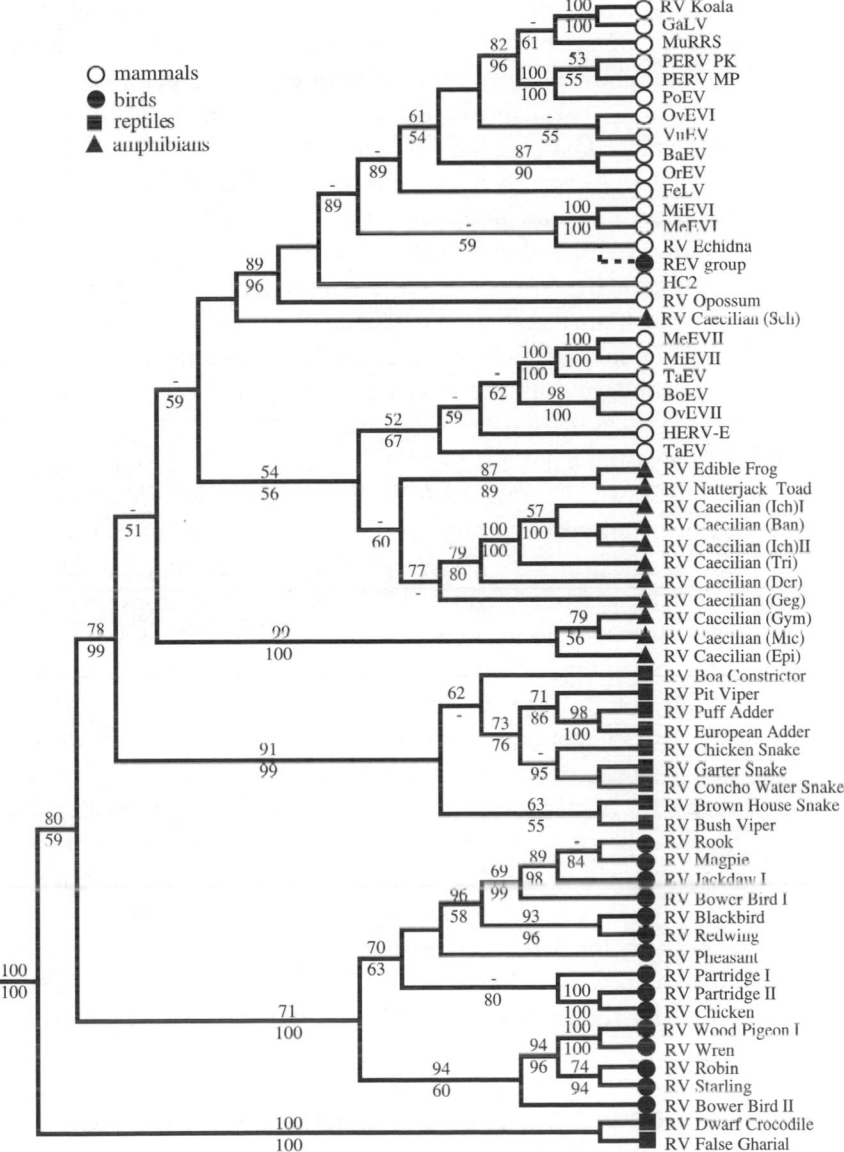

FIGURE 7.1. Rooted majority rule consensus of four maximum parsimony trees constructed from a 260-amino-acid alignment of retroviral polymerase proteins, showing the host class from which the virus was isolated. The figure above each branch represents bootstrap support from 100 maximum parsimony replicates, whereas the lower figure represents bootstrap support from 1,000 neighbor-joining replicates. The dotted line shows the location of the REV group (polymerase sequence information is currently available from only one member of this group).

up this clade in the majority of analyses. Most maximum parsimony trees placed one group of amphibian viruses with the type 2 mammalian MLVs and a second amphibian clade basal to both of the mammalian clades. In addition, a single caecilian sequence (derived from *Schistometopum thor*) was often placed basal to the type 1 mammalian clade (fig. 7.1). Most neighbor-joining trees weakly supported inclusion of all the amphibian-derived sequences into one group clustering with the type 2 mammalian MLVs (not shown), although some placed the *Schistometopum thor* virus in the same position (at the base of the type 1 mammalian subgroup) as seen in the maximum parsimony analyses. No bootstrap support was ever achieved with either maximum parsimony or neighbor joining for the positioning of the caecilian sequence with the type 1 mammalian MLVs, so the exact relationship of this clone remains unresolved.

All but one of the avian MLV–related viruses were placed in a well-supported cluster, which appeared to be divided into two groups: one of 10 isolates (termed the Rook group after the prototype member) and one of five (including the prototype RV-Wren). The average amino acid identity shown between all 15 avian MLVs was 45%, while calculations separating the two groups resulted in average identities of 62% (Rook group) and 59% (Wren group). One avian MLV (spleen necrosis virus of chickens and ducks) did not cluster with viruses derived from other avian hosts in any of the trees generated. While three MLVs from other fowl were placed within the avian clade, SNV consistently and robustly grouped with a novel echidna virus (95% bootstrap support) among the type 1 mammalian MLVs. This was the only example of a virus from one vertebrate class clustering robustly within a clade of isolates from another. As SNV is very closely related to several other reticuloendotheliosis viruses, the group as a whole is probably located in this position. However, with the exception of SNV, no sequence information from the appropriate region of *pol* is available for other REVs, and hence they are indicated with a dotted line in figure 7.1.

The clustering of viruses derived from hosts of the same vertebrate class is quite striking, and there are two scenarios which could explain the observed patterns: (1) the MLVs integrated or evolved within an ancestral vertebrate, and endogenous elements have been cospeciating with their hosts ever since; or (2) the MLVs arose at some point during the evolution of the vertebrates and have occasionally transmitted infectiously into a new vertebrate class and radiated within that class. Alternatively, a combination of these two scenarios may best explain the observed patterns of MLV distribution.

The phylogeny presented in figure 7.1 demonstrates some evidence for the first scenario—that MLVs have been passaged vertically for long periods of time, and have therefore tracked the phylogenies of their hosts. Not only do MLVs cluster into groups of elements derived from a single vertebrate class, but also some of the clustering within those groups reflects the phylogenies of the hosts. The avian MLVs show a separate clustering together of sequences from the Corvidae (rook, jackdaw, and magpie), Muscicapidae (blackbird and redwing) and Phasianidae (chicken, partridge, and pheasant). Furthermore, within the snake clade, several of the viruses derived from the Viperidae (pit viper, puff adder, and European adder) cluster together, separate from the colubrids and boas. However, the number of snake taxa is small at present, and further sequences are clearly required. Although we identified numerous MLVs within caecilians, there is, as yet, no clear phylogeny for the hosts; they have therefore been excluded from the other analyses presented here.

In contrast with the viruses within birds and snakes, there appeared, in general, to be less of a correlation between the mammalian viruses and their hosts. For example, viruses isolated from rabbits (OrEV) and baboons (BaEV) appear relatively closely related, although their hosts are not, and this was also the case for other viral sister taxa such as those derived from koala and echidna hosts. One possible reason for the viral topology is that viruses are possibly being transmitted between mammals more frequently when compared with their avian or reptilian counterparts (i.e., viruses with mammalian hosts may have a higher incidence of intraclass transmission than the viruses in other vertebrate classes). To investigate this hypothesis in more detail, we constructed host versus virus phylogenies of the avian, reptilian, and mammalian viruses and determined to what extent the viruses in each class were cospeciating with their hosts.

Are MLVs Cospeciating with Their Hosts?

Figure 7.2a shows a mammalian host-versus-virus phylogeny, with the host phylogeny being derived from the literature (as were all the other host phylogenies presented here). The two mammalian virus clades were combined in this particular analysis. The number of cospeciating and noncospeciating nodes was determined using the heuristic search option in TREEMAP (Page, 1994); there were 11 cospeciating nodes and 11 noncospeciating nodes. To determine whether the level of cospeciation was significantly more than would be expected given the number and relationships of the host taxa, 100 random viral phylogenies were constructed using the shuffle option in MacClade (Maddison and Maddison, 1992). The randomized

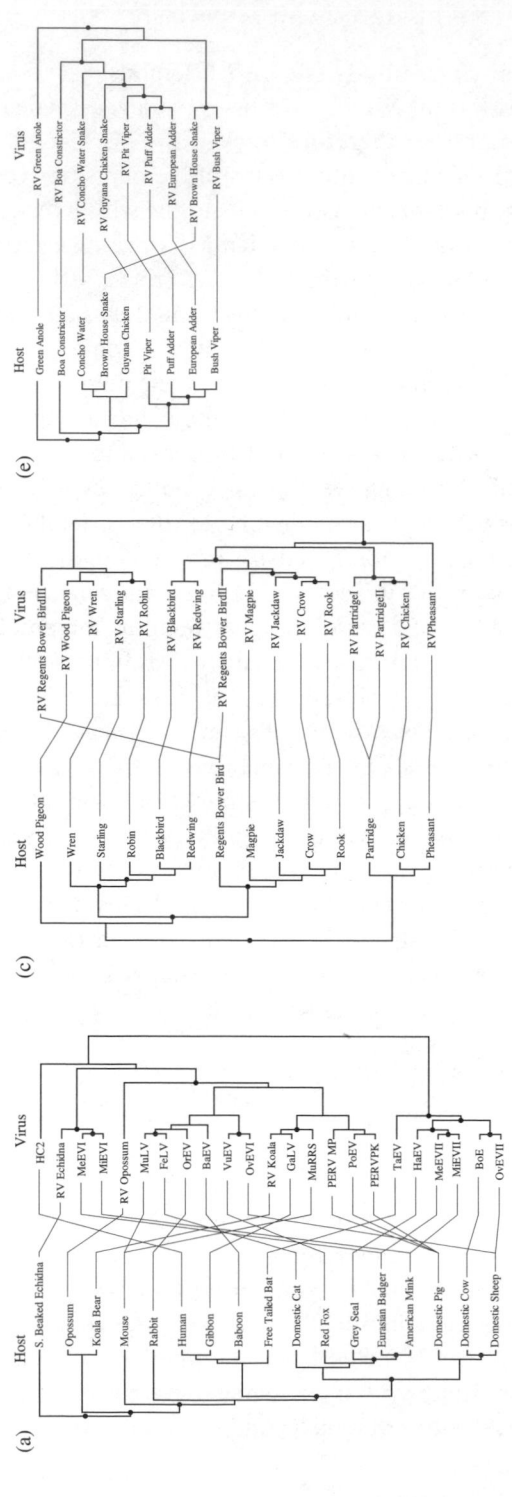

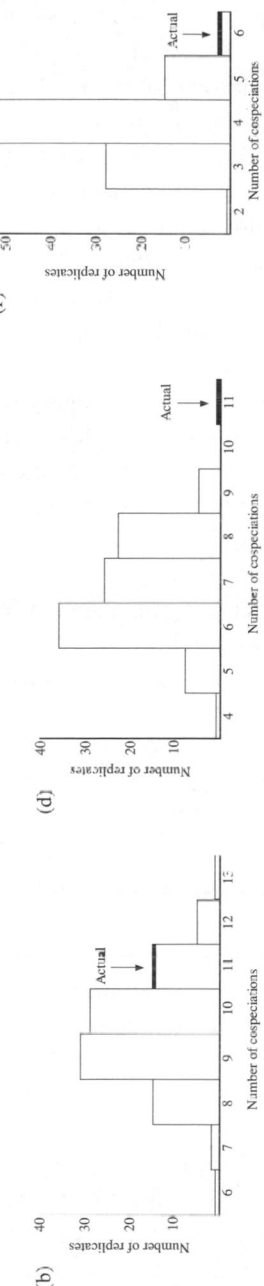

FIGURE 7.2. Cospeciation frequencies of mammalian (*a* and *b*), avian (*c* and *d*), and reptilian (*e* and *f*) MLVs and their hosts. For each set of viruses and their hosts a tanglegram (*a*, *c*, *e*) was constructed in TREEMAP (Page, 1994). Filled circles represent cospeciating nodes, with the lines between the two phylogenies connecting the viruses with the host taxon from which they were derived. The viral phylogeny was generated in PAUP* (Swofford, 1999) using the maximum parsimony approach; the host phylogeny was obtained from the literature. The bar charts (*b*, *d*, *f*) compare the number of cospeciating nodes occurring in the tanglegrams to those occurring in 100 randomly generated viral phylogenies. The random phylogenies were generated in MacClade (Maddison and Maddison, 1992), and the number of cospeciating nodes were then calculated in TREEMAP.

viral phylogenies were imported into TREEMAP and each was then subjected to a heuristic search. Figure 7.2b shows a bar chart of the number of cospeciating nodes in each of the 100 random viral phylogenies as well as the actual phylogeny. Sixteen of the 100 random phylogenies had the same or a greater number of cospeciating nodes than the actual phylogeny (range 6–13, mean 9.6), indicating that the level of cospeciation between the mammalian viruses and their hosts was not significant.

The same approach was then pursued in order to determine whether the level of cospeciation in the avian (fig. 7.2c and d) and reptilian (fig. 7.2e and f) data sets was significant. In the case of the avian viruses the randomly generated phylogenies all showed a lower level of cospeciating nodes (range 4–9, mean 6.8) than the 11 in the actual phylogeny ($p < .01$). Although the number of taxa in the reptilian data set was lower, the level of cospeciation was still significant (2–6 cospeciating nodes with a mean of 3.9 in the randomly generated phylogenies, compared with 6 in the actual phylogeny), with a p value of $<.05$.

To test our hypothesis that mammalian viruses may have a higher level of horizontal transmission (or at least a lower level of cospeciation) than the viruses in reptiles or birds, we examined whether the numbers of cospeciating and noncospeciating nodes in each of the three viral phylogenies were significantly different from each other. Two different methods were employed: the G-test (Crawley, 1993) and the Fisher exact test (Siegel, 1956). All the comparisons between the cospeciation rates of the viruses in the different vertebrate classes (e.g., mammalian viruses versus avian viruses) gave nonsignificant results (see table 7.3). The overall lack of significant differences revealed by these tests imply that host life-history factors cannot, as yet, be implicated in causing elevated levels of retroviral horizontal transmission. The fact that a virus is present within a mammalian host does not, in itself, suggest that it has a greater potential for horizontal transmission than if it were present within an avian or reptilian host. However, as discussed above, it is possible that viral genotypes, rather than host life-history traits, could be important in determining transmission frequencies.

To test this second hypothesis we made use of the observation that there are two lineages of mammalian viruses in figure 7.1: the MLV subgroup (type 1) and the HERV.E subgroup (type 2). These two lineages can be investigated independently of each other, and any significant differences between them would suggest that the rate of cospeciation is being influenced at least partly by viral genotype. Host versus virus phylogenies of the two lineages indicated that the level of cospeciation is not significant in the

TABLE 7.3 Significance of cospeciation to noncospeciation frequencies between mammalian, avian, and reptilian viruses

	Cospeciations	Noncospeciations	Row Totals	G Test	Fisher Exact Test
Mammalian versus avian					
Mammalian	11	11	22		
Avian	11	4	15	NS	NS
Column totals	22	15	37		
Mammalian versus reptilian					
Mammalian	11	11	22		
Reptilian	6	2	8	NS	NS
Column totals	17	13	30		
Avian versus reptilian					
Avian	11	4	15		
Reptilian	6	2	8	NS	NS
Column totals	17	6	23		

*NS = not significant

TABLE 7.4 Significance of cospeciation to noncospeciation frequencies between type 1 and 2 mammalian viruses

	Cospeciations	Noncospeciations	G Test	Fisher Exact Test
Type 1 (MLV subgroup)	6	10		
			N/A	$p < 0.05$
Type 2 (HERV.E subgroup)	5	0		

type 1 mammalian viruses (fig. 7.3a), whereas it is significant within the type 2 clade (fig. 7.3b). In contrast with the previous statistical analyses, there was a significant difference in the rates of cospeciation in the two viral groups. A comparison of cospeciation to noncospeciation frequencies using the Fisher exact test gave a p value of .035, as shown in table 7.4 (the G test could not be used, as the number of noncospeciations in the type 2 clade was zero).

Splitting the mammalian viruses into two clades has another effect, as it allows the cospeciation rates in four viral groups (the type 1 and 2 mammalian viruses, the avian viruses, and the reptilian viruses) to be compared and contrasted. Contingency tables constructed in STATVIEW (table 7.5) demonstrated that there were no significant differences in cospeciation rates between the last three of the four groups indicated above ($\chi^2 = 1.67, p = .43$), but there was a significant difference when the type 1 viruses were included in the analysis ($\chi^2 = 8.64, p = .035$). It was therefore

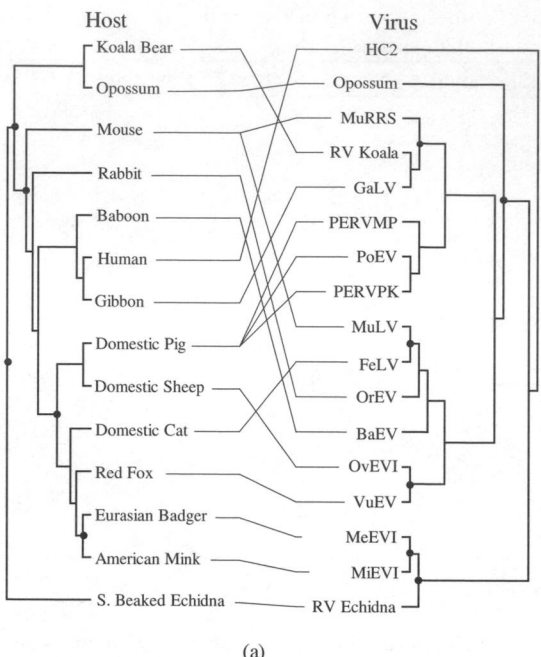

(a)

FIGURE 7.3. Cospeciation rates between the mammalian type 1 *(a)* and type 2 *(b)* MLVs and their hosts. In these analyses the mammalian viruses have been split into two groups, as represented by the two lineages indicated in figure 7.1. Host versus virus phylogenies and the levels of cospeciation were calculated as in figure 7.2.

TABLE 7.5 3-by-2 contingency table of cospeciation and noncospeciation frequencies

	Cospeciations	Noncospeciations
Type 2 mammalian viruses	5	0
Avian viruses	11	4
Reptilian viruses	6	2

$\chi^2 = 1.669, p = .43$

4-by-2 contingency table of cospeciation and noncospeciation frequencies

	Cospeciations	Noncospeciations
Type 1 mammalian viruses	6	10
Type 2 mammalian viruses	5	0
Avian viruses	11	4
Reptilian viruses	6	2

$\chi^2 = 8.636, p = .035$

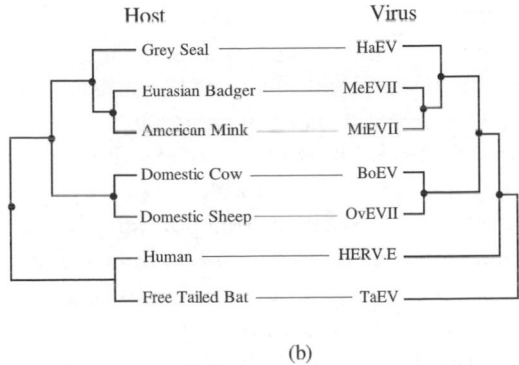

(b)

FIGURE 7.3 (*continued*).

TABLE 7.6 Significance of cospeciation to noncospeciation frequencies between type 1 mammalian viruses and other vertebrate MLVs

	Cospeciations	Noncospeciations	G Test	Fisher Exact Test
Type 1 mammalian viruses	6	10		
			$p < .01$	$p < .01$
Type 2 mammalian, avian, and reptilian viruses	22	6		

appropriate to compare cospeciation rates in the type 1 mammalian viruses against cospeciation rates in the other three viral groups combined (table 7.6). This comparison was highly significant using both the G test ($p = .007$) and Fisher exact test ($p = .009$), strongly suggesting that the genotype of the type 1 mammalian viruses is causing reduced rates of cospeciation in this viral lineage. It is our opinion that the most likely explanation for this is that the type 1 mammalian viruses have elevated horizontal transmission frequencies when compared with the other viruses in the phylogeny.

Discussion

Although there appears to be a relatively high level of intraclass transmission among the type 1 mammalian MLVs, the sheer number of both the potential host taxa and viruses means that it has been extremely difficult to determine the exact pattern and timing of individual horizontal transmission events, although there are one or two exceptions (Benveniste and Todaro, 1974; Van Der Kuyl et al., 1995; Mang et al., 2000). Studies of several endogenous retroviruses in the genomes of Old World monkeys have revealed a picture of repeated horizontal transmission events over the past 9 million years, with many of the transmission events occurring between species that still share the same habitats in the present day (Van Der Kuyl et al., 1995; Mang et al., 2000). Studies such as these are important, as they may eventually enable us to track the migration and transmission patterns of viruses over long periods of time.

Another group of viruses where we are gaining insights into these processes is the Gibbon ape leukemia–related viruses (GaLV) (Delassus et al., 1989). GaLV, an exogenous virus, is very closely related to an endogenous retroviral fragment we identified in a koala bear (approximately 85% nucleotide identity across 900 bp of *pol*), suggesting that at least one of this group of viruses has been horizontally transmitted between placental and marsupial mammals in the recent past (Martin et al., 1999). However, as the two host taxa do not share the same geographic host range (and have not done so recently), it is probable that one or more alternative hosts have vectored the virus between the two species (Martin et al., 1999).

GaLV-like viruses have previously been shown (generally via Southern hybridization) to be present in the genomes of several Southeast Asian rodents, including *Mus caroli, Mus cervicolor,* and *Vandeleuria oleracea,* although no sequence information has been reported (Benveniste and Todaro, 1973, 1974; Lieber et al., 1975; Callahan et al., 1979). Furthermore, there is evidence to suggest that rodents have migrated from Southeast Asia to Australia on several occasions in the past, indicating that they may have vectored GaLV-like viruses between the two continents. For the above reasons we screened a number of Southeast Asian rodents for the presence of endogenous GaLV-like viruses, and detected them in the genomes of several species of rats and feral mice (including *M. caroli* and *M. cervicolor*). Figure 7.4 shows a maximum parsimony phylogeny of this viral group, which also includes the recently described endogenous virus MDEV, isolated from *Mus dunni,* another Asian rodent (Wolgamot et al., 1998). Although none of the rodent viruses split GaLV and RV-Koala, the phylogeny is consistent with the proposal that rodents may well be the vectoring species in this case. The recent transmission of members of this viral group is supported by the nondefective nature of most of the viral fragments shown in figure 7.4. All had a single, uninterrupted ORF across the amplified region (with the exception of the *M. caroli* virus, which had a single frame shift), suggesting they have only recently integrated into the genomes of their hosts. Furthermore, in addition to GaLV (which is exogenous), a C-type virus has been implicated in spontaneous lymphoid

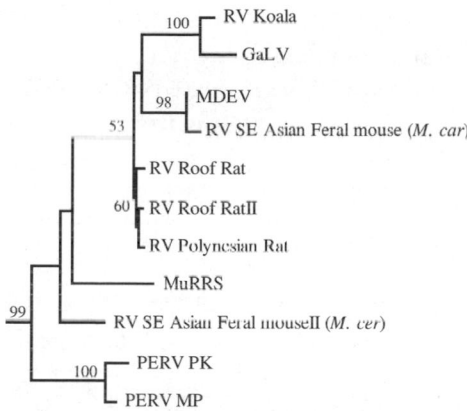

FIGURE 7.4. Maximum parsimony phylogeny of a subgroup of type 1 mammalian MLVs demonstrating that the horizontal transmission of these viruses may be mediated by rodent vectors. Figures on each branch represent bootstrap support from 100 replicates. All sequences were obtained by us except PERV PK and MP (porcine endogenous retrovirus: Le Tissier et al., 1997), MDEV (*Mus dunni* endogenous virus: Wolgamot et al., 1998), and MuRRS (Murine retrovirus–related sequence: Schmidt et al., 1985).

neoplasia in the koala (Canfield et al., 1988), raising the possibility that the RV-koala isolate may still be capable of producing infectious viral particles.

Future Directions

We intend to continue our efforts to elucidate the patterns and processes of the dynamics of retroviral horizontal transmission. As the MLV data set expands we expect to gain further insights into individual transmission events such as those described within the GaLV-related viruses above, and to increase our understanding into the more general patterns of transmission. For example, we have not yet examined whether viruses present within carnivores have a greater potential for transmission than those viruses present within other hosts, and there may be many other such host life-history factors that can be investigated. Furthermore, although we have yet to find an effect of the host class of origin in influencing the viral transmission rate (as opposed to viral genotype), this may still become apparent once the data set is expanded significantly.

We also intend to construct a second data set, based on a group of viruses related to the MLVs termed the HERV.I-related viruses (after their prototypic member human endogenous retrovirus type I: Maeda and Kim, 1990). These viruses have the largest host range of any viral group (Martin et al., 1997), and preliminary analysis suggests they may be more capable of interclass transmission than the MLV-related viruses. Thus there are likely to be many lineages of HERV.I-related viruses within each vertebrate class in which they are present, and this will enable factors associated with infectious horizontal transmission to be more easily investigated using independent contrast analysis (Purvis and Rambaut, 1995).

Acknowledgments

We thank A. Purvis, J. Taylor, and J. Cook for comments and discussion, A. Burt for help with STATVIEW, and Robin Lawson for providing many of the DNA and tissue samples. This work was supported by the NERC Initiative in Taxonomy, and by the Wellcome Trust.

REFERENCES

Benveniste, R. E., and G. J. Todaro. 1973. Homology between type C viruses of various species as determined by molecular hybridisation. *Proceedings of the National Academy of Science of the USA* 70:3316–20.

———. 1974. Evolution of C-type viral genes: Inheritance of exogenously acquired viral genes. *Nature* 252:456–59.

Callahan, R., C. Meade, and G. J. Todaro. 1979. Isolation of an endogenous type C virus related to the infectious primate type C viruses from the Asian rodent *Vandeleuria oleracea. Journal of Virology* 30:124–31.

Canfield, P. J., J. M. Sabine, and D. N. Love. 1988. Virus particles associated with leukemia in a koala. *Australian Vetinary Journal* 65:327–28.

Coffin, J. M. 1992. Structure and classification of retroviruses. In *The retroviridae,* edited by J. A. Levy, 19–49. New York: Plenum Press.

Crawley, M. J. 1993. *GLIM for ecologists.* Oxford, U.K.: Blackwell Scientific.

Delassus, S., P. Sonigo, and S. Wain-Hobson. 1989. Genetic organization of gibbon ape leukemia virus. *Virology* 173:205–13.

Gross, L. 1951. "Spontaneous" leukemia developing in C3H mice following inoculation in infancy, with A-K leukemic extracts, or A-K embryos. *Proceedings of the Society of Experimental Biology and Medicine* 76:27–32.

Herniou, E., J. Martin, K. Miller, J. Cook, M. Wilkinson, and M. Tristem. 1998. Retroviral diversity and distribution in vertebrates. *Journal of Virology* 72:5955–66.

Le Tissier, P., J. P. Stoye, Y. Takeuchi, C. Patience, and R. A. Weiss. 1997. Two sets of human tropic pig retrovirus. *Nature* 389:681–82.

Levy, J. A. 1973. Xenotropic viruses: Murine leukaemia viruses associated with NIH Swiss, NZB, and other mouse strains. *Science* 182:1151–3.

Lieber, M. M., C. J. Sherr, G. J. Todaro, R. E. Benveniste, R. Callahan, and H. G. Coon. 1975. Isolation from the Asian mouse *Mus caroli* of an endogenous type C virus related to infectious primate type C viruses. *Proceedings of the National Academy of Sciences of the USA* 72:2315–19.

Lunger, P. D., W. D. Hardy, and H. F. Clark. 1974. C-type particles in a reptilian tumor. *Journal of the National Cancer Institute* 52:1231–35.

Maddison, W. P., and D. R. Maddison. 1992. MacClade: Analysis of phylogeny and character evolution, version 3. Sunderland, Mass.: Sinauer Associates.

Maeda, R. E., and H. S. Kim. 1990. Three independent insertions of retrovirus-like sequences in the haptoglobin gene cluster of primates. *Genomics* 8:671–83.

Mang, R., J. Maas, A. C. van der Kuyl, and J. Goudsmit. 2000. *Papio cynocephalus* endogenous retrovirus among Old World monkeys: Evidence for coevolution and ancient cross-species transmissions. *Journal of Virology* 74:1578–86.

Martin, J., E. Herniou, J. Cook, R. Waugh O'Neill, and M. Tristem. 1997. Human endogenous retrovirus type I-related viruses have an apparently widespread distribution within vertebrates. *Journal of Virology* 71:437–43.

———. 1999. Interclass transmission and phyletic host tracking in murine leukaemia virus related retroviruses. *Journal of Virology* 73:2442–49.

Overbaugh, J., P. R. Donahue, S. L. Quackenbush, E. A. Hoover, and J. I. Mullins. 1988. Molecular cloning of a feline leukaemia virus that induces fatal immunodeficiency disease in cats. *Science* 239:906–10.

Page, R. D. M. 1994. Parallel phylogenies: Reconstructing the history of host-parasite assemblages. *Cladistics* 10:155–73.

Payne, L. N. 1992. Biology of avian retroviruses. In *The Retroviridae,* edited by J. A. Levy, 299–389. New York: Plenum Press.

Purvis, A. 1995. A composite estimate of primate phylogeny. *Philosophical Transactions of the Royal Society of London,* ser. B, 348:405–21.

Purvis, A., and A. Rambaut. 1995. Comparative analysis by independent contrasts (CAIC): An Apple Macintosh application for analysing comparative data. *Computer Applications in the Biological Sciences* 11:247–51.

Repaske, R., P. E. Steele, R. R. O'Neill, A. B. Rabson, and M. A. Martin. 1985. Nucleotide sequence of a full-length human endogenous retroviral segment. *Journal of Virology* 54:764–72.

Schmidt, M., T. Wirth, B. Kroger, and I. Horak. 1985. Structure and genomic organization of a new family of murine retrovirus–related DNA sequences (MuRRS). *Nucleic Acids Research* 13:3461 70.

Shih, A., E. E. Couvavas, and M. G. Rush. 1991. Evolutionary implications of primate endogenous retroviruses. *Virology* 182:495–501.

Shinnick, T. M., R. A. Lerner, and J. G. Sutcliffe. 1981. Nucleotide sequence of Moloney murine leukaemia virus. *Nature* 293:543–48.

Siegel, S. 1956. *Nonparametric statistics for the behavioural sciences.* New York: McGraw Hill.

Swofford, D. L. 1999. *PAUP*. Phylogenetic Analysis Using Parsimony (*and other methods).* Sunderland, Mass.: Sinauer Associates.

Tristem, M. 1996. Amplification of divergent retroelements by PCR. *Biotechniques* 20:608–12.

Tristem, M., P. Kabat, L. Lieberman, S. Linde, A. Karpas, and F. Hill. 1996. Characterization of a novel murine leukemia virus–related subgroup within mammals. *Journal of Virology* 70:8241–46.

Van Der Kuyl, A. C., J. T. Dekker, and J. Goudsmit. 1995. Distribution of Baboon endogenous virus among species of African green monkeys suggests multiple ancient cross-species transmissions in shared habitats. *Journal of Virology* 69:7877–87.

Weaver, T. A., K. J. Talbot, and A. T. Panganiban. 1990. Spleen necrosis virus gag polyprotein is necessary for particle assembly and release but not for proteolytic processing. *Journal of Virology* 64:2642–52.

Wilhelmsen, K. C., K. Eggleton, and H. T. Temin. 1984. Nucleic acid sequences of the oncogene v-rel in reticuloendotheliosis virus strain T and its cellular homolog, the proto-oncogene c-rel. *Journal of Virology* 52:172–82.

Wilkinson, D. A., D. L. Mager, and J. C. Leong. 1994. Endogenous human retroviruses. In *The Retroviridae,* edited by J. A. Levy, 465–535. New York: Plenum Press.

Wolgamot, G., L. Bonham, and A. D. Miller. 1998. Sequence analysis of *Mus dunni* endogenous virus reveals a hybrid VL30/gibbon ape leukemia virus-like structure and a distinct envelope. *Journal of Virology* 72:7459–66.

Xiong, Y., and T. H. Eickbush. 1990. Origin and evolution of retroelements based upon their reverse transcriptase sequences. *EMBO Journal* 9:3353–62.

Zeigel, R. F., and H. F. Clark. 1971. Histologic and electron microscopic observations on a tumor-bearing viper: Establishment of a C type virus-producing cell line. *Journal of the National Cancer Institute* 46:309–21.

8

COPHYLOGENY BETWEEN POCKET GOPHERS

AND CHEWING LICE

Mark S. Hafner, James W. Demastes,
Theresa A. Spradling, and David L. Reed

Introduction

Since first reported by Timm (1979, 1983), evidence of cophylogeny between pocket gophers (rodents of the family Geomyidae: box 8.1) and chewing lice (insects of the order Phthiraptera: box 8.2) has become so conclusive that the gopher-louse system is now perhaps the preeminent "textbook example" of cophylogeny (e.g., Noble et al., 1989; Esch and Fernández, 1993; Ridley, 1993; Brown and Lomolino, 1998; Page and Holmes, 1998). This pattern of cophylogeny between gophers and lice is now known to extend to the very deepest branches in the gopher tree (Hafner et al., 1994), suggesting that this host-parasite association has remained intact for several million years (based on known fossil dates for the gopher phylogeny: Russell, 1968). The gopher-louse assemblage is among the few mammal-parasite systems in which cospeciation has been documented (see chapter 10), and it is, by far, the most intensively investigated case study of cospeciation to date. The intense interest in this system and the central role it has played in development of methods for analysis of cophylogeny, leads one to ask, Why pocket gophers and chewing lice?

Herein, we argue that pocket gophers and chewing lice represent an unusual—but probably not unique—symbiosis between creatures whose life histories are highly conducive to parallel speciation. We will review these life-history attributes of gophers and lice, followed by a general discussion of major findings that have emerged from study of this unusual system. Finally, we will highlight several potential directions for future research on the gopher-louse assemblage, emphasizing how knowledge of this system has the potential to transcend gophers and lice to elucidate our general understanding of speciation, coevolution, rates of evolution, and host-parasite ecology.

Box 8.1

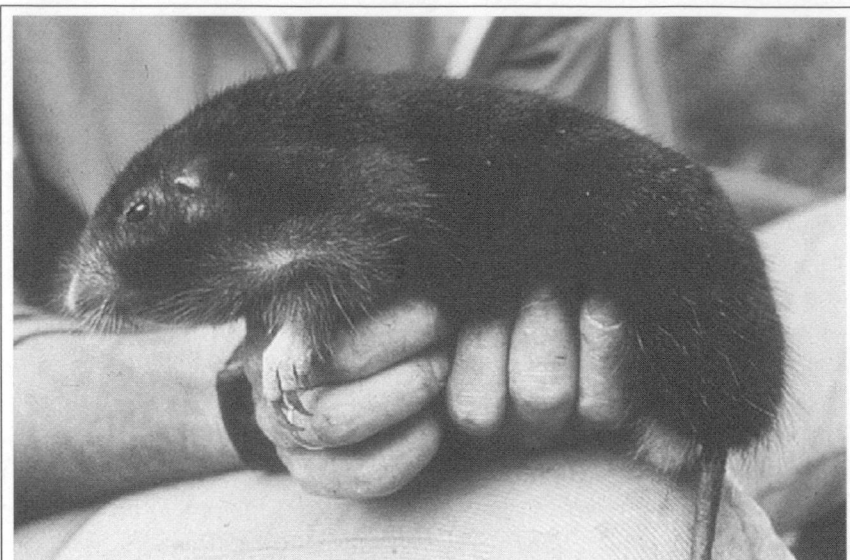

Pocket gophers (family Geomyidae) are herbivorous rodents that spend nearly their entire lives in subterranean tunnels, which they dig while foraging through the earth for food. Most specialists recognize six genera and approximately 40 species of pocket gophers. The family Geomyidae (Gk. "earth mouse family") is found exclusively in the New World, with a geographic distribution ranging from southwestern Canada to northwestern Colombia.

Fossorial Mammals: A Potential Wellspring of Coevolution

Of the thousands of species of mammals that find shelter in subterranean burrow systems, only a few species spend most of their lives beneath the ground, where they tunnel through the soil in search of food. This highly specialized subterranean lifestyle is termed "fossorial" (*fossor* L. "digger"), and fossorial species share numerous adaptations for subterranean life, including a fusiform body shape, enlarged incisors and claws for digging, and valvular ears and nostrils that can be closed to prevent the entry of soil. Fossoriality has evolved independently in multiple lineages of mammals, including marsupial moles of the family Notoryctidae (1 species), moles and golden moles of the insectivoran families Talpidae (42 species) and Chrysochloridae (19 species), and rodents (many species in several families). By far, the greatest diversity of fossorial mammals is seen in the order Rodentia, which includes the African mole-rats of the family

Box 8.2

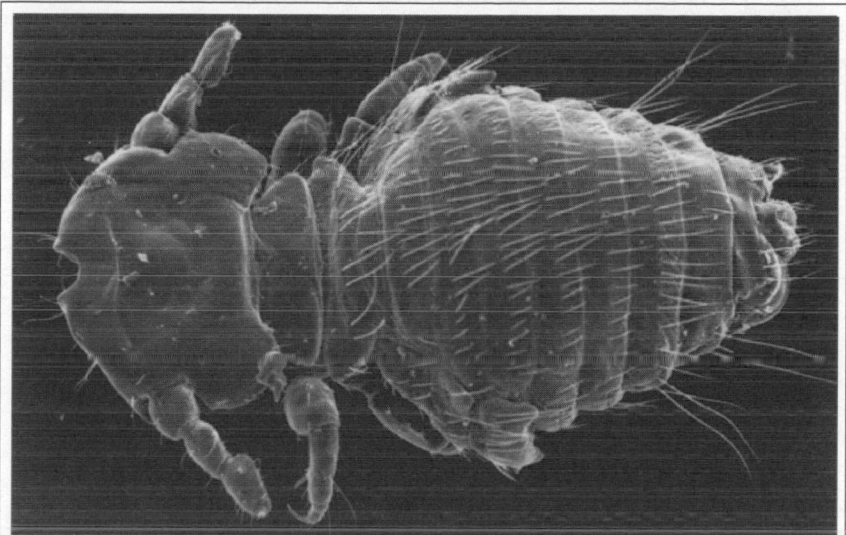

Chewing lice of the genera *Geomydoecus* and *Thomomydoecus* (family Trichodec-tidae) are wingless insects that are found exclusively on pocket gophers (see box 8.1). The louse's entire life cycle takes place on its host, and lice cannot live for extended periods of time off their host. Unlike sucking lice, chewing lice do not pierce the skin of their host, but rather feed on sloughed skin and hair detritus. To date, more than 120 species of chewing lice have been described from pocket gophers (Hellenthal and Price, 1991).

Bathyergidae (14 species); the South American tuco-tucos and octodonts of the families Ctenomyidae (48 species) and Octodontidae (1 fossorial species); the fossorial voles, blind mole-rats, and bamboo rats of the family Muridae (27 fossorial species); and pocket gophers of the New World family Geomyidae (40 species; data from Hall, 1981; Nowak, 1999).

These fossorial mammals share more than superficial morphological similarities. For a number of reasons, the fossorial mode of existence usually involves a solitary lifestyle, with individuals often living in single-occupant burrow systems within small, isolated populations (Pearson, 1959; Patton, 1972; Nevo, 1979). This patchy distribution of populations usually reflects the patchy distribution of friable and flood-free soils, compounded by the often dispersed nature of suitable food resources. Within these isolated populations, individual subterranean mammals often live alone in burrow systems that are defended vigorously against intruders, including conspecifics. Presumably, the enormous energy invested in the excavation of a complex system of tunnels argues against sharing of burrows in most subterranean mammals. Notable exceptions include the highly social naked mole-rats *(Heterocephalus glaber)* of Africa (Jarvis and Bennett, 1991) and the colonial tuco-tuco *(Ctenomys sociabilis)* of South America (Lacey et al., 1997).

In the act of foraging through the soil, subterranean mammals create a cave environment that is ecologically very different from the typical surface environment of most mammals. Through evolutionary time, many creatures have invaded this cave environment (fig. 8.1), and some of these

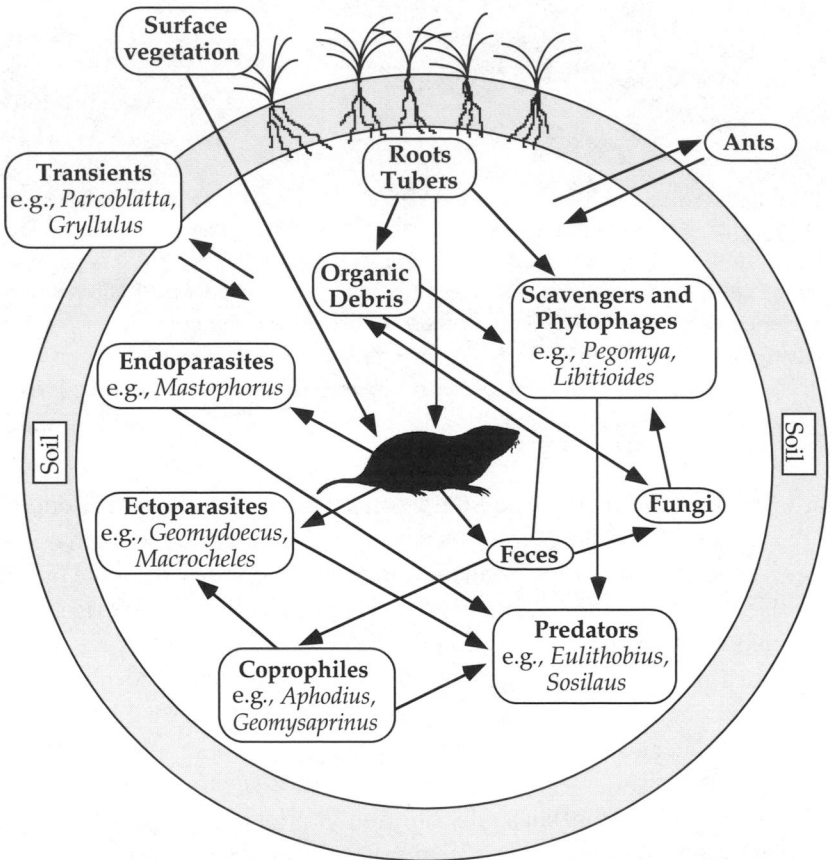

FIGURE 8.1. A diagrammatic (cross-section) view of a pocket gopher in its burrow. Arrows indicate trophic relationships among burrow associates (modified from Hubbell and Goff, 1940). Each of the organisms that coexists with a fossorial mammal is a promising candidate for study of coevolution and cophylogeny.

cave occupants have evolved specialized adaptations unique to the unusual biotic and abiotic regime in which they live (Kim, 1985a, b). Often, cave-specific adaptations, such as reduction of eyes and limbs, render the organism poorly adapted for life on the surface, meaning that subterranean mammals and many of the organisms that coexist with them may be locked into a long-term ecological association. If so, the likelihood of eventual biological interaction between these obligate cave dwellers is high—and biological interaction is, of course, the necessary precursor to coevolution. Clearly, the distribution of a species of subterranean mammal will influence (if not determine) the distribution of organisms that live in its burrow, which means that the latter creatures often will show the same patchy distribution characteristic of so many species of subterranean mammals.

In most species of fossorial mammals, the above factors work in concert to isolate individual mammals and their burrow cohabitants from conspecific individuals, from other subterranean organisms, and from surface-dwelling creatures, in general. Geographic isolation and small population sizes characteristic of subterranean organisms are powerful catalysts for evolutionary change and provide ample opportunity for the evolution of long-term symbioses between fossorial mammals and the many organisms that coexist with them. A recent review of subterranean rodents and their burrow associates (Hafner et al., 2000) identified more than 200 associations that are likely to show some evidence of coevolution. Of these, only two host-parasite systems—ctenomyid rodents and their nematode parasites (Gardner, 1991) and geomyid rodents and their chewing lice (Hafner et al., 1994)—have been investigated in any detail.

The Case for Cophylogeny between Fossorial Mammals and Their Parasites

Perhaps the most conclusive evidence for coevolution comes from the study of host-parasite relationships (Stone and Hawksworth, 1986). In most cases, the parasite's dependence on the host for some or all of its required resources, and the resultant energetic drain on the host, combine to generate an evolutionary "arms race" as the host attempts to defend its energy resources and the parasite responds by countering the host's defenses. Although there are many well-studied host-parasite assemblages, few studies have examined the association from a historical perspective, and fewer yet are unequivocal examples of strict coevolution (defined here as reciprocal adaptation in the hosts and parasites: Janzen, 1980; Hafner et al., 2000). Most studies document an adaptive response on the part of the parasite but fail to show a reciprocal response on the part of the host (Jermy,

1984). Those that show apparent evolutionary responses in both the host and parasite tend to lack evidence of a long-term relationship between the two, which leaves open the possibility of a recent association between the symbionts (termed pseudo-coevolution: Hafner et al., 2000).

Endoparasites of fossorial mammals would seem to be ideal candidates for coevolutionary studies because of their unusually intimate symbiosis with their hosts. Although endoparasites have been reported from several groups of fossorial mammals (see references in Hafner et al., 2000), only a few of these studies have explored the relationship from a coevolutionary perspective (e.g., Gardner, 1991). Our experience with pocket gophers suggests that they harbor many fewer metazoan parasites (especially helminth parasites) than do typical surface-dwelling rodents. Although more study is needed, it is possible that the solitary existence of individuals of many species of fossorial mammals makes them unlikely candidates for acquisition and spread of endoparasites.

Ectoparasitic arthropods provide outstanding opportunities for study of coevolution in fossorial mammals because many ectoparasitic arthropods are restricted to a single host taxon, are geographically widespread, and show high prevalence and abundance on their hosts. For example, chewing lice (Amblycera and Ischnocera) and sucking lice (Anoplura) have limited dispersal abilities and cannot survive for long periods of time off their host (Kellog, 1913; Marshall, 1981). As a result, lice generally rely on host-to-host contact for dispersal, unlike other, more vagile, arthropods. Because host-to-host contact among fossorial mammals is almost exclusively intraspecific, there are few opportunities for lice to colonize new host species; hence we see a high degree of host specificity among lice (Hopkins, 1957; Marshall, 1981). High host specificity, in turn, makes lice ideal candidates for study of host-parasite coevolution. Although current taxonomy suggests that chewing lice are generally species-specific and sucking lice are only genus-specific, this difference may be more an artifact of taxonomy, rather than biological reality (Hopkins, 1957).

Host-parasite systems are intrinsically interesting to evolutionary biologists because they signal a long and intimate association between two or more groups of organisms that are distantly related and quite dissimilar biologically. This long history of association may lead to reciprocal adaptations in the hosts and their parasites (strict coevolution) as well as contemporaneous cladogenic events in the two lineages (variously termed parallel cladogenesis, cospeciation, or cophylogeny). The phenomenon of cophylogeny is of particular interest to comparative phylogeneticists because coincident branching points in the host and parasite trees represent temporal links that facilitate comparative studies of rates of evolution in the two groups. Evidence of cophylogeny also can be used to test hypotheses of coadaptation in the hosts and parasites (see Studies of Coadaptation). Given the unusual natural history of many species of fossorial mammals—including their asocial behavior, patchy geographic distributions, and relative isolation from parasites of terrestrial animals—future studies of cophylogeny involving fossorial mammals and their parasites should prove unusually interesting and rewarding.

Study of Cophylogeny in Pocket Gophers and Chewing Lice
Historical Perspective

Recent advances in our knowledge of pocket gopher and chewing louse cophylogeny would not have been possible without the pioneering taxonomic studies of chewing lice by K. C. Emerson, R. A. Hellenthal, R. D. Price, and R. M. Timm (e.g., see Emerson and Price, 1981, and included references). In particular, Price and Hellenthal undertook an exhaustive survey of geographic variation in the two genera of chewing lice that parasitize pocket gophers (*Geomydoecus* and *Thomomydoecus*), which required examination of countless ectoparasites from many thousands of study skins contained in museum collections in the United States (Hellenthal and Price, 1991). These critical studies identified major lineages within *Geomydoecus* and *Thomomydoecus* (e.g., Price and Hellenthal, 1981; Hellenthal and Price, 1984), and revealed the now well-established pattern of host specificity among these lice and their hosts. In 1983, R. M. Timm, a former student of R. D. Price's, was the first to propose an explicit hypothesis of cophylogeny between pocket gophers and chewing lice (Timm, 1983).

Studies of Gopher-Louse Cophylogeny Based on Allozymes

R. D. Price visited Hafner's laboratory in 1984 to brush ectoparasites from study skins of pocket gophers that Hafner and his colleagues had collected in Central America. Discussions between Price and Hafner concerning relationships among louse species stimulated Hafner (who was then using allozymes to investigate pocket gopher relationships) to attempt starch-gel electrophoresis of individual chewing lice to obtain independent, molecular-based phylogenies for the lice, which he could then compare to the gopher phylogenies. Over the next few years, techniques for protein electrophoresis of chewing lice were developed in Hafner's lab by postdoctoral associate S. A. Nadler. During this same time, C. H. C. Lyal (1985, 1986, 1987) published important empirical analyses and theoretical discussions of cophylogeny between chewing lice and their mammalian hosts based primarily on cladistic analyses of morphological data. Also during this time, the landmark book *Coevolution and Systematics* (Stone and Hawksworth, 1986) was published. In their review of the Stone and Hawksworth book, Futuyma and Kim (1987, p. 441) offered this less-than-optimistic summary of the state of knowledge of cophylogeny:

> [T]he message nonetheless emerges that the phylogenies of hosts and parasites show little congruence at any taxonomic level. Cospeciation is far from universal, and host lineages seem often to have lost their parasites.

In 1988, the allozyme studies in Hafner's lab culminated in publication of the first molecular-based gopher and louse phylogenies (Hafner and Nadler, 1988). This study documented significant congruence between independently derived gopher and louse phylogenies (fig. 8.2), thereby supporting Timm's (1983) hypothesis of cophylogeny in this host-parasite assemblage.

By late 1988, empirical studies of gopher and louse relationships were far ahead of available methods for comparative analysis of host and parasite trees. Although statistical methods for comparing trees were available (e.g., Nelson and Platnick, 1981; Simberloff, 1987), most were developed for use in biogeographical studies, and each had serious limitations for analysis of cophylogeny. Recognizing this need, R. D. M. Page (then a graduate student in New Zealand) contacted Hafner in 1989 to obtain the allozyme data sets for gophers and lice, which he used to develop his nascent program for comparative analysis of trees (the forerunner of his program COMPONENT: Page, 1993a). Thereafter, a synergistic relationship developed between the Hafner and Page laboratories because of Hafner's need for tools to analyze his gopher and louse phylogenies and Page's need for empirical data to develop his analytical tools for investigation of cophylogeny.

Publication of the allozyme-based phylogenies for gophers and lice (Hafner and Nadler, 1988) stimulated a series of publications on methods and theory of cophylogeny analysis (e.g., Hafner and Nadler, 1990; Nadler et al., 1990; Page, 1990; Demastes and Hafner, 1993; Page, 1993a, 1993b, 1993c, 1994b). These contributions emphasized the need for increased rigor in phylogenetic studies, including the need for independent host and parasite phylogenies, improved statistical tests of tree similarity, and widespread (if not exhaustive) sampling of parasite taxa.

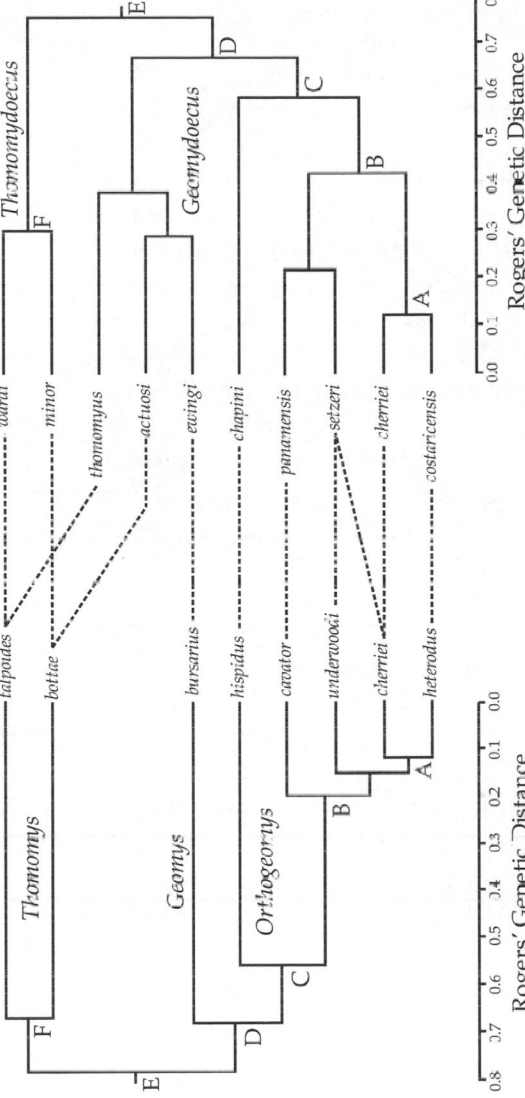

FIGURE 8.2. Phenogram of pocket gopher and chewing louse relationships inferred by Hafner and Nadler (1988) from protein electrophoretic data. Data were clustered using UPGMA analysis (Sneath and Sokal, 1973) of Rogers's genetic distance values (Rogers, 1972). This was the first molecular-based study of gopher-louse cophylogeny, and the first to document a significant level of branching similarity between the gopher and louse trees. Rates of protein evolution in gophers and lice could not be compared using these data because different suites of proteins were examined in the two groups (Hafner and Nadler, 1988). Letters identify analogous nodes in the two trees.

Studies of Cophylogeny Based on DNA Sequences

By the mid-1990s, cophylogeny of pocket gophers and their chewing lice had been investigated based on morphology (e.g., Timm and Price, 1980; Page et al., 1995) and allozymes (e.g., Hafner and Nadler, 1988; Demastes and Hafner, 1993; Highland, 1996). Each of these studies reinforced previous evidence of cophylogeny in this host-parasite assemblage. In the first DNA-based study of the gopher-louse system, Hafner et al. (1994) obtained nucleotide sequences from a 379-bp (base pair) region of the cytochrome c oxidase subunit I (COI) gene from the mitochondria of 15 taxa of pocket gophers and 17 taxa of lice that parasitize these gophers. Hafner et al. (1994) used four tree-building methods to reconstruct gopher and louse relationships and showed that major portions of the phylogenies (fig. 8.3) were recovered consistently regardless of the method of analysis.

Given robust and well-resolved phylogenies for gophers and lice, Hafner et al. (1994) employed a simple test to determine whether the structure of the parasite tree was independent of that of its host. If so, one would expect the amount of cophylogeny observed between the hosts and parasites (i.e., the number of cospeciation events in the two phylogenies) to be no greater than that expected between the host tree and random parasite trees (Page, 1994b). Applying this test to the phylogenies in figure 8.3 (using COMPONENT: Page, 1993a), Hafner et al. (1994) rejected the hypothesis that the louse phylogeny was independent of the gopher phylogeny ($p = .004$, computed using 1,000 random trees). Thus, the DNA sequence data corroborated earlier (morphological and allozyme) evidence for cophylogeny in the gopher-louse assemblage. An intensive DNA-based investigation of cophylogeny involving the gopher genera *Cratogeomys* and *Pappogeomys* and their lice is currently under way and nearing completion (Demastes et al., in prep.). This study will be the first exhaustive (or nearly so) survey of multiple lineages of chewing lice collected from a geographically widespread host taxon.

Studies of Evolutionary Rates

Perhaps the most exciting aspect of cospeciation analysis has been its recent application to comparative studies of evolutionary rates (e.g., Hafner et al., 1994; Hafner and Page, 1995; Page, 1996; Page and Hafner, 1996; Huelsenbeck and Rannala, 1997; Huelsenbeck et al., 2000). There are many ways to convert molecular data (including data from allozymes, restriction-fragment patterns, and protein and DNA sequences) into

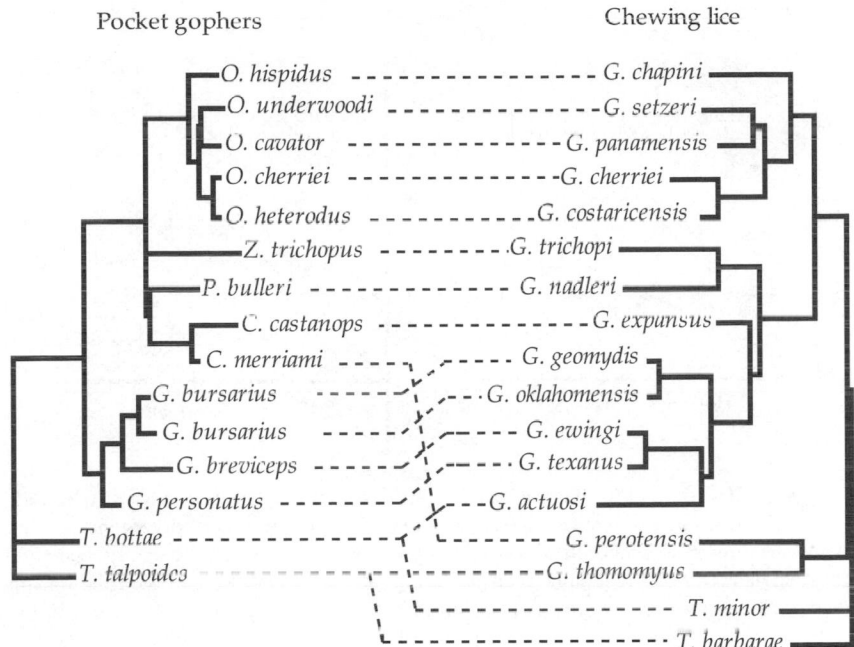

FIGURE 8.3. Phylogenies of pocket gophers and their chewing lice based on nucleotide sequence data analyzed by Hafner et al. (1994). Shown are composite trees based on multiple methods of phylogenetic analysis. Branch lengths are proportional to inferred amounts of genetic change. Pocket gopher genera are *Orthogeomys, Zygogeomys, Pappogeomys, Cratogeomys, Geomys,* and *Thomomys. Geomys bursarius* is represented by two subspecies (a = *G. b. halli;* b = *G. b. majusculus*). Chewing louse genera are *Geomydoecus* and *Thomomydoecus.* The program COMPONENT (Page, 1993a) was used to document significant similarity in branching structure between these trees. Because the host and parasite trees were based on DNA sequences from the same gene (cytochrome *c* oxidase subunit I), rates of DNA evolution could be compared in the two groups. Based on these data, Hafner et al. (1994) estimated that chewing lice were evolving approximately 10 times faster than pocket gophers in this gene region. Page (1996) reanalyzed these same data using a different estimate of transition bias and reported only a twofold rate difference between gophers and lice.

estimates of genetic divergence (Swofford et al., 1996). Each method has inherent advantages and limitations, and each involves assumptions about the nature of evolutionary change at the molecular level. Most comparative studies of genetic differentiation in hosts and parasites have used either pairwise estimates of genetic distance (e.g., Hafner and Nadler, 1990; Page, 1990) or estimates of length of homologous branches in the host and parasite trees (e.g., Hafner et al., 1994; Page, 1996).

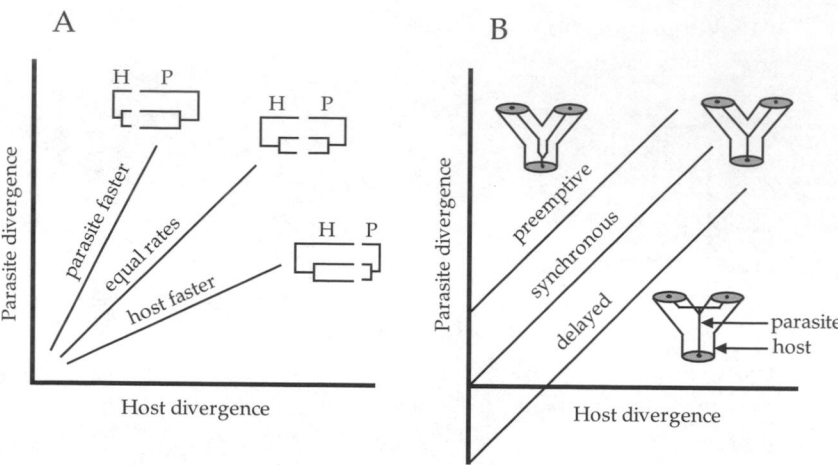

FIGURE 8.4. Bivariate plots illustrating the graphical method developed by Hafner and Nadler (1990) to investigate relative rates of evolution and timing of divergence events in cospeciating hosts and their parasites. Both panels compare genetic divergence between pairs of hosts and pairs of cospeciating parasites. The slope of the relationship (panel *A*) indicates relative rates of evolution in the two groups. The trees (inset in panel *A*) are drawn with branch lengths proportional to amount of genetic change in the hosts (H) and parasites (P). The *y*-intercept (panel *B*) indicates the relative timing of divergence events in the two groups. A positive *y*-intercept is evidence of "preemptive cospeciation," in which case the parasites are assumed to have diverged in advance of their hosts. A negative *y*-intercept signals "delayed cospeciation" (hosts diverge in advance of their parasites), and a *y*-intercept of zero is evidence of "synchronous cospeciation" in the two groups. The inset figures in panel *B* illustrate relative timing of divergence events in the hosts (outer portion of figure) and their parasites (thin line within each figure). Figure modified from Hafner and Nadler (1990, fig. 2) and Hafner and Page (1995, fig. 4).

Hafner and Nadler (1990) proposed a theoretical framework for comparing host and parasite genetic divergence, whether measured as genetic distance or relative length of branches on a phylogenetic tree. The regression of parasite divergence against host divergence (fig. 8.4) allows us to describe simultaneously two aspects of host-parasite divergence. The slope of the line (fig. 8.4A) is an estimate of the relative rate of genetic change in the two groups. The *y*-intercept of the line (fig. 8.4B) measures genetic divergence in the parasites at the time of host speciation. For example, an intercept of zero indicates synchronous cospeciation, wherein hosts and parasites diverge simultaneously. A negative intercept suggests delayed cospeciation, in which case the parasites tend to diverge consistently after their hosts. Finally, a positive intercept signals preemptive cospeciation, in which case the parasites diverge prior to their hosts.

Although it is widely acknowledged that estimates of DNA sequence divergence should be adjusted for the effects of saturation (repeated nucleotide substitutions at a single site), there is no general consensus as to how this should be done. For example, Hafner et al. (1994) attempted to correct for transition bias in the gopher and louse COI data by using the largest observed pairwise transition bias in a maximum-likelihood phylogeny reconstruction. They reasoned that this value, which is usually measured between the most recently diverged taxa, is least likely to be affected by saturation and, therefore, is the most reasonable estimate of the actual transition bias for this gene region. In contrast, Page (1996) recommended use of the transition bias estimate that maximizes the likelihood of the phylogeny. Use of these different correction factors can have profound influence on estimates of branch length. For example, Hafner et al.'s analysis suggested that lice are evolving 10 to 11 times more rapidly than pocket gophers at selectively neutral sites. In contrast, Page's reanalysis of Hafner et al.'s data suggested that lice are evolving only two to three times as fast as gophers, and a similar result was obtained by Huelsenbeck et al. (1997). Research into the effects of transitional saturation (and evolutionary models, in general) is now moving at a rapid pace, and we expect that some degree of consensus will be reached in the near future.

Studies of Gopher-Louse Cophylogeny at Lower Taxonomic Levels

The large-scale pattern of cophylogeny observed between genera of pocket gophers and their chewing lice (e.g., Hafner and Nadler, 1988; Hafner et al., 1994) must ultimately emanate from biological interactions at lower taxonomic levels and on smaller geographic scales. With this view in mind, Demastes and Hafner (1993) and Spradling (1997) investigated gopher-louse cophylogeny within a single genus of pocket gophers (*Geomys* and *Thomomys*, respectively) in the southwestern United States. Although the predicted pattern of cophylogeny was evident in both studies, there were numerous inconsistencies between the host and parasite trees that were variously attributed to host switching by the parasites, retention of ancestral parasite taxa, and poorly delineated taxonomic boundaries in the hosts and parasites. Demastes and Hafner (1993) and Spradling (1997) concluded that studies of cophylogeny at lower taxonomic levels are likely to be confounded by population-level phenomena, including reticulate evolution of host taxa (i.e., merging of taxa previously thought to be distinct) and retention of ancestral parasite taxa on recently evolved host lineages. This topic of spatial and temporal scales, and their effects on cophylogeny analysis, is explored in greater detail in the next chapter, and by Rannala and Michalakis (chap. 5).

Studies of Gopher-Louse Cophylogeny on Small Geographic Scales

Studies of cophylogeny that are restricted both taxonomically (e.g., to a single species of host) and geographically (to a small portion of the host's distribution) offer the greatest potential for elucidating fundamental host-parasite interactions that ultimately generate the pattern of cophylogeny. For example, Nadler et al. (1990) compared population structure in several populations of *Thomomys bottae* and their lice to determine if structuring of parasite populations on individual hosts (which likely is influenced by founder events as new hosts are colonized) tends to accelerate parasite evolution relative to that of their hosts. Indeed, Nadler and his colleagues found significant levels of genetic differentiation among louse populations collected from hosts living only meters apart, reinforcing previous evidence that louse transfer among host individuals is severely restricted (Nadler and Hafner, 1989). Nadler et al. (1990) also showed that among-population differentiation in lice (i.e., lice from hosts collected at different geographic localities) was similar to that measured among the host populations themselves, suggesting a close association between gene flow in pocket gophers and gene flow in their lice.

Working at an even smaller geographic scale, Demastes (1996) examined gopher and louse genetics in a single alfalfa field in central New Mexico. Operating under the then-widespread assumption that cophylogeny is the inevitable result of chewing lice being passed from mother to offspring within lineages of related hosts (see Studies of Parasite Transmission), Demastes used nuclear DNA fingerprint data to generate host pedigrees ("family trees") in order to trace louse transmission patterns. Surprisingly, this microspatial genetic analysis revealed little or no relationship between genetic relatedness among gophers and the composition of their louse populations. Instead, Demastes showed that the genetic composition of louse populations on individual pocket gophers shows significant spatial autocorrelation at this locality, meaning that louse transmission depends more on spatial proximity of the hosts (i.e., the louse composition on nearest neighbors) than on host mating regimes. This topic is explored in more detail in the next chapter.

Studies of Contact Zones

Studies of the distribution of chewing lice at zones of hybridization between gopher taxa have yielded important information about the history of the zone (e.g., Patton et al., 1984; Nadler et al., 1990; Hafner et al., 1998). If genetic introgression is present in both the hosts and parasites (as in the study by Nadler et al., 1990), then rate and pattern of introgression can be compared to reveal common demographic patterns. In other cases, parasites can be treated as "genes" of their hosts to serve as an independent measure of extent of host introgression (e.g., Bohin and Zimmerman, 1982; Patton et al., 1984). Along these lines, Hafner et al. (1998) studied movement of a contact zone between two species of chewing lice in the Rio Grande Valley of New Mexico and showed that the midpoint of the contact zone had moved approximately 1 km during a five-year period (fig. 8.5).

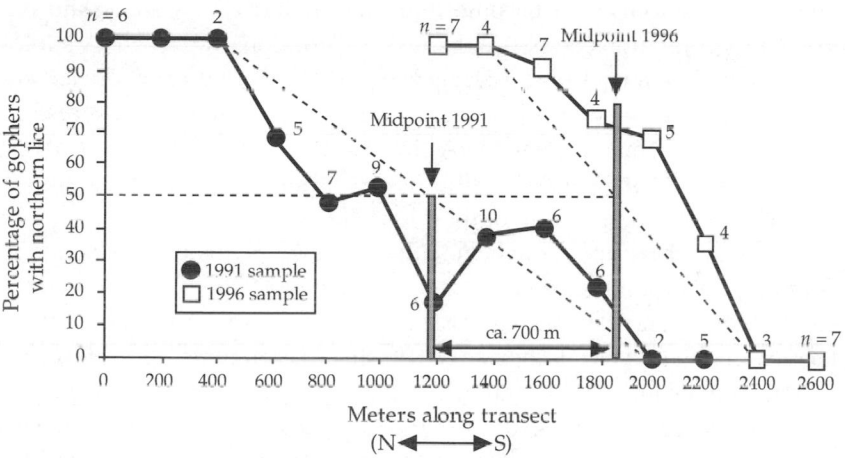

FIGURE 8.5. Transects (sampled in 1991 and 1996) through the zone of overlap between two species of lice that meet in the Rio Grande Valley of central New Mexico. For each 200-m interval along the transect, Hafner et al. (1998) calculated the percentage of pocket gophers that hosted the "northern" (*Geomydoecus aurei*) versus "southern" (*G. centralis*) louse species (*n* = number of pocket gophers surveyed at each interval along the transect). The midpoint of each zone (calculated as one-half of the distance between the northernmost occurrence of a southern louse and the southernmost occurrence of a northern louse) moved approximately 700 m southward between 1991 and 1996. This discovery had important implications for rate of louse dispersal (estimated from these data to be approximately 150 m/yr) and age of this contact zone (estimated to be approximately 50 years old). Hafner and his colleagues returned to the contact zone in 2001 and determined that zone movement continues in the predicted direction (southward) and at the predicted rate (150 m/yr) (unpub. data).

This rate of zone movement, coupled with knowledge of the genetics of the hosts, had important implications, not only for rate of louse dispersal, but also for estimating the age of the contact zone—these new data suggested that the zone was only about 50 years old, rather than >10,000 years old, as previously hypothesized (Smith et al., 1983).

Studies of Parasite Transmission

As mentioned earlier, researchers have long believed that chewing lice are transmitted primarily—if not exclusively—from a mother host to her offspring (e.g., Newson and Holmes, 1968; Rust, 1974), thereby generating the lineage-specific transmission pattern that, when viewed at large scales, is termed cophylogeny. A study by Demastes et al. (1998) tested this "maternal transmission hypothesis" using an indirect approach that compared the distribution of louse populations with the distribution of mitochondrial DNA (mtDNA) haplotypes in the pocket gophers (mtDNA haplotypes are known to be inherited maternally). Their study showed no significant relationship between louse distributions and mtDNA haplotypes for the gophers, thus falsifying the hypothesis that chewing lice are transmitted exclusively from mother to offspring. These results were consistent with previous nuclear DNA fingerprint studies (Demastes, 1996) in showing little or no relationship between genetic relatedness among gophers and the composition of their louse fauna.

Given that there appear to be no intrinsic barriers to louse transmission within a population of pocket gophers, Reed and Hafner (1997) investigated whether such barriers may exist between host populations of varying degrees of relatedness. Laboratory transfer experiments were designed to test whether lice could establish successful colonies when transferred between two subspecies of the same host species, between two species of the same host genus, and between two genera of pocket gopher hosts. Although lice established successful colonies at each level, rate of colonization of new hosts (i.e., percentage of transfers that were successful) diminished with increasing phylogenetic distance from the natural host of each louse. Reed and Hafner (1997) concluded that the pattern of cophylogeny may result primarily from lack of opportunity for lice to colonize new hosts. However, in rare cases in which lice disperse to new hosts, survival may be difficult on hosts that are not closely related to the natural host, which would reinforce the pattern of cophylogeny. Reed et al. (2000a) implicate hair diameter as an important factor that may determine suitability of new hosts for dispersing lice.

Studies of Coadaptation

Component analysis (Page, 1993a) identifies pairs of equivalent nodes in host and parasite trees that reflect the same historical event. Hypotheses of coadaptation in the hosts and parasites can be tested using these nodes. For example, Harvey and Keymer (1991) used simplified phylogenies of gophers and lice taken from Hafner and Nadler (1988) to show that evolution of body size in lice and their hosts is highly correlated. Similarly, Morand et al. (2000) and Reed et al. (2000a) used independent contrasts (which require knowledge of gopher and louse phylogenies) to investigate gopher and louse body size relationships and to analyze the relationship between louse body size and hair diameter of the host. Reed et al. (2000a) documented a significant, positive relationship between hair diameter in pocket gophers and rostral groove dimensions of their chewing lice (lice use the rostral groove to grasp the hair of their host: fig. 8.6). Coupled with previous evidence of a strong allometric relationship between louse body size and rostral groove width (Morand et al., 2000), this finding supports the contention that hair diameter of the host may be an important determinant of body size in chewing lice. Numerous other morphological, physiological, and ecological attributes of hosts and parasites can be compared using the cophylogeny framework.

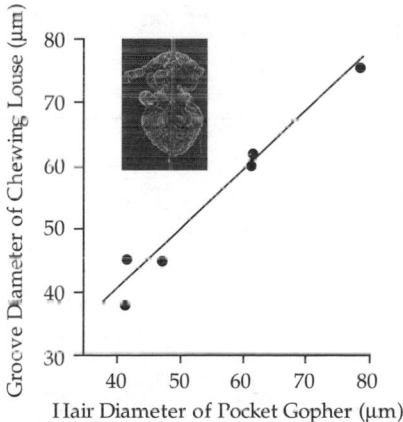

FIGURE 8.6. Linear regression of chewing louse groove width and pocket gopher hair diameter from Reed et al. (2000a). Chewing lice use the head groove to grasp the hair of their host (inset). This significant relationship suggests that hair diameter of the host is an important element of the louse's physical environment and may be an important determinant of whether a louse is able to colonize a new species of host.

Studies of Parasite Ecology

In efforts to understand the fundamental biological basis of cophylogeny between pocket gophers and chewing lice, our research has moved progressively from the study of large-scale patterns (e.g., studies at the genus level in hosts and parasites) to microgeographic studies at the level of the host and parasite population. These latter studies have lain to rest the intuitively appealing—but, apparently, incorrect—notion that the pattern of cophylogeny emanates solely from mother-to-offspring transfer of chewing lice by pocket gophers (Demastes et al., 1998). Although maternal transmission of chewing lice is known to occur (Rust, 1974) and certainly contributes to the emergent pattern of cophylogeny (Timm, 1983), it alone cannot explain the pattern. Reed and Hafner (1997) have suggested that cophylogeny may result from a combination of factors, including patchy distribution of hosts, low dispersal abilities of the parasites, and ecological constraints on the part of the parasites that reduce their abilities to establish successful colonies on new host taxa. Reed et al. (2000a) showed that one of these constraints may relate to the louse's ecological requirement for host hairs of a certain diameter or range of diameters.

Another ecological factor that may constrain or prevent louse colonization of new host taxa is simple competition between the dispersing louse and the louse population already in residence on the new host. It is reasonably well documented that two congeneric species of lice (whether of the genus *Geomydoecus* or *Thomomydoecus*) do not—and probably cannot—coexist for any meaningful period of time on a single host individual (Timm, 1983; Demastes, 1996; Hafner et al., 1998). This suggests that any louse that manages to disperse to a new host taxon (an event that, itself, will be rare, given the patchy distribution of host taxa and the poor dispersal abilities of lice) will find it difficult to establish a successful colony on the new host in the face of competition from the more numerous resident lice. The exact nature of this louse competition is presently unknown, but ongoing and planned studies in the Hafner laboratory using high-resolution videomicroscopy should reveal whether species of lice engage in direct (interference) competition—such as fighting or destroying the eggs of the competitor—when experimentally introduced onto a single host individual.

Although congeneric species of chewing lice appear unable to coexist on a single host individual, several species of pocket gopher are known to host mixed louse populations consisting of one species of *Geomydoecus* and one species of *Thomomydoecus*. These lice coexist in what appears to be stable, long-term equilibrium on single individual hosts. Reed et al. (2000b) explored the nature of this coexistence and discovered that the chewing lice appear to partition available host resources spatially, with the *Geomydoecus* species living primarily on the lateral and dorsal regions of the host, and the *Thomomydoecus* species living primarily on the lateral and ventral regions (fig. 8.7). This dorsoventral habitat partitioning does not appear to be explained by hair diameter, which means that the lice may be responding to some other factor that varies dorsoventrally, such as temperature or humidity gradients of the host's body (Reed et al., 2000b).

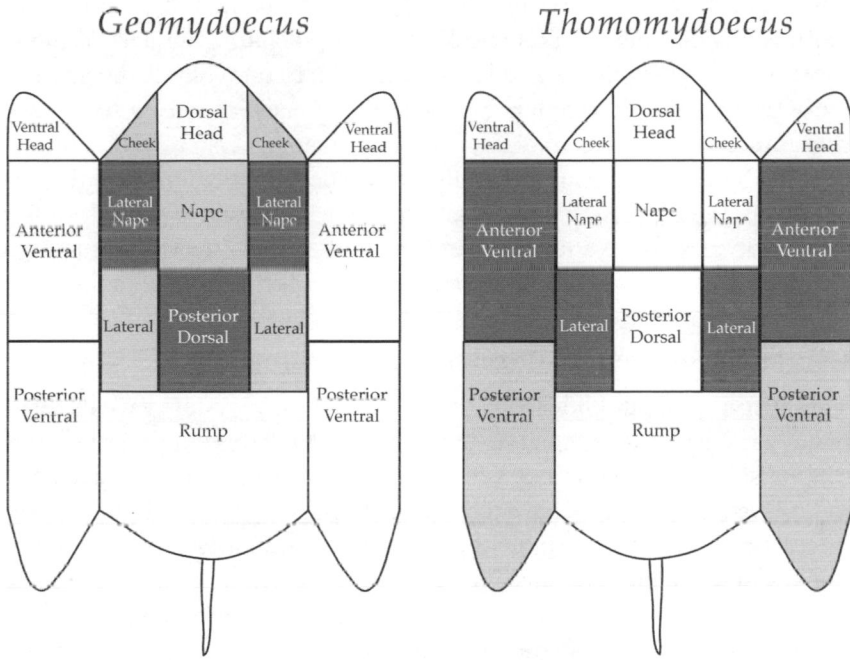

FIGURE 8.7. Reed et al. (2000b) studied the point-in-time distribution of two species of chewing lice that coexist on individual pocket gophers of the genus *Thomomys*. This is a diagrammatic view of the external surface of a pocket gopher (the skin is incised mid-ventrally, then laid flat) showing the simultaneous distributions of *Geomydoecus aurei* and *Thomomydoecus minor*. Distributions were analyzed on three host individuals, and pooled results are presented here. Darkly shaded regions contained more lice than expected in all three pocket gopher specimens examined. Stippled regions contained more lice than expected in two of the three hosts examined. Unshaded regions contained fewer lice than expected on at least two of the hosts examined. These results suggest that the two species of lice tend to partition their habitat dorsoventrally.

Development of Evolutionary Models

The remarkable level of cophylogeny documented for gophers and lice, and the size and scope of the gopher-louse DNA data set (Hafner et al., 1994; Hafner and Page, 1995), have made this system especially useful for development and testing of models in the areas of phylogenetics and molecular evolution. Studies using the gopher-louse data have focused on use of differential equations for modeling in biology (e.g., Taubes, 2001), testing of similarity between evolutionary trees (e.g., Page, 1993c; Huelsenbeck et al., 1997; Huelsenbeck et al., 2000), measurement of evolutionary rate differences in hosts and parasites (e.g., Hafner et al., 1994; Page, 1996; Huelsenbeck et al., 1997), and development of likelihood-ratio tests to address a wide range of biological questions, including models of DNA substitution (Huelsenbeck and Rannala, 1997). In this volume the gopher-louse data feature in chapters 3 and 4. There is every reason to believe that future DNA-based studies of the gopher-louse system will continue to contribute, both directly and indirectly, to development of evolutionary models in phylogenetics, molecular evolution, and related fields. Studies that involve exhaustive sampling of host and parasite taxa within a large and diverse assemblage (discussed below) offer special promise in this area.

Conclusions and Future Directions for Research

This overview of our knowledge of the gopher-louse system reveals three major gaps in our understanding of the evolution of this unusual symbiotic relationship. First, we do not yet know the root cause of host-specificity in this system, nor do we have a clear understanding of how louse species interact ecologically on an individual host. Second, all studies of the gopher-louse system published thus far have suffered from sparse taxonomic sampling, and we have yet to explore the potentially confounding effects of inadequate phylogenetic sampling on studies of cophylogeny. Finally, although we have produced clear and convincing evidence of cophylogeny between gophers and lice, we have yet to investigate whether other participants in this system (e.g., other parasites of the gophers or parasites of the lice) also show cophylogeny with their symbionts.

Our ongoing and future studies of the gopher-louse system will explore these major gaps in our knowledge and will necessarily require studies on three very different evolutionary and geographic scales. The first of these will employ controlled laboratory and field transfer experiments and videomicroscopy techniques to further our understanding of the intimate ecological relationship between gophers and lice and between coexisting louse species. Our working hypothesis is that these ecological factors

reinforce the overall pattern of cophylogeny, which is determined ultimately by the patchy distribution of the hosts and the low dispersal ability
of the lice.

Our second line of inquiry will focus on completion of the first exhaustive taxonomic survey of a single lineage of gophers and their lice.
For years, researchers have stressed the critical importance of thorough
phylogenetic sampling in evolutionary studies (e.g., Moore et al., 1976;
Fitch and Bruschi, 1987; Sanderson, 1990). More recently, Page (1996)
documented the adverse effects of incomplete phylogenetic sampling on
host-parasite studies. We are currently completing an exhaustive survey of
pocket gophers of the sister taxa *Cratogeomys* and *Pappogeomys* and their
chewing lice. This has involved several months of fieldwork in Mexico and
the United States and will result in a complete systematic revision of the
host genera. This data set promises to have multiple applications that transcend the basic question of cophylogeny between these particular hosts and
parasites, including development of models for testing similarity between
phylogenetic trees, detection of multiple parasite lineages, identification of
host-switching events, and study of rates of molecular evolution.

Our third major research effort involves expansion of our ongoing studies of gopher and louse cophylogeny to include selected lineages of endosymbiotic bacteria hosted by the chewing lice. We are currently collab-
orating with systematic microbiologists F. A. Rainey and N. L. Ward to
investigate phylogenetic patterns in this three-tiered system. To date, we
have amplified, cloned, and sequenced between 500 and 1,534 base pairs
of the 16S ribosomal RNA gene for 234 unique clones representing 35 bacterial lineages associated with the gopher-louse system (Reed and Hafner,
in press).

Phylogenetic analysis of sequence data for *Staphylococcus* (one of the
most widespread bacterial lineages in chewing lice) has revealed significant
phylogenetic structure and suggests some level of cophylogeny between
the bacteria and their hosts (Reed and Hafner, in press). However, the
preliminary analysis also suggests the possibility of multiple lineages of
Staphylococcus like bacteria in chewing lice, which may reflect multiple,
independent invasions. This possibility will be tested by examination of
additional clones, and it is conceivable that cophylogeny may have occurred
subsequent to each independent bacterial invasion. More sophisticated
phylogenetic analyses using complex maximum-likelihood models may
reveal relationships heretofore unresolved in our preliminary analyses.

Although, at present, there are few published studies of cophylogeny
explored from a molecular perspective (e.g., Moran et al., 1993; Hafner
et al., 1994; Moran et al., 1995; Page et al., 1998, see also chapters 9,

11–13), we anticipate rapid growth in this research area as molecular tools become more widely available and the advantages of this approach better known. We believe that fossorial mammals and their parasites, especially the gopher-louse assemblage, will continue to play a prominent role in cophylogeny research, largely because of the unusual life-history characteristics of subterranean mammals, which are particularly conducive to parallel speciation. Unfortunately, many host-parasite systems will show little or no evidence of cophylogeny (e.g., Baverstock et al., 1985), which will preclude comparative studies of higher-order phenomena, such as evolutionary rates. However, in those systems with appreciable cospeciation, the researcher will have the unparalleled opportunity to compare evolution in the same genes, and over the same period of time, in distantly related organisms. Within this framework, the potential is great for discovery of large-scale evolutionary patterns that apply to diverse groups of organisms.

REFERENCES

Baverstock, P. R., M. Adams, and I. Beveridge. 1985. Biochemical differentiation in bile duct cestodes and their marsupial hosts. *Molecular Biology and Evolution* 2:321–37.

Bohin, R. G., and E. G. Zimmerman. 1982. Genic differentiation of two chromosome races of the *Geomys bursarius* complex. *Journal of Mammalogy* 63:218–28.

Brown, J. H., and M. V. Lomolino. 1998. *Biogeography*. Sunderland, Mass.: Sinauer Associates.

Demastes, J. W. 1996. Analysis of host-parasite cospeciation: Effects of spatial and temporal scale. Ph.D. diss., Louisiana State University.

Demastes, J. W., and M. S. Hafner. 1993. Cospeciation of pocket gophers *(Geomys)* and their chewing lice *(Geomydoecus)*. *Journal of Mammalogy* 74:521–30.

Demastes, J. W., M. S. Hafner, D. J. Hafner, and T. A. Spradling. 1998. Pocket gophers and chewing lice: A test of the maternal transmission hypothesis. *Molecular Ecology* 7:1065–69.

Emerson, K. C., and R. D. Price. 1981. A host-parasite list of the Mallophaga on mammals. *Miscellaneous Publications of the Entomological Society of America* 12:1–72.

Esch, G. W., and J. C. Fernández. 1993. *A functional biology of parasitism*. London: Chapman and Hall.

Fitch, W. M., and M. Bruschi. 1987. The evolution of prokaryotic ferredoxins—with a general method correcting for unobserved substitutions in less branched lineages. *Molecular Biology and Evolution* 4:381–94.

Futuyma, D. J., and J. Kim. 1987. Phylogeny and coevolution. *Review of* Coevolution and systematics, edited by A. R. Stone and D. L. Hawksworth. 1986 (New York: Oxford University Press). *Science* 237:441–42.

Gardner, S. L. 1991. Phyletic coevolution between subterranean rodents of the genus *Ctenomys* (Rodentia: Hystricognathi) and nematodes of the genus *Paraspididera* (Heterakoidea: Aspidoderidae) in the neotropics: Temporal and evolutionary implications. *Zoological Journal of the Linnean Society* 102:169–201.

Hafner, M. S., J. W. Demastes, D. J. Hafner, T. A. Spradling, P. D. Sudman, and S. A. Nadler. 1998. Age and movement of a hybrid zone: Implications for dispersal distance in pocket gophers and their chewing lice. *Evolution* 52:278–82.

Hafner, M. S., J. W. Demastes, and T. A. Spradling. 2000. Coevolution and subterranean rodents. In *Life underground: The biology of subterranean rodents*, edited by E. A. Lacey, J. L. Patton, and G. N. Cameron, 370–88. Chicago: University of Chicago Press.

Hafner, M. S., and S. A. Nadler. 1988. Phylogenetic trees support the coevolution of parasites and their hosts. *Nature* 332:258–59.

———. 1990. Cospeciation in host-parasite assemblages: Comparative analysis of rates of evolution and timing of cospeciation. *Systematic Zoology* 39:192–204.

Hafner, M. S., and R. D. M. Page. 1995. Molecular phylogenies and host-parasite cospeciation: Gophers and lice as a model system. *Philosophical Transactions of the Royal Society of London,* ser. B, 349:77–83.

Hafner, M. S., P. D. Sudman, F. X. Villablanca, T. A. Spradling, J. W. Demastes, and S. A. Nadler. 1994. Disparate rates of molecular evolution in cospeciating hosts and parasites. *Science* 265.1087–90.

Hall, E. R. 1981. *The mammals of North America.* New York: John Wiley and Sons.

Harvey, P. H., and A. E. Keymer. 1991. Comparing life histories using phylogenies. *Philosophical Transactions of the Royal Society of London,* ser. B, 332:31–39.

Hellenthal, R. A., and R. D. Price. 1984. Distributional associations among *Geomydoecus* and *Thomomydoecus* lice (Mallophaga: Trichodectidae) and pocket gopher hosts of the *Thomomys bottae* group (Rodentia: Geomyidae). *Journal of Medical Entomology* 21:432–46.

———. 1991. Biosystematics of the chewing lice of pocket gophers. *Annual Review of Entomology* 36:185–203.

Highland, R. 1996. Cospeciation of Midwestern pocket gophers and their ectoparasitic chewing lice. M.S. thesis, Northern Illinois University.

Hopkins, G. H. E. 1957. The distribution of Pthiraptera on mammals. In *Premier symposium sur la spécificité parasitaire des parasites de Vertébrés*, edited by J. G. Baer, 88–119. Neuchatel: Institut de Zoologie, Université de Neuchatel.

Hubbell, T. H., and C. C. Goff. 1940. Florida pocket gophers and their arthropod inhabitants. *Proceedings of the Florida Academy of Science* 4:127–66.

Huelsenbeck, J. P., and B. Rannala. 1997. Phylogenetic methods come of age: Testing hypotheses in an evolutionary context. *Science* 276:227–32.

Huelsenbeck, J. P., B. Rannala, and B. Larget. 2000. A Bayesian framework for the analysis of cospeciation. *Evolution* 54:352–64.

Huelsenbeck, J. P., B. Rannala, and Z. Yang. 1997. Statistical tests of host-parasite cospeciation. *Evolution* 51:410–19.

Janzen, D. H. 1980. When is it coevolution? *Evolution* 34:611–12.

Jarvis, J. U. M., and N. C. Bennett. 1991. Ecology and behavior of the family Bathyergidae. In *The biology of the naked mole-rat*, edited by P. W. Sherman,

J. U. M. Jarvis, and R. D. Alexander, 66–96. Princeton, N.J.: Princeton University Press.

Jermy, T. 1984. Evolution of insect/host plant relationships. *American Naturalist* 124:609–30.

Kellog, V. L. 1913. Distribution and species forming of ectoparasites. *American Naturalist* 47:129–58.

Kim, K. C. 1985a. Coevolution of fleas and mammals. In *Coevolution of parasitic arthropods and mammals,* edited by K. C. Kim, 295–437. New York: John Wiley and Sons.

———. 1985b. Evolution and host associations of Anoplura. In *Coevolution of parasitic arthropods and mammals,* edited by K. C. Kim, 197–231. New York: John Wiley and Sons.

Lacey, E. A., S. H. Braude, and J. R. Wieczorek. 1997. Burrow sharing by colonial tuco-tucos *(Ctenomys sociabilis). Journal of Mammalogy* 78:556–62.

Lyal, C. H. C. 1985. A cladistic analysis and classification of trichodectid mammal lice (Phthiraptera: Ischnocera). *Bulletin of the British Museum (Natural History), Entomology* 51:187–346.

———. 1986. Coevolutionary relationships of lice and their hosts: A test of Farenholz's Rule. In *Coevolution and systematics,* edited by A. R. Stone and D. L. Hawksworth, 77–91. Oxford: Clarendon Press.

———. 1987. Co-evolution of trichodectid lice (Insecta: Phthiraptera) and their mammalian hosts. *Journal of Natural History* 21:1–28.

Marshall, A. G. 1981. *The ecology of ectoparasitic insects.* London: Academic Press.

Moore, G. W., M. Goodman, C. Callahan, R. Holmquist, and H. Moise. 1976. Stochastic versus augmented maximum parsimony method for estimating superimposed mutations in the divergent evolution of protein sequences: Methods tested on cytochrome *c* amino acid sequences. *Journal of Molecular Biology* 105:15–37.

Moran, N. A., M. A. Munson, P. Baumann, and H. Ishikawa. 1993. A molecular clock in endosymbiotic bacteria is calibrated using the insect hosts. *Proceedings of the Royal Society of London,* ser. B, 253:167–71.

Moran, N. A., C. D. van Dohlen, and P. Baumann. 1995. Faster evolutionary rates in endosymbiotic bacteria than in cospeciating insect hosts. *Journal of Molecular Evolution* 41:727–31.

Morand, S., M. S. Hafner, R. D. M. Page, and D. L. Reed. 2000. Comparative body size relationships in pocket gophers and their chewing lice. *Biological Journal of the Linnean Society* 70:239–49.

Nadler, S. A., and M. S. Hafner. 1989. Genetic differentiation in sympatric species of chewing lice (Mallophaga: Trichodectidae). *Annals of the Entomological Society of America* 82:109–13.

Nadler, S. A., M. S. Hafner, J. C. Hafner, and D. J. Hafner. 1990. Genetic differentiation among chewing louse populations (Mallophaga: Trichodectidae) in a pocket gopher contact zone (Rodentia: Geomyidae). *Evolution* 44:942–51.

Nelson, G., and N. I. Platnick. 1981. *Systematics and biogeography: Cladistics and vicariance.* New York: Columbia University Press.

Nevo, E. 1979. Adaptive convergence and divergence of subterranean mammals. *Annual Review of Ecology and Systematics* 10:269–308.

Newson, R. M., and R. G. Holmes. 1968. Some ectoparasites of the coypu *(Myocastor coypus)* in eastern England. *Journal of Animal Ecology* 37:471–81.

Noble, E. R., G. A. Noble, G. A. Schad, and A. J. MacInnes. 1989. *Parasitology. The biology of animal parasites.* Philadelphia: Lea and Febiger.

Nowak, R. M. 1999. *Walker's mammals of the world.* Baltimore: Johns Hopkins University Press.

Page, R. D. M. 1990. Temporal congruence and cladistic analysis of biogeography and cospeciation. *Systematic Zoology* 39:205–26.

———. 1993a. COMPONENT, tree comparison software for Microsoft®Windows™, version 2.0. London: The Natural History Museum.

———. 1993b. Genes, organisms, and areas: The problem of multiple lineages. *Systematic Biology* 42:77–84.

———. 1993c. Parasites, phylogeny and cospeciation. *International Journal for Parasitology* 23:499–506.

———. 1994a. Maps between trees and cladistic analysis of historical associations among genes, organisms, and areas. *Systematic Biology* 43:58–77.

———. 1994b. Parallel phylogenies: Reconstructing the history of host-parasite assemblages. *Cladistics* 10:155–73.

———. 1996. Temporal congruence revisited: Comparison of mitochondrial DNA sequence divergence in cospeciating pocket gophers and their chewing lice. *Systematic Biology* 45:151–67.

Page, R. D. M., and M. S. Hafner. 1996. Molecular phylogenies and host-parasite cospeciation: Gophers and lice as a model system. In *New uses for new phylogenies,* edited by P. H. Harvey, A. J. Leigh Brown, J. Maynard Smith, and S. Nee, 255–70. Oxford: Oxford University Press.

Page, R. D. M., and E. C. Holmes. 1998. *Molecular evolution: A phylogenetic approach.* Oxford: Blackwell Science.

Page, R. D. M., P. L. M. Lee, S. A. Becher, R. Griffiths, and D. H. Clayton. 1998. A different tempo of mitochondrial DNA evolution in birds and their parasitic lice. *Molecular Phylogenetics and Evolution* 9:276–93.

Page, R. D. M., R. D. Price, and R. A. Hellenthal. 1995. Phylogeny of *Geomydoecus* and *Thomomydoecus* pocket gopher lice (Phthiraptera: Trichodectidae) inferred from cladistic analysis of adult and first instar morphology. *Systematic Entomology* 20:129–43.

Patton, J. L. 1972. Patterns of geographic variation in karyotype in the pocket gopher, *Thomomys bottae* (Eydoux and Gervais). *Evolution* 26:574–86.

Patton, J. L., M. F. Smith, R. D. Price, and R. A. Hellenthal. 1984. Genetics of hybridization between the pocket gophers *Thomomys bottae* and *Thomomys townsendii* in northeastern California. *Great Basin Naturalist* 44:431–40.

Pearson, O. P. 1959. Biology of the subterranean rodent, *Ctenomys,* in Peru. *Memorias del Museo de Historia Natural "Javier Prado"* 9:1–56.

Price, R. D., and R. A. Hellenthal. 1981. A review of the *Geomydoecus californicus* complex (Mallophaga: Trichodectidae) from *Thomomys* (Rodentia: Geomyidae). *Journal of Medical Entomology* 18:1–23.

Reed, D. L., and M. S. Hafner. 1997. Host specificity of chewing lice on pocket gophers: A potential mechanism for cospeciation. *Journal of Mammalogy:* 78:655–60.

————. In press. Phylogenetic analysis of bacterial communities associated with ectoparasitic chewing lice of pocket gophers: A culture-independent approach. *Microbial Ecology.*

Reed, D. L., M. S. Hafner, and S. K. Allen. 2000a. Mammal hair diameter as a possible mechanism for host specialization in chewing lice. *Journal of Mammalogy* 81:999–1007.

Reed, D. L., M. S. Hafner, S. K. Allen, and M. B. Smith. 2000b. Spatial partitioning of host habitat by chewing lice of the genera *Geomydoecus* and *Thomomydoecus* (Phthiraptera: Trichodectidae). *Journal of Parasitology* 86:951–55.

Ridley, M. 1993. *Evolution.* Oxford: Blackwell Scientific.

Rogers, J. S. 1972. Measures of genetic similarity and genetic distance. *Studies in Genetics VII,* The University of Texas Publication no. 7213:145–53.

Russell, R. J. 1968. Evolution and classification of the pocket gophers of the subfamily Geomyinae. University of Kansas Publications of the Museum of Natural History 16:473–579.

Rust, R. W. 1974. The population dynamics and host utilization of *Geomydoecus oregonus,* a parasite of *Thomomys bottae. Oecologia* 15:287–304.

Sanderson, M. J. 1990. Estimating rates of speciation and evolution: A bias due to homoplasy. *Cladistics* 6:387–92.

Simberloff, D. 1987. Calculating the probabilities that cladograms match: A method of biogeographic inference. *Systematic Zoology* 36:175–95.

Smith, M. F., J. L. Patton, J. C. Hafner, and D. J. Hafner. 1983. *Thomomys bottae* pocket gophers of the central Rio Grande Valley, New Mexico: Local differentiation, gene flow, and historical biogeography. *Occassional Papers, Museum of Southwest Biology, University of New Mexico* 2:1–16.

Sneath, P. H. A., and R. R. Sokal. 1973. *Numerical taxonomy.* San Francisco: Freeman and Co.

Spradling, T. A. 1997. Relative rates of molecular evolution in rodents and their symbionts. Ph.D. diss., Louisiana State University.

Stone, A. R., and D. L. Hawksworth. 1986. *Coevolution and systematics.* Oxford: Clarendon Press.

Swofford, D. L., G. J. Olsen, P. J. Waddell, and D. M. Hillis. 1996. Phylogenetic inference. In *Molecular systematics,* edited by D. M. Hillis, C. Moritz, and B. K. Mable, 407–514. Sunderland, Mass.: Sinauer.

Taubes, C. H. 2001. *Modeling differential equations in biology.* Upper Saddle River, N.J.: Prentice Hall.

Timm, R. M. 1979. The *Geomydoecus* (Mallophaga: Trichodectidae) parasitizing pocket gophers of the *Geomys* complex (Rodentia: Geomyidae). Ph.D. diss., University of Minnesota.

————. 1983. Farenholz's rule and resource tracking: A study of host-parasite coevolution. In *Coevolution,* edited by M. H. Nitecki, 225–66. Chicago: University of Chicago Press.

Timm, R. M., and R. D. Price. 1980. The taxonomy of *Geomydoecus* (Mallophaga: Trichodectidae) from the *Geomys bursarius* complex (Rodentia: Geomyidae). *Journal of Medical Entomology* 17:126–45.

9

THE EFFECTS OF SPATIAL AND TEMPORAL
SCALE ON ANALYSES OF COPHYLOGENY

*James W. Demastes, Theresa A. Spradling,
and Mark S. Hafner*

Introduction

Host-parasite cophylogeny can be studied at multiple levels depending upon the evolutionary scale of the question being addressed and
the resolving power of the techniques used. Prior to development of DNA
sequencing techniques, researchers were limited to comparisons of host
and parasite phylogenies and a general comparison of the overall tempo
of evolution in the associated lineages (Hafner and Nadler, 1988). With
the advent of DNA sequencing techniques, researchers are no longer limited to documenting a history of cophylogeny, but can now examine rates
of change in orthologous genes among vastly different organisms, such as
hosts and their parasites (Hafner et al., 1994; Page et al., 1998).

If host and parasite lineages are compared to gene and organismal lineages, several similarities come to light (Hafner and Page, 1995). In this
comparison, the parasites may be thought of as genes that are passed from
generation to generation and are subject to the same biological phenomena
as other heritable genes. For example, parasite lineages may be affected
by reticulate evolution of host taxa, resulting in apparent host switches by
the parasites. The same stochastic processes that affect gene trees may also
affect parasites in the form of loss or fixation of parasites through drift or
incomplete lineage sorting (Avise et al., 1984). Hence, it is possible that
studies of cophylogeny at lower taxonomic (microevolutionary) levels—
generally below the species level—are likely to encounter one or more
of these confounding biological processes. Conversely, studies at higher
levels—generally at or above the species level—are more likely to find evidence of cophylogeny (Demastes and Hafner, 1993). This increased likelihood of demonstrating cophylogeny stems from host lineages having been
genetically (and geographically) isolated for longer periods of time and

because chance extinction of parasite lineages will inevitably lead to reciprocal monophyly (analogous to lineage sorting: Avise et al., 1984; Demastes
and Hafner, 1993). Therefore, in this chapter, we explore the relationship
between the temporal scale of lineages studied and the degree of cophylogeny that is detectable using current techniques.

Several studies have introduced the possibility that population demo-
graphics influence rate of molecular evolution (DeSalle and Templeton,
1988; Crozier et al., 1989; Easteal and Collet, 1994; Spradling et al., 2001).
Population dynamics could influence the outcome of examinations of relative rates of genetic change in hosts and parasites such as those performed
by Hafner et al. (1994) and Page et al. (1998). Therefore, by comparing
studies that utilize DNA sequence data, we examine any possible effects
that spatial scale (population-level parameters) may have on studies of
tempo and mode of evolution in cospeciating lineages. We rely heavily on
past studies of pocket gophers (Rodentia: Geomyidae) and their ectoparasitic chewing lice (Phthiraptera: Trichodectidae) in examining the effects
of both temporal and spatial scale because of the availability of multiple
studies examining associations at different scales.

Comparison One: Potential Effects of Temporal Scale

One of the earliest studies to use statistical tests to document presence
of host-parasite cophylogeny was that of Hafner and Nadler (1988). In
that study, the authors used allozymes to construct independently derived
phylogenetic trees for 8 taxa of pocket gophers and 10 taxa of chewing lice hosted by the gophers (fig. 9.1A). The pocket gophers included
in this study represented eight species in three genera (five subgenera).
In contrast, an allozyme study by Demastes and Hafner (1993) included
10 gopher taxa (excluding the outgroup), all of which belong to a single
genus *(Geomys)* and span four closely related species (fig. 9.1B). Both of
these studies used statistical tests to document presence of a history of
widespread cophylogeny between the gophers and lice. However, a comparison of the two tanglegrams (fig. 9.1) illustrates some of the problems
that are more likely to be encountered as studies move from large scale
(fig. 9.1A) to the microevolutionary scale (fig. 9.1B).

Insufficient Time of Isolation

The study dealing with older lineages (fig. 9.1A) has equal numbers of
host and parasite taxa, with the exception of the presence of two genera of
chewing lice *(Geomydoecus* and *Thomomydoecus)* on some of the pocket

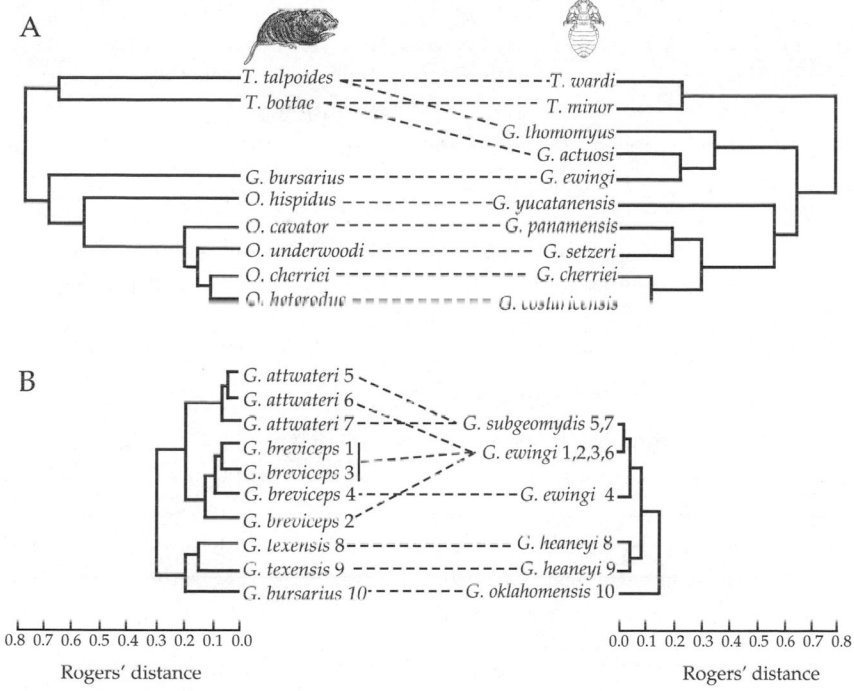

FIGURE 9.1. UPGMA phenograms based on Rogers's (1972) Genetic Distances for *A*, the allozyme based cospeciation studies of Hafner and Nadler (1988), and *B*, Demastes and Hafner (1993). Pocket gopher *(Geomys, Orthogeomys,* and *Thomomys)* relationships (left) are compared to chewing louse *(Geomydoecus* and *Thomomydoecus)* relationships (right). Dashed lines connect each parasite to its host.

gophers *(Thomomys).* The presence of two species of lice on one host is analogous to a duplication event in a gene tree. In contrast, the host and parasite trees in the study focusing on the genus *Geomys* (fig. 9.1B) are unbalanced, with 10 host lineages and only 6 parasite lineages, as indicated by the allozyme data. This discrepancy could be due to the lower number of allozyme loci surveyed for the chewing lice (precluding discovery of cryptic species), but it could also be due to incomplete geographic isolation between several subspecies of *Geomys.* For example, within the *Geomys breviceps* clade, *G. breviceps* populations 1, 2, and 3 are parapatric and host lice that are similar both morphologically (Timm and Price, 1980) and allozymically (Demastes and Hafner, 1993). The only population that is geographically disjunct (*G. breviceps* 4) is host to the only *Geomydoecus ewingi* population that is genetically distinct from its conspecifics (fig. 9.1B). Although the allozyme data fail to reflect the geographic isolation of *G. breviceps* 4 from *G. breviceps* 1, 2, and 3, a more recent study (Demastes, 1994) showed that this geographically isolated population is, indeed, genetically distinct from these other populations of *G. breviceps* (mitochondrial cytochrome-*b* sequence divergence $= 7.9\%$). Similarly, a study by Bohlin and Zimmerman (1982) documented absence of genetic introgression across contact zones between *G. bursarius* and *G. breviceps.* Chewing louse distributions were concordant with this genetic break in the hosts (Bohin and Zimmerman, 1982), which reinforces our suggestion that the imbalance between the trees in figure 9.1B is the result of incomplete geographic and genetic isolation.

Retained Ancestral Lineages

The louse population hosted by *Geomys attwateri* (6; fig. 9.1B) is a possible example of a retained ancestral lineage in the form of the chewing louse *Geomydoecus ewingi.* This population (found just south and west of San Antonio, Texas, USA) is separated from the nearest conspecific population of chewing lice by approximately 260 km, with the intervening area occupied by the closely related species, *Geomydoecus subgeomydis.* Both morphological (Timm and Price, 1980) and allozyme (Demastes and Hafner, 1993) data reflect this disjunct distribution of *G. ewingi;* the most parsimonious biogeographic hypothesis involves a widely distributed ancestral taxon, *G. ewingi,* giving rise to a daughter species *(G. subgeomydis),* with a remnant population of *G. ewingi* persisting in the patchily distributed suitable habitat of southwest Texas. The close genetic relationship of these two species of chewing lice is supported by the allozyme data (Demastes and Hafner, 1993), and is clearly indicated in the extensive phylogenetic analysis of the genus by Page et al. (1995).

Comparison Two: Potential Effects of Spatial Scale

Hafner et al. (1994) documented a history of cophylogeny for pocket gophers and their chewing lice using DNA sequence data from the mitochondrial cytochrome *c* oxidase subunit I gene (COI: see Hafner, Demastes, Spradling, and Reed, chap. 8). This study involved comparisons of hosts at or above the species level, with the highest concentration of cospeciating taxa within the chewing louse genus *Geomydoecus* and the pocket gopher genus *Orthogeomys* (fig. 9.2A). Beyond merely documenting cospeciation, the use of orthologous gene regions allowed for a direct comparison of evolutionary rates in the hosts and parasites (Hafner et al., 1994; Page, 1996). These analyses have documented that chewing lice have a higher rate of COI evolution than do pocket gophers.

To facilitate a comparison with the data of Hafner et al. (1994), we examined previously unpublished data from pocket gophers and chewing lice (Spradling, 1997, app. 1). These data consist of the same 379-nucleotide positions of the mitochondrial COI gene studied by Hafner et al. (1994). Twenty taxa of gophers (seven species) within a single genus *(Thomomys)* and their chewing lice *(Geomydoecus* and *Thomomydoecus)* were examined. Although many gophers of the genus *Thomomys* host both genera of lice, it is unusual for a pocket gopher population to host more than one species of louse of each genus (Hellenthal and Price, 1984a). Because the two genera of chewing lice are distinct both morphologically (Hellenthal and Price, 1984b) and genetically (Nadler and Hafner, 1993), they constitute two independent lineages that were compared separately with the host lineage, *Thomomys*.

DNA isolation, PCR (including primer sequences), and sequencing techniques for *Thomomys* and their lice followed Hafner et al. (1994). Sequences are available under GenBank accession numbers AF284356-284375 and AF284391-284418; data matrices also are available from TreeBASE (http://www.treebase.org/) under accession numbers SN520-521. Phylogenetic hypotheses for the chewing lice and the pocket gophers (fig. 9.2B and C) were generated using maximum-likelihood methods (Olsen et al., 1994). To increase the probability of finding the best maximum-likelihood tree, a variety of transition:transversion ratios were incorporated into the model. In addition, the global rearrangement option was used and the input order of taxa was jumbled until a tree with the best log-likelihood score was found a minimum of four times. A transition bias of 3·1 yielded the highest log-likelihood scores for both the hosts (−2755.64056) and parasites (−4772.46372).

To determine if parallel cladogenesis is the fundamental mechanism governing host-parasite associations in this assemblage, host and parasite phylogenies were compared to determine if topological similarity was greater than would be expected by chance. The TREEMAP program (Page, 1994b) was used to reconcile host and parasite trees (allowing host switching to be postulated), thereby estimating the number of potential cospeciation events between the two groups. Tests of significance were conducted by randomizing the parasite trees 1,000 times. These randomly generated trees then were compared with the actual host tree. The fit between the actual host and parasite trees was significantly better than the fit between the host tree and randomly generated parasite trees ($p < .02$, 15 cospeciation events estimated). The *Fit* command of the COMPONENT program (Page, 1993a) also was used to reconcile the host and parasite cladograms, thereby estimating the minimum number of events that must have occurred

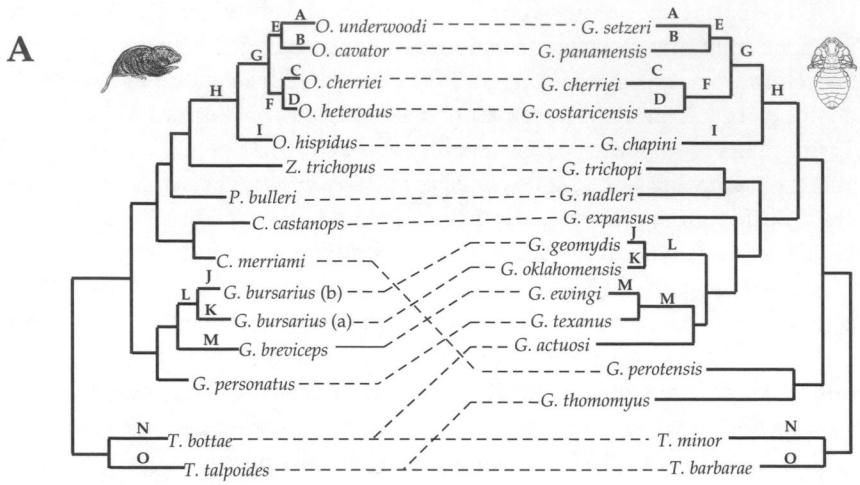

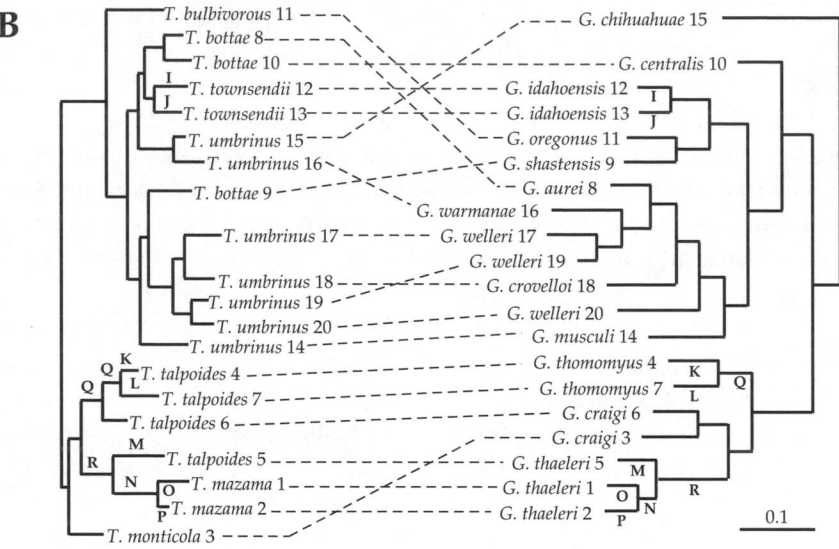

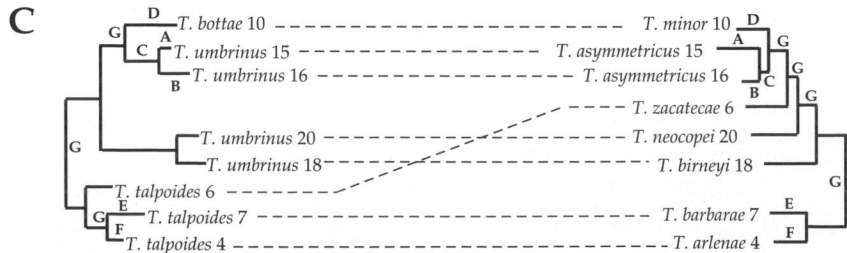

("number of leaves added" and "number of losses": Page, 1994a) to produce the observed level of incongruence in the host and parasite trees if no host switching has occurred. Fit between the host and parasite trees was significantly better than the fit between the actual parasite tree and 1,000 randomly generated host trees ($p < .001$).

Host and parasite genetic data were further tested for significant association by comparing maximum-likelihood sequence-divergence matrices (Mantel, 1967) using the R-Package program (Legendre and Vaudor, 1991). The matrices for the *Geomydoecus* comparisons (fig. 9.2B) exhibited a greater association than expected by chance ($p = .001$), as did the matrices for the *Thomomydoecus* comparisons (fig. 9.2C; $p = 0.02$). Thus, all evidence involving tree topology and genetic divergence points to a history of widespread cospeciation between these pocket gophers and chewing lice.

Hafner et al. (1994) compared branch lengths from phylogenetic trees of pocket gophers and their cospeciating chewing lice to compare rates of molecular evolution in the two groups. Making a similar comparison for the *Thomomys* and lice described here, we investigated rates of evolution in the hosts and parasites through comparisons of maximum-likelihood branch lengths, which were generated using the optimal maximum-likelihood transition:transversion ratio (3:1). Host-parasite associations that are the result of host switching are not relevant to comparisons of relative rates of molecular evolution (Page, 1993b). Accordingly, branch-length comparisons were limited to those branches that are most obviously orthologous (fig. 9.2B and C, table 9.1). The analysis revealed no significant difference in rate of evolution between pocket gophers and their chewing lice (Wilcoxon signed-ranks tests, $p > .50$). In contrast, the results for the comparison of gopher and louse branch lengths in the Hafner et al. (1994) study, reanalyzed here using the above parameters (fig. 9.2A), indicated significantly different rates of evolution between the hosts and their parasites (Wilcoxon signed-ranks test, $p < 0.01$).

FIGURE 9.2. (*Facing page*) Maximum-likelihood trees based on COI-sequence data (transition:transversion ratio of 3:1) with branch lengths for the trees drawn to scale. Pocket gopher (*Cratogeomys, Geomys, Orthogeomys, Pappogeomys, Thomomys,* and *Zygogeomys*) relationships (left) are compared with chewing louse (*Geomydoecus* and *Thomomydoecus*) relationships (right). Dashed lines connect each parasite to its host. Letters indicate corresponding branches between cospeciating hosts and their parasites. *A,* Data from Hafner et al. (1994) reanalyzed using the parameters described here (gopher tree, ln = −2472.61244, louse tree = −3646.89484). This topology corresponds to figure 2B of Hafner et al. (1994). *B, C,* Data from Spradling (1997).

TABLE 9.1 Branch lengths (expected number of substitutions per 100 sites) from maximum-likelihood trees of pocket gopher and chewing louse COI nucleotide sequences. Letters correspond to branches labeled in fig. 9.2B and C, which indicate cospeciating taxa.

Chewing Louse		Pocket Gopher	
Branch	Length	Branch	Length
A	5.6	A	1.5
B	2.0	B	3.9
C	1.5	C	8.3
D	4.3	D	3.6
E	4.8	E	5.1
F	4.6	F	2.3
G	15.5	G	18.2
I	4.4	I	4.4
J	3.9	J	3.6
K	6.6	K	2.3
L	6.2	L	5.1
M	5.3	M	6.6
N	2.5	N	5.9
O	4.1	O	4.0
P	4.3	P	1.5
Q	4.2	Q	5.3
R	12.6	R	4.4

To allow for a more detailed comparison, orthologous host and parasite branches from both studies were plotted, and reduced major-axis regression was used to find the slope of the best-fit line through the origin using the SYSTAT program (version 5.2, SYSTAT, Inc., Evanston, Ill.). Regression analysis of the branch lengths determined here for the Hafner et al. data (1994) indicated that chewing lice are evolving about 1.6 times faster than their hosts (Model II regression, slope = 1.62: fig. 9.3A). In contrast, a plot of the branch lengths for hosts and parasites in the Spradling (1997) study reveals a wide scatter of points (fig. 9.3B). This, coupled with the paucity of information at greater levels of sequence divergence, provides little basis for comparing relative rate of evolution of these gophers and lice. More important, these points show no obvious linear structure consistent with a pattern of cophylogeny (Hafner and Nadler, 1990) and suggest a lack of any predictable relationship between genetic divergence in these host and parasite lineages. The major difference between these two

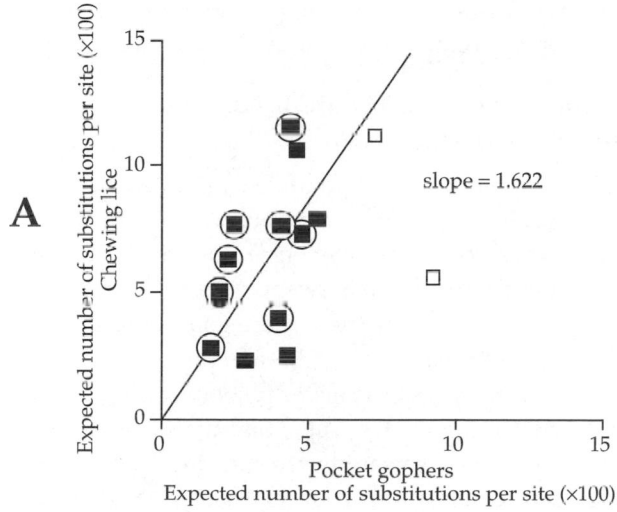

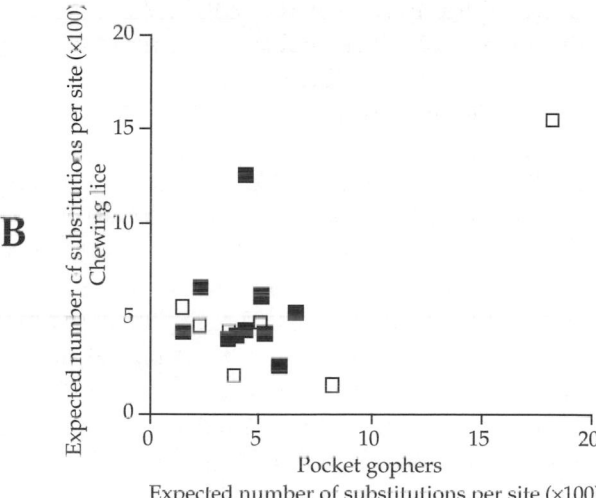

FIGURE 9.3. Bivariate plots of pocket gopher and chewing louse maximum-likelihood branch lengths (*Geomydoecus* comparisons = closed squares, *Thomomydoecus* comparisons = open squares) from figure 9.2. A transition:transversion bias of 3:1, which yielded the best maximum-likelihood scores for trees shown in figure 9.2B and C, was assumed for pocket gophers and chewing lice of Hafner et al. (1994) to allow a more direct comparison. *A,* Reduced major-axis regression (through the origin) of reanalyzed data from Hafner et al. (1994; points represent lettered branches in figure 9.2A). Circled points represent *Orthogeomys* comparisons based on the best maximum-likelihood trees for pocket gophers and chewing lice. *B,* Points represent lettered branches in figure 9.2B and C. Note the absence of an obvious linear relationship.

gopher-louse COI data sets, therefore, involves presence or absence of a linear relationship between branch lengths of hosts and parasites.

Reasons for a Nonlinear Relationship in the Divergence of *Thomomys* and Their Lice

It is possible that the scatter of points seen in figure 9.3B is the result of inclusion of several host and parasite pairs that did not diverge contemporaneously (i.e., pseudocospeciation events), despite the fact that a number of tests indicated sufficient match between trees to indicate a history of cophylogeny over much of the history of these gophers and lice. Many of the apparently cospeciating sister taxa of both gophers and lice (fig. 9.2B and C) involve intraspecific comparisons (e.g., between *Thomomys mazama* populations 1 and 2 in fig. 9.2B). Thus, this analysis may be confounded by some of the problems associated with studying more recently diverged lineages (discussed earlier in this chapter), such as host switching and incomplete lineage sorting.

It also is possible, however, that the scatter of points in fig. 9.3B is the result of the absence of a local molecular clock within the pocket gophers, the chewing lice, or both of these groups. In order for the regression approach to yield meaningful results, a molecular clock must be present (Page and Hafner, 1996). Therefore, several relative-rate tests (Mindell and Honeycutt, 1990) were conducted both within and between subgenera of pocket gophers (using both *Pappogeomys* and *Thomomys* taxa as outgroups) and within and between subgenera of chewing lice (using *Geomydoecus* as the outgroup for *Thomomydoecus,* and vice versa, and also using congeneric lice as outgroups). These relative-rate tests included all possible comparisons between sister taxa of pocket gophers and between sister taxa of chewing lice thought to have cospeciated with their hosts (lettered branches in fig. 9.2B and C). Among pocket gopher taxa, overall rate of substitution did not differ significantly in eight independent comparisons, regardless of which outgroup taxon was used (binomial test, $p > .26$). Similarly, among chewing lice, no significant difference in rate of substitution was detected, regardless of which outgroup taxon was used (binomial test, $p > .26$ in ten comparisons).

Despite the absence of a significant difference in rate of evolution as determined by relative-rate tests, comparisons of maximum-likelihood branch lengths show evidence of heterogeneity in rate of evolution (fig. 9.4). Because sister taxa, by definition, have diverged for the same length of time, the lengths of branches leading to sister taxa from their common ancestor

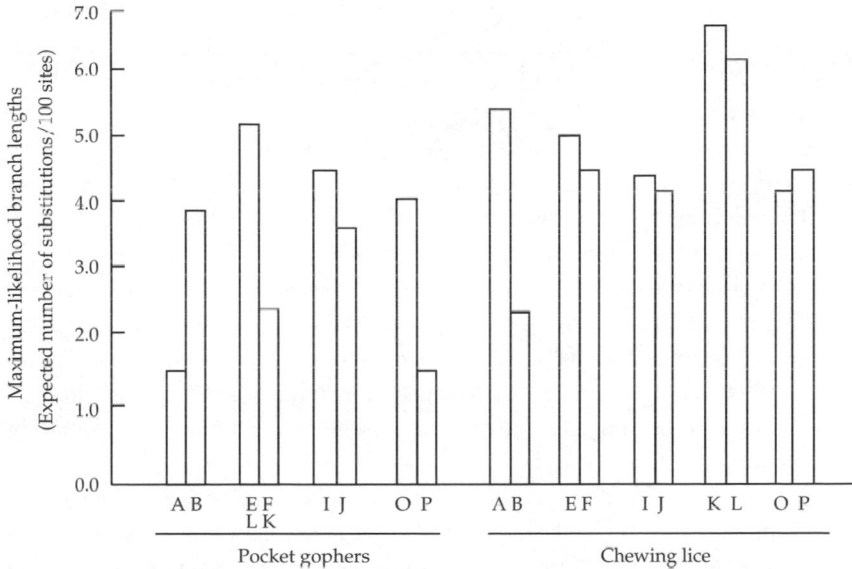

FIGURE 9.4. Comparison of maximum-likelihood branch lengths (table 9.1) of cospeciating sister taxa of pocket gophers and chewing lice. Letters correspond to branches lettered in figure 9.2B and C. If rates of evolution are equal between sister taxa, branch lengths should also be roughly equal. Though not statistically significant, there is enough apparent variation in rate of evolution to introduce nonlinearity to a plot of gopher branch lengths vs. louse branch lengths (fig. 9.3).

should be roughly equal if DNA substitutions have accumulated at the same rate in the two taxa. Visual inspection of branches leading to sister taxa of pocket gophers and of chewing lice (fig. 9.4) reveals differences in branch lengths, which seem more prominent in the pocket gopher comparisons than in the chewing louse comparisons. Further examination of rates of evolution between these pairs of taxa was performed using the two-cluster test of Takezaki et al. (1995), which tests for heterogeneity in the average substitution rate of sister lineages. This test also showed no significant differences in rate of evolution among the pairs of gophers ($Z = 0.817 - 1.611, p = .107 - .402$) or lice ($Z = 0.421 - 1.258, p = .211 - .674$) of figure 9.4.

Although the relative-rate tests revealed no significant heterogeneity in overall substitution rate in these groups, analysis of only 379 base pairs may be insufficient to discern any but the most dramatic differences in rate of evolution. Scherer (1989, 440) pointed out that "passing the relative rate test does not guarantee a uniform rate of molecular evolution." Therefore, conclusions of rate homogeneity always should be viewed with caution. Even slight variation in estimated branch length within either or both of two trees being compared will tend to confound comparisons of overall rate of evolution between the two groups. For example, the branch leading to *T. mazama*-1 (branch O, fig. 9.2B) was greater than twice the length of the branch leading to its sister taxon, *T. mazama*-2 (branch P, fig. 9.2B). Differences of this magnitude, although not statistically significant, will introduce some degree of nonlinearity in a bivariate plot, making it difficult to compare rates of evolution of the hosts and their parasites. A similar comparison of branch lengths of cospeciating sister taxa examined by Hafner et al. (1994: see fig. 9.2A) revealed much more homogeneity in estimated rate of evolution, perhaps contributing to the more linear appearance of points in figure 9.3A.

If, as seems likely, the lack of a correlation between genetic divergence in *Thomomys* and their lice stems from the heterogeneous nature of evolution among the pocket gopher hosts (fig. 9.4), then a comparison of these hosts with those in the Hafner et al. (1994) study is appropriate. Rate of evolution appears much more homogeneous in pairs of pocket gopher sister taxa in the Hafner et al. (1994) study (fig. 9.2A) than in the *Thomomys* (fig. 9.2B and C) examined here. This is interesting given that it is widely accepted that species within the genus *Orthogeomys* are valid biological species (Hafner, 1991) and that the subspecies of *Thomomys* are not reproductively isolated (Hafner et al., 1987; Patton and Smith, 1994). These differences in reproductive status are not reflected by amount of COI divergence, however, as the *Orthogeomys* comparisons are at a level of divergence similar to that between the *Thomomys* taxa of interest (fig. 9.2). If the *Thomomys* taxa studied are more recently diverged than the *Orthogeomys* taxa, we might expect there to be more stochastic variation in rate of evolution among the *Thomomys* because of insufficient time for an average molecular clock to emerge. But, because both the *Orthogeomys* and *Thomomys* lineages studied are at roughly the same stage of COI genetic divergence, insufficient time of genetic divergence must not be the source of the observed variation in branch lengths in the *Thomomys* tree (fig. 9.4). Thus, temporal scale may not play as great a role in explaining the differences in these two studies as it did in the two allozyme studies described earlier in this chapter. Furthermore, Smith (1998) used the mitochondrial cytochrome-*b* gene (1,140 base pairs) to construct a phylogenetic tree of *Thomomys* and showed rate heterogeneity among taxa equivalent to that presented here. Thus, differences in branch lengths observed in our study (fig. 9.4) are not simply the result of gene-specific factors or a small number of nucleotides sampled.

Spatial Scale and Rates of Evolution

The question of why there would be disparate rates of evolution among subspecies of *Thomomys* and not among species of *Orthogeomys* potentially may be answered by examining fundamental differences in population structure of these two genera. "Nearly-neutral" theory predicts that populations having small effective sizes should be exposed to weaker purifying selection and thus accumulate slightly deleterious mutations at a faster rate than would larger populations (Ohta, 1973, 1974). Therefore, smaller average population size should tend to promote a higher rate of evolution in a species, and variation in size among populations would result in a variety of evolutionary rates. DeSalle and Templeton (1988) provided empirical evidence supporting this prediction. Further empirical evidence for the possible effect of population dynamics on rate of evolution comes from an earlier study involving *Thomomys*. Patton and Smith (1981) sampled 159 populations of *Thomomys* and compared rates of allozymic change among 35 lineages belonging to five species. Four apparent rate decelerations and one apparent rate acceleration were identified within *T. bottae*, and four additional rate decelerations were identified within *T. umbrinus*. A cursory examination of the branch lengths in their figure 2 (Patton and Smith, 1981, p. 497) reveals several other pairs of sister taxa with disparate rates of change. Patton and Smith (1981, p. 499) concluded that these rate differences likely were caused by "variation in population size, degree and temporal length of isolation, and hybridization."

Of the *Thomomys* examined here, there is great variation in range size and habitat patchiness. A single species, *Thomomys bottae*, has a distribution that is orders of magnitude larger than the combined geographic ranges of four of the *Orthogeomys* species included in the Hafner et al. (1994) study. The 191 described subspecies of *Thomomys bottae* (Hall, 1981) vary greatly in the size of their distributions and must have been subjected to differing levels of climatic, geological, and stochastic perturbations throughout their evolutionary histories. For example, the distribution of *Thomomys mazama-2* (appendix 9.1) is more than four times the size of *Thomomys mazama-*1, and the geographic range of *Thomomys umbrinus-*17 is more than five times larger than that of either *Thomomys umbrinus-*19 or *Thomomys umbrinus-*20. More important, variation in effective population size is expected to result in large differences in heterozygosity among gopher populations. Indeed, *Thomomys talpoides* populations studied by Nevo et al. (1974) show a wide range of heterozygosity values from 0.008 to 0.085, and heterozygosity in *Thomomys bottae* populations studied by Patton and Yang (1977) varied from 0.030 to 0.169. Patton and Yang (1977) found a correlation between heterozygosity within *Thomomys bottae* populations and patchiness of suitable habitat, which varies greatly across the range of *Thomomys*. Although it is impossible to know the past geographic distributions of these taxa and their historical population dynamics, current distributions suggest that population subdivision and effective population size may explain the evolutionary rate differences observed between pairs of *Thomomys* (fig. 9.4).

The genus *Orthogeomys* is very different from *Thomomys* in many aspects of population structure involving spatial scale. The genus *Orthogeomys* has a predominantly Central American distribution, and the pairs of sister taxa included in the Hafner et al. (1994) study have roughly similarly sized ranges (Hafner, 1991). A study of the mitochondrial cytochrome-*b* and 16S rDNA genes (Sudman and Hafner, 1992) and the study by Hafner et al. (1994) both showed roughly equivalent branch lengths between sister taxa of *Orthogeomys*. Furthermore, Demastes et al. (1996) demonstrated that at least one of these species *(O. underwoodi)* fits an isolation-by-distance model of geographic variation, indicating that gene flow is restricted more by distance than by any physical barriers between populations. Isolated populations undoubtedly exist in the genus, but because of the restricted geographic range, these populations of *Orthogeomys* can only differ in size by a relatively small degree of magnitude. Therefore, we expect to see more homogeneity in rate of evolution among *Orthogeomys* populations than among *Thomomys* populations.

Unlike their hosts, most comparisons of the chewing lice of *Thomomys* reveal relatively little heterogeneity in evolutionary rate (fig. 9.4). Given that most pocket gopher individuals host hundreds of lice (Rust, 1974), it is likely that chewing louse populations experience less population fragmentation and bottlenecking than do their hosts. Therefore, population-level phenomena may have less dramatic effects on relative rates of evolution in chewing lice, yielding slightly more homogeneous branch lengths compared with those of their hosts (fig. 9.4).

Conclusions

When conducting analyses of cospeciation among lineages that have diverged relatively recently, one must be aware of the potential confounding effects of incomplete lineage sorting of the parasites, loss of parasite lineages through chance extinction events, retention of ancestral parasite lineages, and recent host-switching events (Demastes and Hafner, 1993; Hafner and Page, 1995). Certain of these effects, most notably incomplete sorting in the form of retention of ancestral parasite lineages, were evident in the study by Demastes and Hafner (1993).

The DNA-sequence studies discussed herein reveal yet another series of potential problems when trying to extend the scope of cospeciation studies to include comparisons of relative rates of evolution between hosts and their associated parasites. Although previous studies have demonstrated disparate rates of evolution between hosts and their associated parasitic lice (Hafner et al., 1994; Page et al., 1998), a study focused on the genus *Thomomys* failed to corroborate this finding. The lack of a significant relationship between host and parasite evolutionary rates in the *Thomomys* data is of interest in that the causal factors, once identified, will aid in the design of future studies. What we have observed, however, is potentially an example of a new and different type of scaling that may play a role in our understanding of host-parasite cophylogeny. Instead of the entire system being too recently evolved to demonstrate cospeciation effectively, the problem of differing spatial scales *within* an assemblage may adversely affect our abilities to compare evolutionary rates between hosts and their cospeciating parasites because of the absence of a local molecular clock. We have shown, therefore, that both temporal and spatial factors may bring to studies of cophylogeny a complex array of influences, including incomplete lineage sorting, chance extinction events, and heterogeneity in rates of evolution within groups resulting from population dynamics.

REFERENCES

Avise, J. C., J. E. Neigel, and J. Arnold. 1984. Demographic influences on mitochondrial DNA lineage survivorship in animal populations. *Journal of Molecular Evolution* 20:99–105.

Bohlin, R. G., and E. G. Zimmerman. 1982. Genic differentiation of two chromosome races of the *Geomys bursarius* complex. *Journal of Mammalogy* 63.218–28.

Crozier, R. H., Y. C. Crozier, and A. G. Mackinlay. 1989. The CO-I and CO-II region of honeybee mitochondrial DNA: Evidence for variation in insect mitochondrial evolutionary rates. *Molecular Biology and Evolution* 6:399–411.

Demastes, J. W. 1994. Systematics and zoogeography of the Mer Rouge pocket gopher *(Geomys breviceps breviceps)* based on cytochrome-*b* sequences. *Southwestern Naturalist* 39:276–80.

Demastes, J. W., and M. S. Hafner. 1993. Cospeciation of pocket gophers *(Geomys)* and their chewing lice *(Geomydoecus)*. *Journal of Mammalogy* 74:521–30.

Demastes, J. W., M. S. Hafner, and D. J. Hafner. 1996. Phylogeographic variation in two Central American pocket gophers *(Orthogeomys)*. *Journal of Mammalogy* 77:917–27.

DeSalle, R., and A. R. Templeton. 1988. Founder effects and the rate of mitochondrial DNA evolution in Hawaiian *Drosophila*. *Evolution* 42:1076–84.

Easteal, S., and C. Collet. 1994. Consistent variation in amino-acid substitution rate, despite uniformity of mutation rate: Protein evolution in mammals is not neutral. *Molecular Biology and Evolution* 11:643–47.

Hafner, M. S. 1991. Evolutionary genetics and zoogeography of Middle American pocket gophers, Genus *Orthogeomys*. *Journal of Mammalogy* 72:1–10.

Hafner, M. S., J. C. Hafner, J. L. Patton, and M. F. Smith. 1987. Macrogeographic patterns of genetic differentiation in the pocket gopher *Thomomys umbrinus*. *Systematic Zoology* 36:18–34.

Hafner, M. S., and S. A. Nadler. 1988. Phylogenetic trees support the coevolution of parasites and their hosts. *Nature* 332:258–59.

———. 1990. Cospeciation in host-parasite assemblages: Comparative analysis of rates of evolution and timing of cospeciation. *Systematic Zoology* 39:192–204.

Hafner, M. S., and R. D. M. Page. 1995. Molecular phylogenies and host-parasite cospeciation: Gophers and lice as a model system. *Philosophical Transactions of the Royal Society of London,* ser. B, 349:77–83.

Hafner, M. S., P. D. Sudman, F. X. Villablanca, T. A. Spradling, J. W. Demastes, and S. A. Nadler. 1994. Disparate rates of molecular evolution in cospeciating hosts and parasites. *Science* 265:1087–90.

Hall, E. R. 1981. *The mammals of North America.* New York: John Wiley and Sons.

Hellenthal, R. A., and R. D. Price. 1984a. Distributional associations among *Geomydoecus* and *Thomomydoecus* lice (Mallophaga: Trichodectidae) and pocket gopher hosts of the *Thomomys bottae* group (Rodentia: Geomyidae). *Journal of Medical Entomology* 21:432–46.

———. 1984b. *Geomydoecus thomomyus* complex (Mallophaga: Trichodectidae) from pocket gophers of the *Thomomys talpoides* complex (Rodentia: Geomyidae) of the United States and Canada. *Annals of the Entomological Society of America* 82:286–97.

Legendre, P., and A. Vaudor. 1991. The R Package: Multidimensional analysis, spatial analysis. Département de Sciences Biologiques, Université de Montréal.

Mantel, N. 1967. The detection of disease clustering and a generalized regression approach. *Cancer Research* 27:209–20.

Mindell, D. P., and R. L. Honeycutt. 1990. Ribosomal RNA in vertebrates: Evolution and phylogenetic applications. *Annual Review of Ecology and Systematics* 21:541–66.

Nadler, S. A., and M. S. Hafner. 1993. Systematic relationships among pocket gopher chewing lice (Phthiraptera: Trichodectidae) inferred from electrophoretic data. *International Journal for Parasitology* 23:191–201.

Nevo, E., Y. C. Kim, C. R. Shaw, and C. S. Thaeler. 1974. Genetic variation, selection and speciation in *Thomomys talpoides* pocket gophers. *Evolution* 28:1–23.

Ohta, T. 1973. Slightly deleterious mutant substitutions in evolution. *Nature* 246:96–98.

————. 1974. Mutational pressure as the main cause of molecular evolution and polymorphism. *Nature* 252:351–54.

Olsen, G. J., H. Matsuda, R. Hagstrom, and R. Overbeck. 1994. fastDNAml: A tool for construction of phylogenetic trees of DNA sequences using maximum likelihood. *Computer Applications in the Biological Sciences* 10:41–48.

Page, R. D. M. 1993a. COMPONENT, tree comparison software for Microsoft® Windows™, version 2.0. London: The Natural History Museum.

————. 1993b. Parasites, phylogeny and cospeciation. *International Journal for Parasitology* 23:499–506.

————. 1994a. Maps between trees and cladistic analysis of historical associations among genes, organisms, and areas. *Systematic Biology* 43:58–77.

————. 1994b. Parallel phylogenies: Reconstructing the history of host-parasite assemblages. *Cladistics* 10:155–73.

————. 1996. Temporal congruence revisited: Comparison of mitochondrial DNA sequence divergence in cospeciating pocket gophers and their chewing lice. *Systematic Biology* 45:151–67.

Page, R. D. M., and M. S. Hafner. 1996. Molecular phylogenies and host-parasite cospeciation: Gophers and lice as a model system. In *New uses for new phylogenies*, edited by P. H. Harvey, A. J. Leigh Brown, J. Maynard Smith, and S. Nee, 255–70. Oxford: Oxford University Press.

Page, R. D. M., P. L. M. Lee, S. A. Becher, R. Griffiths, and D. H. Clayton. 1998. A different tempo of mitochondrial DNA evolution in birds and their parasitic lice. *Molecular Phylogenetics and Evolution* 9:276–93.

Page, R. D. M., R. D. Price, and R. A. Hellenthal. 1995. Phylogeny of *Geomydoecus* and *Thomomydoecus* pocket gopher lice (Phthiraptera: Trichodectidae) inferred from cladistic analysis of adult and first instar morphology. *Systematic Entomology* 20:129–43.

Patton, J. L., and M. F. Smith. 1981. Molecular evolution in *Thomomys:* Phyletic systematics, paraphyly, and rates of evolution. *Journal of Mammalogy* 62:493–500.

————. 1994. Paraphyly, polyphyly, and the nature of species boundaries in pocket gophers (genus *Thomomys*). *Systematic Biology* 43:11–26.

Patton, J. L., and S. Y. Yang. 1977. Genetic variation in *Thomomys bottae* pocket gophers: Macrogeographic patterns. *Evolution* 31:697–720.

Rogers, J. S. 1972. Measures of genetic similarity and genetic distance. *Studies in Genetics VII,* The University of Texas Publication no. 7213:145–53.

Rust, R. W. 1974. The population dynamics and host utilization of *Geomydoecus oregonus*, a parasite of *Thomomys bottae*. *Oecologia* 15:287–304.

Scherer, S. 1989. The relative-rate test of the molecular clock hypothesis: A note of caution. *Molecular Biology and Evolution* 6:436–41.

Smith, M. F. 1998. Phylogenetic relationships and geographic structure in pocket gophers in the genus *Thomomys*. *Molecular Phylogenetics and Evolution* 9:1–14.

Spradling, T. A. 1997. Relative rates of molecular evolution in rodents and their symbionts. Ph.D. diss., Louisiana State University.

Spradling, T. A., M. S. Hafner, and J. W. Demastes. 2001. Differences in rate of cytochrome-*b* evolution among species of rodents. *Journal of Mammalogy* 82:65–80.

Sudman, P. D. and M. S. Hafner. 1992. Phylogenetic relationships among Middle American pocket gophers (genus *Orthogeomys*) based on mitochondrial DNA sequences. *Molecular Phylogenetics and Evolution* 1:17–25.

Takezaki, N., A. Rzhetsky, and M. Nei. 1995. Phylogenetic test of the molecular clock and linearized trees. *Molecular Biology and Evolution* 12:823–33.

Timm, R. M., and R. D. Price. 1980. The taxonomy of *Geomydoecus* (Mallophaga: Trichodectidae) from the *Geomys bursarius* complex (Rodentia: Geomyidae). *Journal of Medical Entomology* 17:126–45.

Appendix 9.1

Collection localities for pocket gophers reported in this study. Locality numbers correspond with figures 9.2B and C. Host and parasite vouchers were deposited in the Louisiana State University Museum of Natural Science mammal (LSUMZ) or tissue (M) collection, the New Mexico Museum of Natural History (NMMNH), or the University of South Dakota Natural History Collection (USDNHC). Other initials are those of individual collectors.

Washington: Mason Co.; 2 mi. N Shelton on Hwy. 101, Shelton Airport (LSUMZ 34383; TSD 567)

Oregon: Lane Co.; 1.1 mi. N, 5.5 mi. E McKenzie Bridge (LSUMZ 31398; TSD 504)

Nevada: Washoe Co.; 2.2 mi. N, 2 mi. E intersection Hwy. 28 and Hwy. 431 (LSUMZ 31411; TSD 493)

New Mexico: Sandoval Co.; 11 mi. E, 3 mi. N Jemez Springs (LSUMZ 34031; DLR 265)

Washington: Klickitat Co.; 5 mi. S Trout Lake, on Forest Rd. 051 (LSUMZ 34387; TSD 571)

Utah: Sevier Co.; 5 mi. S, 2 mi. E Monroe (Mud Springs Flat) (LSUMZ 31301; TSD 454)

South Dakota: Lawrence Co.; Timon Campground (USDNHC 1091 and 1094; PDS 453 and SCM 4)

New Mexico: Socorro Co.; 3.5 mi. S La Joya, W side Rio Grande (LSUMZ 29548; JWD 39)

Oregon: Jackson Co.; Ashland, near intersection of Hwy. 66 and Interstate 5 (LSUMZ 34331; TSD 576)

New Mexico: Socorro Co.; 2.0 mi. N, 0.5 mi. E Polvadera (NMMNH 1260; JWD 68)

Oregon: Benton Co.; 4 mi. N Corvallis (Benton Co. Courthouse) (SW 1/4, NE 1/4, sec. 13, T11S, R5W) (LSUMZ 31306; TSD 535)

Nevada: Lander Co.; 9.8 mi. N, 0.3 mi. E Battle Mountain (LSUMZ 31264; TSD 459)

Oregon: Malheur Co.; 15 mi. W Vale (by road) on Hwy. 20 (LSUMZ 34330; TSD 583)

MEXICO: Sinaloa; 1 km N Siqueros, 50 m elev. (M 3042; MSH 1452)

MEXICO: Durango; 12 km E El Salto, 2490 m elev. (M 3040; MSH 1450)

MEXICO: Chihuahua; 13 km E Tomochic, 2100 m elev. (LSUMZ 34346; MSH 1442)

MEXICO: Durango; 50 km N, 20 km W Bermejillo, 1140 m elev. (M 3034; MSH 1444)

Arizona: Santa Cruz Co.; 10.5 mi. S, 0.6 mi. E Patagonia (0.4 mi. N, 0.6 mi. E Crescent Spring) (LSUMZ 33869; TSD 546)

MEXICO: Michoacán; 6.5 km S Pátzcuaro, 2200 m elev. (M 3044; MSH 1459)

MEXICO: México; 9 km N Valle de Bravo, 2370 m elev. (M 3047; MSH 1463)

IO

HAVE MAMMALS AND THEIR CHEWING LICE DIVERSIFIED IN PARALLEL?

Jason Taylor and Andy Purvis

Introduction

Chewing lice (Phthiraptera: Ischnocera: Trichodectidae) are wingless, permanent ectoparasitic insects found exclusively on mammals. The speciation of specialized parasites such as chewing lice is often thought to be driven by the speciation of their hosts. This assumption has been formalised as Fahrenholz's Rule, which states that host and parasite phylogenies should be mirror images due to cospeciation. Fahrenholz's Rule has long been applied to the problems of inferring host or parasite phylogenies, and has been used in particular by many louse taxonomists and systematists (see Lyal, 1983). In fact, modern systematics uses an assumption similar to Fahrenholz's Rule whereby gene phylogenies are interpreted as species phylogenies (see Page and Holmes, 1998).

Previous studies of cospeciation (Brooks, 1988; Moran et al., 1995; Peek et al., 1998; Hugot, 1999), many involving lice and their hosts (Lyal, 1986, 1987; Hafner and Nadler, 1988; Demastes and Hafner, 1993; Paterson et al., 1993; Hafner and Page, 1995; Page et al., 1998), suggest that cospeciation may be a major factor in determining the current host-parasite associations. However, due to the small scale of previous studies, it is difficult to determine if cospeciation is the trend or whether there has been a reporting of particularly notable examples.

Only three studies of cospeciation have investigated large clades (Lyal, 1986, 1987; Hugot, 1999). Hugot's study demonstrated a high degree of cospeciation between primates and their pinworms. Lyal's studies of cospeciation between mammals and their chewing lice were hampered by the lack of host phylogenies and suitable tools for the detection of cospeciation. However, he was able to conclude that a minimum of between 20.7% and 25.7% of louse speciation events are clearly incompatible with the scenario of cospeciation. Lyal also predicted that this figure would rise with greater knowledge of mammal and louse phylogenies (Lyal, 1987).

Here, we test the same clade of chewing lice for cospeciation with their mammalian hosts. However, this study goes beyond that of Lyal in several important ways. When constructing the louse phylogeny we have been able to use all 164 characters used by Lyal in his manual cladistic analysis, plus 95 of his remaining 115 unanalyzed characters (Taylor, 2000). Also, we have performed a much more rigorous phylogenetic analysis using a new and efficient tree-finding strategy (Quicke et al., 2001). Furthermore, due to the availability of many more mammal phylogenies and tools for the investigation of cospeciation (Page, 1994b), this study is the largest study to explicitly map parasite phylogeny to host phylogeny. The scale of this study has allowed us to investigate not only whether cospeciation is the trend, but also two previously unaddressed questions: (1) Do louse clades vary in their level of cospeciation? (2) Do lice found on different families of mammals show different degrees of cospeciation? This investigation may provide insight into the host factors that promote or break down cospeciation.

Host and Parasite Phylogenies

Using a new and efficient tree-finding strategy (Quicke et al., 2001), we had constructed a phylogeny of 234 taxa of chewing lice based on 259 morphological characters collected by Lyal (1983) (Taylor, 2000; this matrix has been deposited in TreeBASE as study number SN 576). While searching for the most parsimonious estimate of chewing louse phylogeny, due to memory constraints, no more than 30,000 equally parsimonious trees were saved.

To test all 30,000 equally parsimonious trees for cospeciation would be prohibitively time consuming, so consensus trees were produced. Therefore, faced with the tradeoff between conservatism and including more taxa, we have chosen to use a semistrict consensus. This contains features that are resolved in all the initial trees, or are resolved in some of the initial trees and not contradicted in others. However, the semistrict consensus tree was very unresolved in places, so we also produced a majority rule consensus with the largest possible majority that would provide enough resolution to increase the scope of the cospeciation analysis. The lowest majority in the consensus tree used is 78%; however, most nodes are represented by a majority exceeding 90%. It should be noted that, because only 30,000 trees could be retained in memory during the estimation of the louse phylogeny, the semistrict consensus produced is from only a subset of all the equally parsimonious trees. Therefore, our semistrict consensus has similarities with a majority rule consensus.

To test for cospeciation, the semistrict and majority rule consensus trees were compared with mammal phylogenies taken from the literature. The phylogenies used in this study are a composite phylogeny of all 271 extant carnivores (Bininda-Emonds et al., 1999), a composite phylogeny of 76 species of African bovids (Seymour, 1997), and a molecular phylogeny of 57 bovids (Gatesy et al., 1997). We constructed phylogenies of the Equidae (horses) and Cervidae (deer) using Matrix Representation with Parsimony (MRP: Sanderson et al., 1998) from studies marked with a † in the reference section. Unfortunately, no Hyracoidae or Erethizontidae phylogenies were found; therefore, the 61 taxa of lice found on these hosts cannot be examined for cospeciation. Although the lack of hyrax phylogenies results in the exclusion of 51 louse taxa, the scope of the study, with regards to the mammals, is not greatly reduced, as there are only 9 species of hyrax.

Testing for Cospeciation

To test for cospeciation, we used the TREEMAP package (Page, 1994b). TREEMAP maps a fully bifurcating parasite phylogeny onto the phylogeny of their hosts in order to maximize the number of cospeciation events while allowing for in situ speciation, host switches, and lineage sorting. The probability of detecting the observed number of cospeciations is calculated by randomizing the parasite tree many times and fitting each random tree to the host tree. The maximum number of cospeciations is calculated for each random tree and a distribution of the frequency of the maximum number of cospeciation events against the number of times it occurred during randomization. From this distribution a p-value can be calculated. However, the TREEMAP program contains a bug that causes the probability distribution to be biased toward accepting the alternate hypothesis that there has been significant cospeciation. To bypass this bug, 100 random parasite trees were produced using the "random joining/splitting" option in MacClade (Maddison and Maddison, 1992). We then used the heuristic search option in TREEMAP to map each one of the 100 random parasite trees to the host in order to maximize cospeciation. The maximum number of cospeciation events was noted for each random tree and a probability distribution thereby produced. This distribution was then used to calculate the probability of the maximum observed number of cospeciations by chance. Care should be taken while bypassing this bug, as the default mapping— a reconciled tree (a mapping that does not include host switches: Page, 1994a)—occasionally finds more cospeciations than the heuristic search.

Therefore, if the reconciled tree contained a greater number of cospeciation events than the heuristic search, then this number of cospeciations was noted and used to produce the probability distribution to test for significance testing. The difference between the default reconciled tree and heuristic search does not account for the biased probability distribution produced by TREEMAP.

Not surprisingly, TREEMAP has difficulty mapping large and very incongruent phylogenies together. Therefore, the louse and mammal phylogenies are too large to be analyzed as complete phylogenies using TREEMAP. Furthermore, due to a lack of resolution in both louse and mammal phylogenies, and the complete lack of some mammal phylogenies (e.g., hyraxes), mapping the two trees together would be impossible. In order to avoid these problems the louse phylogeny was broken into terminal clades (fig. 10.1). Using the terminal clades is the most pragmatic way to break up the tree for several reasons. First, the terminal clades are the most robust parts of the tree (they contain the highest jackknife scores: Farris et al., 1996). The jackknife scores are reported in Taylor (2000). Second, each clade can be tested to determine the maximum amount of cospeciation and whether there is more than expected by chance. Finally, using terminal clades makes it possible to determine whether there is variation among clades.

The terminal clades, if found on a single family of mammals, were also used to examine whether lice found on different families show different levels of cospeciation. However, most terminal clades were found on more than one family of mammals. These clades had the minority of lice found on a different family of mammals removed from the analysis.

In testing the terminal clades for cospeciation, we have not attempted to map the parasite phylogeny to the host phylogeny with the aim of accurately locating events of cospeciation. To pinpoint cospeciation events would be impossible. This is because for most of the terminal clades there were many equally optimal reconstructions. That is, all the optimal reconstructions would have an equal number of cospeciations but vary in their number of host switches, lineage sorting, and in situ speciation events. With additional information it is possible to choose a preferred optimal reconstruction. For example, if the branch lengths of the host and parasite phylogenies can be used as a surrogate for time, it is possible to choose the optimal reconstruction in which the cospeciation events in the two clades are contemporaneous (Page, 1994b, 1996). However, due to the lack of other data in this study, such as branch lengths, it is impossible to choose among optimal reconstructions. Further, it should be noted that, if the optimality criterion or the phylogenies were inaccurate, none of the reconstructions would be correct.

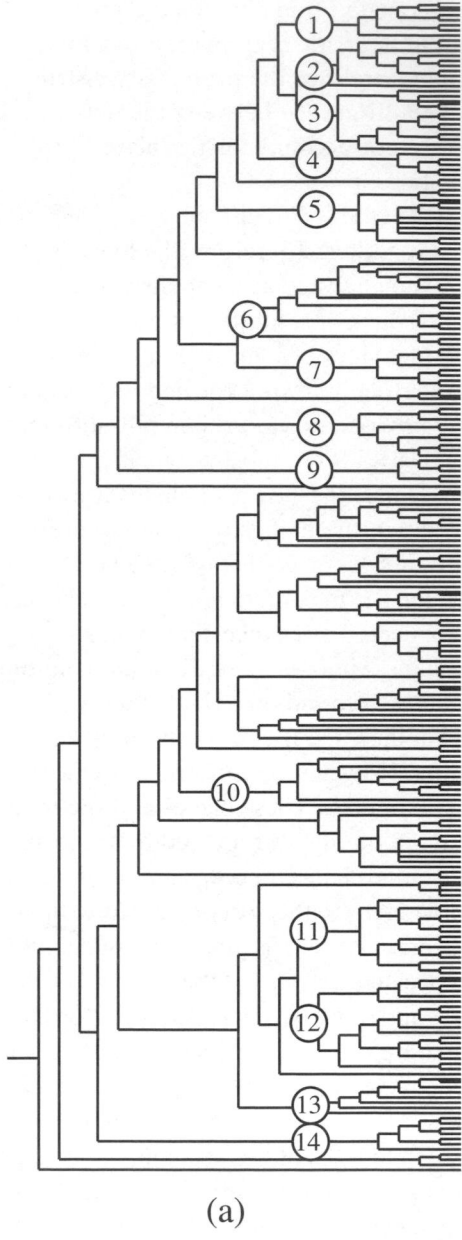

(a)

FIGURE 10.1 *(a)*. The louse majority rule consensus *(a)* and semistrict consensus *(b)* showing the terminal clades (numbered 1–14) used to test for cospeciation. Clades 6, 7, and 11 in the semistrict tree *(b)* were not tested for cospeciation due to the poor resolution of these clades. The large unnumbered clade in the center of the

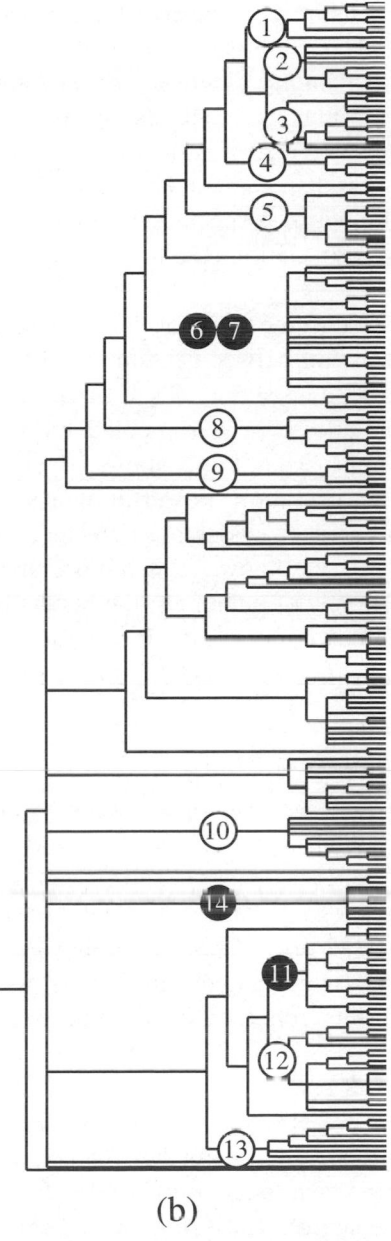

(b)

FIGURE 10.1 *(continued)* phylogeny could not be tested due to a lack of phylogenies for the hosts (see text for details). The names of the terminal taxa are not shown to conserve space. The original trees with the terminal taxa labeled, and the data matrix used to generate the trees, are available from TreeBASE (study number SN 576).

Therefore, rather than attempt to accurately reconstruct past host-parasite evolution, we have simply tested each terminal clade for maximum cospeciation. The louse and mammal phylogenies were mapped together to maximize cospeciation using the exact search option in TREEMAP. The number of cospeciations detected was then tested to determine whether there is more cospeciation than expected by chance (as described above). We also present the number of cospeciations detected as a percentage of the maximum possible cospeciation (the number of nodes in the parasite tree).

We used G-tests to determine whether there is a significant difference in the degree of cospeciation between the terminal louse clades. G-tests were also performed between sister clades. If there was no significant difference between sister clades, they were pooled and then compared with their sister clade. This approach not only allows larger samples to be compared, possibly resulting in a more powerful analysis, but when significant differences between sister clades are detected, the position of this difference on the phylogeny is known. When more than one clade of lice was found on a single family of mammals, the results from each clade, if not significantly different, were also combined and used to test if the lice found on different families of mammals had experienced different levels of cospeciation.

Phylogeny Excluded from Analysis

Of the 234 taxa in the louse phylogeny, pocket gopher lice were removed from the analysis of both the semistrict and majority rule trees because many of the gopher lice have been reduced to a few taxa during the estimation of the phylogeny. Page et al. (1995) have produced a complete phylogeny of the 122 taxa of pocket gopher lice. Unfortunately, a complete host phylogeny is not available; therefore, this clade could not be tested. However, the results from a study of 17 taxa of pocket gophers and their hosts (Hafner et al., 1994) as analyzed using TREEMAP (Page, 1994b) will be compared with our results.

A single taxon—the hypothetical outgroup—was removed from the louse phylogenies before analyses. A further 168 taxa were removed from the semistrict analysis (65 taxa remaining) and 116 from the majority rule analyses (117 taxa remaining). These taxa were removed due to either a lack of host phylogenies, a lack of resolution in host and/or parasite phylogenies, or because they formed a terminal clade of 4 or fewer taxa. Clades of 4 or fewer taxa were not examined, as it would be impossible to detect significant cospeciation. This is because with such small clades even the maximum possible amount of cospeciation will not be significant:

with 4 taxa, the maximum number of cospeciation events (3 events) is expected to occur by chance with a frequency of 6.6%, giving a p-value of $p = .066$.

When testing the terminal louse clades for cospeciation, the deep structure is neglected. The deep structure is very difficult to analyze because of a lack of resolution in the phylogenies. Even with full resolution at the deep structure the analysis would still be difficult to analyze because removing the terminal clades, or reducing them to make the analysis possible, is problematic. The difficulty arises from the need to treat inferences of ancestral hosts from the analysis of terminal clades as data in the deep structure analysis. Therefore, investigation of the deep structure is descriptive rather than analytical.

As mentioned above, due to the lack of some host phylogenies and the need to break the louse phylogeny into its terminal clades, a proportion of the phylogeny remains unexamined. However, it is possible to examine the phylogenetic clustering of the louse-host associations using MacClade (Maddison and Maddison, 1992) across the complete louse phylogeny (see fig. 10.2). This is performed by scoring each louse taxon with a multistate character representing its host family. The minimum number of steps required for these characters on the louse phylogeny is then calculated and compared to 100 replicates of shuffled characters. A smaller number of steps required by the actual associations compared with the shuffled data indicates that the louse-mammal associations are phylogenetically correlated (Herniou et al., 1998).

Results

Main Caveat

Like most phylogenies, the ones used in this study are simply estimates. We do not pretend that they are definitive. The results presented below will, at least to an extent, have been influenced by the accuracy of the phylogenies used. Due to the scale of this project, it could be a long time before enough new data could be accumulated to produce new estimates of phylogeny that could then be used to check the validity of our results. Therefore, rather than wait an undefined period of time, we have proceeded with the data at hand to obtain provisional answers to the questions listed below. A way to investigate the validity of the results presented here is described below in Future Work.

Is Cospeciation the Trend?

Using TreeMap to map the terminal clades of the louse semistrict consensus to their corresponding host phylogenies in order to maximize

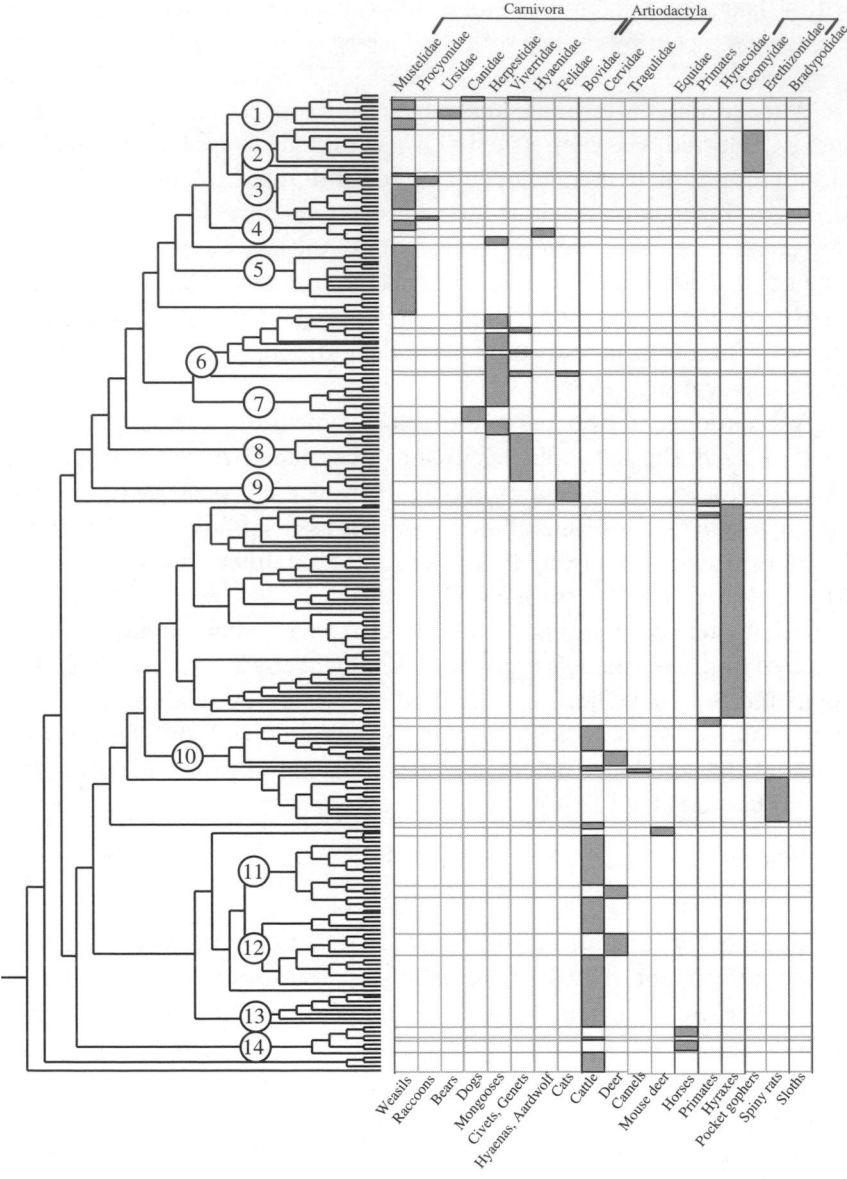

FIGURE 10.2 The phylogenetic clustering of the louse-mammal associations based on the majority rule consensus tree shown in figure 10.1A. The numbered clades are the terminal clades tested for cospeciation.

cospeciation showed that, for the semistrict consensus, from a possible maximum of 59 cospeciation events only 33 were detected (56%). The level of cospeciation detected is slightly less when the majority rule consensus is used (50% cospeciation: 52 out of a maximum 104) (table 10.1). With Hafner et al.'s (1994) gopher-louse study included, the amount of cospeciation detected rises by just over 1% when using the majority rule tree or the semistrict tree. It is difficult to view this level of cospeciation as a trend. In fact, the level of cospeciation is surprisingly low considering that cospeciation has been maximized—so the estimate biased upward—and the obligate parasitic lifestyle and high level of specialization lead to the prediction that cospeciation should be common.

Do Clades Vary in Their Level of Cospeciation?

The level of cospeciation differs slightly among the terminal clades (table 10.1). However, the G-tests show that this variation is not statistically significant: clades (results from gopher lice included) (semistrict consensus: $G = 6.05$, d.f. $= 9$, $p > .1$; majority rule consensus: $G = 10.41$, d.f. $= 13$, $p > .1$). Likewise, despite not being corrected for small sample size, which biases the test toward detecting significant differences, the G-tests also show that the clades are not statistically different when sister clades are compared and pooled (fig. 10.3). It is interesting to note that, despite the generally low level of cospeciation, the gopher-louse study, which has become a textbook example of cospeciation (Page and Holmes, 1998), is not significantly different from the terminal clades examined in this study.

One of the clades examined (clade 4) shows 100% cospeciation (fig. 10.4). Compared with the other louse clades, this is a very high level of cospeciation. If these lice have cospeciated with their hosts, then the basal node of this clade maps as a cospeciation event to the basal node of the carnivore phylogeny. This would mean that this clade of lice is as old as the most recent common ancestor of the extant carnivores (over 50 M yrs; Bininda-Emonds et al., 1999). The scenario of 100% cospeciation also requires that this clade has experienced many lineage sorting events. It is difficult to estimate the number of sorting events due to a lack of resolution in some parts of the carnivore phylogeny. Furthermore, if this clade has experienced 100% cospeciation, most other louse lineages found on carnivores must be at least as old as the extant carnivores. Thus, the 100% cospeciation shown by clade 4 could be more fortuitous than real.

TABLE 10.1 The results of the cospeciation analysis for the majority rule consensus and the semistrict consensus trees shown in figure 10.1. The first columns in the table contain the number of the clade tested. The other columns contain the number of taxa in each clade (No. of taxa); the number of taxa used when testing for cospeciation analysis (No. of taxa used); the maximum possible number of cospeciation events (Max.); the number of cospeciation events detected (Cosp.); the number of cospeciations detected expressed as a percentage (%); and the p-value. The significant results are highlighted in bold. The totals of each column do not include the results from Hafner et al.'s (1994) gopher louse study (Clade 2).

Clade	No. of Taxa	Majority Rule					Semistrict				
		No. of Taxa Used	Max.	Cosp.	%	p-Value	No. of Taxa Used	Max.	Cosp.	%	p-Value
1	8	8	7	4	57	0.45	8	7	4	57	0.45
2	**122**	**17**	**16**	**10**	**63**	**0.03**	**17**	**16**	**10**	**63**	**0.03**
3	13	11	10	5	50	0.38	11	10	5	50	0.38
4	**5**	**5**	**4**	**4**	**100**	**>0.001**	**5**	**4**	**4**	**100**	**>0.001**
5	12	9	8	5	55	0.10	8	7	5	71	0.10
6	17	13	12	4	33	0.42	-	-	-	-	-
7	8	8	7	4	57	0.26	-	-	-	-	-
8	5	5	4	1	25	0.17	10	9	4	44	0.17
9	10	10	9	4	44	0.81	5	4	1	25	0.81
10	**11**	**9**	**8**	**6**	**75**	**0.05**	**6**	**5**	**4**	**67**	**0.05**
11	15	11	10	5	50	0.37	-	-	-	-	-
12	19	14	13	6	46	0.26	8	7	4	57	0.23
13	8	8	7	3	42	0.83	7	6	2	33	0.64
14	7	6	5	1	20	1.00	-	-	-	-	-
Totals	138	117	104	52	50	-	68	59	33	56	-

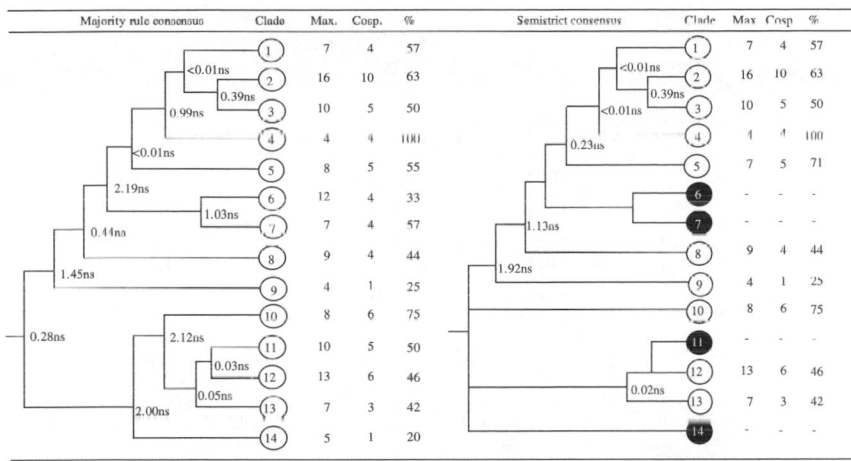

FIGURE 10.3 The results of the G-tests between sister clades for the majority rule consensus and the semistrict consensus trees shown in figure 10.1. The G-values are shown at each node between the sister clades tested. The "ns" highlights that the G- values are not above 3.84, and that the clades are therefore not significantly different. The numbered circles at the tips of each tree under the heading Clade refer to the terminal clades in figure 10.1. The columns next to each tree show the maximum possible cospeciation events (Max.), the number of cospeciation events detected (Cosp.), and the number of cospeciations detected expressed as a percentage (%). The clades shown in gray have not been tested due to a lack of resolution in the louse phylogeny. Nodes without G-values in the semistrict tree have not been tested due to a lack of resolution obscuring sister-clade relationships.

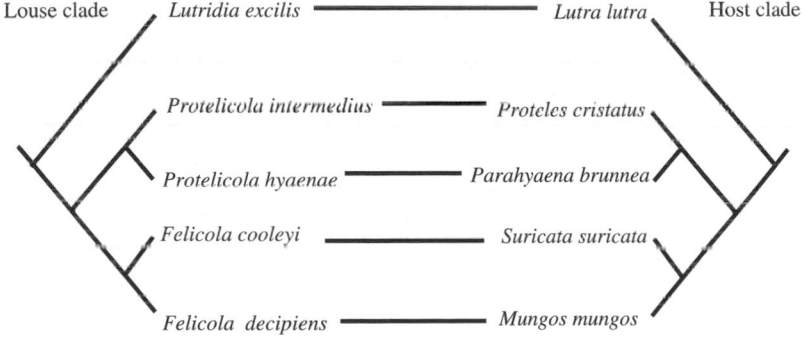

FIGURE 10.4. Clade 4, an example of 100% cospeciation.

Have Lice Found on Different Mammals Experienced Different Degrees of Cospeciation?

Hafner et al.'s (1994) study shows the highest level of cospeciation (table 10.2). Of the remaining louse clades that we have analyzed, the lice found on mustelids (weasels) showed the highest level of cospeciation,

TABLE 10.2 The results of the cospeciation analysis comparing terminal clades restricted to single families of mammals. The first column refers to the terminal clades in figure 10.1. The next column contains the family of mammals with which the louse clade was compared. The other columns contain the number of louse taxa tested (No. of taxa used); the maximum possible number of cospeciation events (Max.); the number of cospeciation events detected (Cosp.); the number of cospeciations detected expressed as a percentage (%); and the *p*-value. Clades that show a significant level of cospeciation are highlighted in bold. The totals of each column do not include the results from Hafner et al.'s (1994) gopher louse study.

| | | Majority Rules | | | | | Semistrict | | | | |
| | | No. of Taxa | | | | | No. of Taxa | | | | |
Clade	Host	Used	Max.	Cosp.	%	*p*-Value	Used	Max.	Cosp.	%	*p*-Value
2	**Geomyidae**	**17**	**16**	**10**	**63**	**0.03**	**17**	**16**	**10**	**63**	**0.03**
3	Mustelidae	8	7	4	57	0.10	8	5	2	57	0.88
5	Mustelidae	9	8	5	55	0.18	9	8	5	55	0.18
6+7	Herpestidae	15	14	6	42	0.56	-	-	-	-	-
8	Viverridae	10	9	4	44	0.17	10	9	4	44	0.17
9	Felidae	5	4	1	25	0.81	5	4	1	25	0.81
11	Bovidae	9	8	4	50	0.44	-	-	-	-	-
12	Bovidae	11	10	5	50	0.37	-	-	-	-	-
13	Bovidae	8	7	3	42	0.83	8	7	3	42	0.83
14	Equidae	6	5	1	20	1.00	6	5	1	20	1.00
Totals	-	81	72	33	46	-	81	38	16	42	-

and those found on felids (cats) and equids (horses) show the lowest. The G-tests reveal that the variation among clades (gopher lice included) is not statistically significant (semistrict consensus: G = 5.14, d.f. = 6, $p > .1$; majority rule consensus: G = 5.13, d.f. = 6, $p > .1$). This is also true when multiple clades of lice found on a single mammal family are pooled (semistrict consensus: G = 4.24, d.f. = 5, p > 0.1; majority rule consensus: G = 4.90, d.f. = 6, $p > .1$).

The Deep Structure

When examining the deep structure of the phylogenies, the most obvious observation is that the phylogenies are not mirror images (fig. 10.5). The incongruence between the phylogenies indicates that pure and faithful cospeciation has not occurred. Furthermore, as pointed out by Lyal (1986), if faithful cospeciation has occurred, then each parasite should have one host and each host should have one parasite (for a full test of this prediction for these lice and their hosts, see Lyal, 1986). It can be clearly seen (fig. 10.5) that several groups of mammals completely lack lice while others are host to several, often distantly related louse clades. Furthermore, several clades of lice are not restricted to single clades of mammals.

Without suitable techniques for analyzing the deep structure, combined with the lack of resolution in the phylogenies, it is possible to postulate many different scenarios, but difficult to determine which is most likely, or if all possibilities have been explored. It is possible that a level of cospeciation has occurred and the incongruence has been caused by a combination of other events such as host switching, in situ speciation, and lineage sorting. Therefore, all that we can determine is that faithful cospeciation has not occurred and that many alternative scenarios are possible. Due to the difficulties in mapping the deep structure of the host and parasite phylogenies together, it is also impossible to determine the origin and age of the Trichodectidae.

Unusual Host Associations

From inspection of the deep structure (see figs. 10.1 and 10.5), some particularly unusual louse-mammal associations can be seen, in that some closely related lice are found on distantly related hosts. Three examples are particularly striking and merit further discussion: the sloth lice (whose sister clade is on carnivores), the pocket gopher lice (sister to the aforementioned clade of sloth and carnivore lice), and a hyrax louse that is also found on a primate.

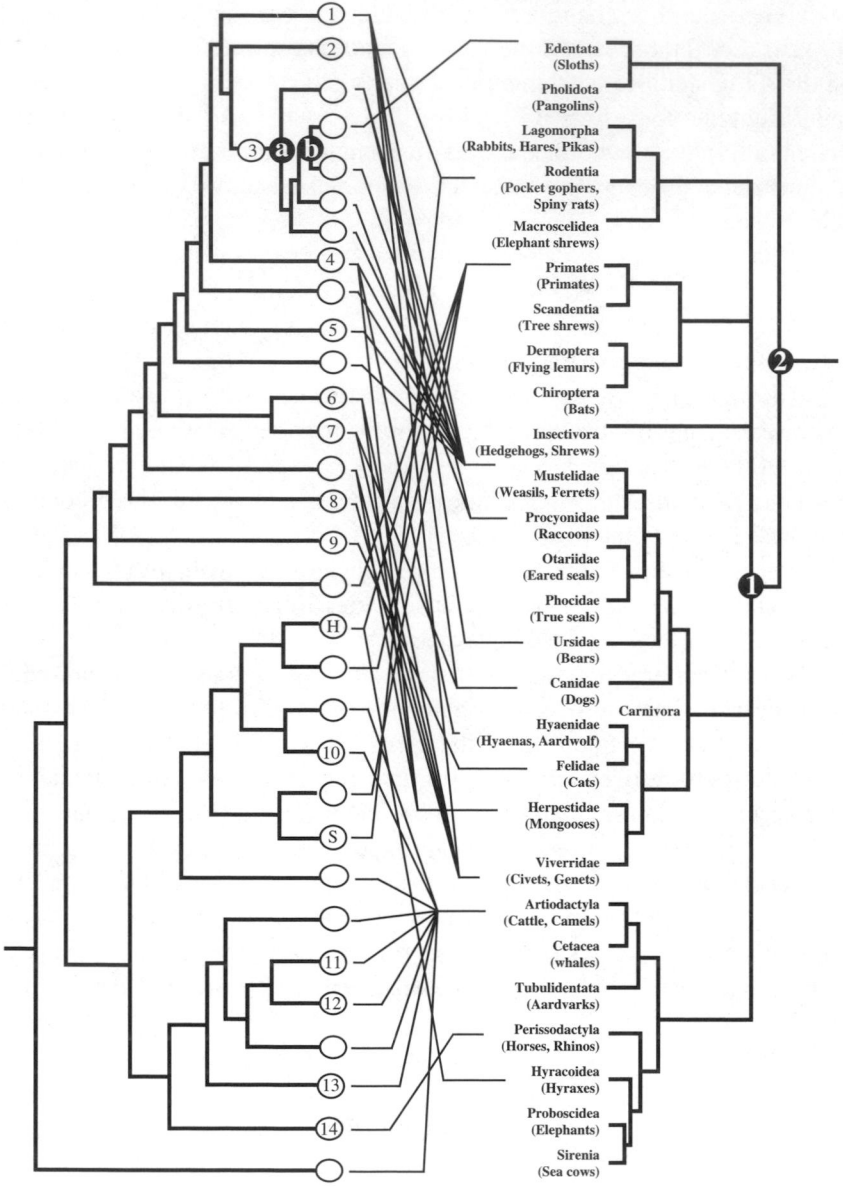

If the association between the sloth lice and carnivore lice was formed by cospeciation (node *b* mapped onto node 2 in fig. 10.5), this would imply that the split between these clades is roughly 100 M yrs old (Novacek, 1992; Kumar and Hedges, 1998). However, this scenario seems unlikely, as it would require all nodes found between this split and the root of the tree to be at least as old as most of the extant mammals and invokes a very large number of sorting events. Therefore, it would seem likely that the association between sloths and their lice has not been formed by cospeciation. A simple alternative scenario is that of a host switch from carnivores to sloths. This is the same conclusion as Lyal (1987), who states that the suggestion that the association between these sloths and their lice is "primary and ancient" is unfounded and argues that a secondary infestation (a host switch) from carnivores seems to be the most likely explanation. The fact that the two louse species, *Neotrichodectes (Nasuicola) pallidus* and *N. (Trigonodectes) barbarae,* which are basal to the sloth lice are also found on hosts (*Nasua nasua* and *Eira barabara,* respectively) that inhabit the same continent as the sloths—South America—further supports the scenario of host switching.

The pocket gopher lice are the sister group of a clade containing carnivore and sloth lice. Again a simple scenario would be a host switch from carnivores to an ancestor of the pocket gophers. The alternative situation of cospeciation (node *a* mapped onto node 1 in fig. 10.5) again seems unlikely for reasons similar to those given for the sloth lice. Again this conclusion is in agreement with Lyal (1987), who also presents additional support for the host switching scenario by pointing out that gopher lice have occasionally been found on the gophers' mustelid predators. However, we are not aware of any carnivore lice being found on gophers.

FIGURE 10.5. (*Facing page*) The louse majority rule consensus tree (fig. 10.1) compared with a mammalian tree based on Novacek (1992) and Bininda-Emonds et al. (1999) to illustrate the high degree of incongruence between louse and mammal phylogenies. The numbered tips of the louse phylogeny are those tested for cospeciation (see fig. 10.1), with clade 3 expanded to highlight the unusual gopher-louse and sloth-louse associations. The tip of the louse phylogeny labeled with an *H* represents the 51 louse taxa found on hyraxes (with one louse also found on a primate). The tip labeled S represents the 11 taxa of lice found on spiny rats. These two clades could not be tested for cospeciation due to a lack of host phylogenies. The unlabeled tips represent clades that were too small (containing an average of 2 taxa) to test for cospeciation. The nodes labeled *a, b,* 1, and 2 are referred to in the discussion of sloth-louse and gopher-louse associations (see text).

It is impossible for the split between the sloth lice and carnivore lice, and the split between the pocket gopher lice and their sister clade to have both been cospeciation events. From inspection of figure 10.5, it is obvious that it is impossible to map node *a* onto node 1 and also to map node *b* onto node 2. It is possible that either node *a* or node *b* could have been a cospeciation event as discussed above, but we prefer an alternative scenario, such as two individual host switches, that would seem more probable than cospeciation.

Another unusual association is the presence of an African hyrax louse *(Procaviphilus [Meganarionoides] colobi)* on an African primate *(Colobus guereza)* (Lyal, 1985). As all other members of this large louse clade are found on hyraxes, it seems reasonable to presume that this association has been formed by a host switch from hyraxes to primates(Lyal, 1987).

There are some other unusual louse-mammal associations, for example, two South American primate lice that are the sister clade to the large clade of African hyrax lice (fig. 10.1). However, the other associations are difficult to interpret, as there are many possible scenarios and it is difficult to determine which could be the most likely. Therefore, we have only considered the three particularly notable examples discussed above. Nevertheless, simple examination of the deep structure has shown three possible host switches across large phylogenetic distances, for example, from carnivores to sloths, for parasites that have often been thought to show high levels of cospeciation.

At the beginning of this section we stressed that our results are dependent upon the accuracy of the phylogenies. Here we would like to add that the deep structure of the louse phylogeny of the louse tree is the least supported part of the tree (most nodes have a jackknife score below 50%). However, the conclusions drawn regarding the deep structure do not vary with different mammal phylogenies (for example, that of de Jong, 1998).

Although the cospeciation analysis of the terminal clades and inspection of the deep structure suggest cospeciation is not the trend, the louse-mammal associations shown in figure 10.2 are not random. The MacClade (Maddison and Maddison, 1992) analysis reveals that the number of steps required to generate the louse-mammal associations on the majority rule consensus (33 steps) and semistrict consensus (31 steps) is significantly smaller than the number of steps required by 100 replicates in which the host-parasite associations have been shuffled. The range for majority rule consensus was 150–167 steps ($p < .01$) and the range for semistrict consensus was 144–161 ($p < .01$). Therefore, although cospeciation does not appear to be the trend, the associations are not random: some other process, such as resource tracking (Kethley and Johnston, 1975), has been operating.

Discussion

Our results have shown a low level of cospeciation between mammals and their chewing lice. This is surprising because the methods used are designed to detect the maximum amount of cospeciation; theory predicts that highly specialized parasites with low powers of independent dispersal, such as chewing lice, should cospeciate; and a previous study between gophers and their lice has shown a high degree of cospeciation. However, the results have also shown that the gopher-louse example is not statistically different from the other clades we have tested. Furthermore, although our results show a generally low level of cospeciation, they also show clades that have more cospeciation than the gopher louse example. Our results also fit Lyal's prediction that with further study the level of cospeciation detected between mammals and their chewing lice would decrease (Lyal, 1987).

Inspection of the deep structure also revealed three putative host switches across large phylogenetic distances, for example, from carnivores to sloths. Therefore, not only do our results show that cospeciation is not the trend, but it would also appear that lice are capable of making host switches across large phylogenetic distances.

Despite the low level of cospeciation, the louse-mammal association is not random, suggesting that some other process has been operating. It is possible, for example, that the lice have been tracking a host resource (Kethley and Johnston, 1975) rather than tracking the phylogeny.

Our analysis of the louse clades showed that only 3 clades showed significant levels of cospeciation (including Hafner et al.'s 1994 gopher louse study). These results also showed a range of levels of cospeciation greater than any other previous study (ranging from 20% to 75%, excluding a clade that is thought to have shown 100% cospeciation fortuitously). However, the differences between clades were shown not to be statistically significant. Nevertheless, there is still considerable variation to explain, and the results from these clades will be used in further work to examine the host factors associated with the level of cospeciation.

Our study also examines whether the level of cospeciation varies among families of host. The results showed that Hafner et al.'s (1994) gopher louse example showed the highest level of cospeciation and was the only clade to show a significant level of cospeciation. Of the clades we tested, the lice found on mustelids (weasels) showed the highest level of cospeciation and those found on felids (cats) and equids (horses) showed the lowest.

However, the difference among the lice of different mammal families was shown to not be significant. Had a significant difference been detected, this would have been useful in formulating hypotheses as to the host factors that affect the level of cospeciation.

As previously mentioned, the conclusions drawn from this study will, at least to an extent, be affected by the accuracy of the phylogenies (see Main Caveat above). However, the results will also depend upon the ability of TREEMAP to accurately map the parasite tree to the host tree.

Future Work

The low level of cospeciation detected in this study also raises the question: how difficult is cospeciation to detect? In other words, even with a high level of cospeciation, can combinations of "other events"(host switching, in situ speciation, lineage sorting, and host and/or parasite extinction) occurring at low levels make cospeciation difficult to detect? A promising avenue might be to simulate various levels and combinations of cospeciation and the other events listed above to determine what levels of cospeciation can be accurately detected.

Summary

Previous studies (Hafner and Nadler, 1988; Demastes and Hafner, 1993; Hafner et al., 1994; Hafner and Page, 1995) of cospeciation between lice and their mammalian hosts (excluding Lyal, 1986, 1987) have focused on a small sample of pocket gophers and their lice. These studies have shown a high level of cospeciation, in line with the prediction that the speciation of highly specialized parasites with low powers of independent dispersal is driven by the speciation of their hosts.

The study presented here is the largest test of cospeciation to date, and the only statistical broad-scale study of cospeciation between mammals and their chewing lice. The results of this study have shown a surprisingly low level of cospeciation (between 50% and 56% of louse speciation events). Furthermore, we have shown that the results of our study are not significantly different from the textbook gopher louse study of Hafner et al. (1994). In light of these results, and contrary to most expectations, it would appear that cospeciation between mammals and their chewing lice is not the trend. However, there are alternative explanations for this low level of cospeciation and this raises several questions for future investigation, for instance; Will new estimates of louse phylogeny validate our results? Is cospeciation harder to detect than originally thought?

Bearing in mind the difficulties of determining whether we have accurately estimated the level of cospeciation, our next stage of work will begin to investigate the host factors associated with the level of cospeciation. We hope that this will provide new evidence for the role of several biological traits in promoting or breaking down cospeciation.

Acknowledgments

We thank M. Charleston and R. Page for their patience and advice. We thank M. Tristem and J. Harral for comments during the preparation of this manuscript and R. Seymour for permission to use his bovid phylogeny. This work was funded by a NERC Ph.D. studentship to J.T.

REFERENCES

†Bennett, D. K. 1980. Stripes do not a zebra make, part 1: A cladistic analysis of *Equus*. *Systematic Zoology* 29:272–87.

Bininda-Emonds, O. R. P., J. L. Gittleman, and A. Purvis. 1999. Building large trees by combining phylogenetic information: A complete phylogeny of the extant Carnivora (Mammalia). *Biological Reviews* 74:143–75.

†Bogenberger, J. M., H. Neitzel, and F. Fittler. 1987. A highly repetitive DNA common to all Cervidae: Its organization and chromosomal distribution during evolution. *Chromosoma* 95:154–61.

Brooks, D. R. 1988. Macroevolutionary comparisons of host and parasite phylogenies. *Annual Review of Ecology and Systematics* 19:235–59.

†Cronin, M. A. 1991. Mitochondrial-DNA phylogeny of deer (Cervidae). *Journal of Mammalogy* 72:533–66.

†————1996. κ-casein phylogeny of the higher ruminants (Pecora, Artiodactyla). *Molecular Phylogenetics and Evolution* 6:295–311.

de Jong, W. W. 1998. Molecules remodel the mammalian tree. *Trends in Ecology and Evolution* 13:270–74.

Demastes, J. W., and M. S. Hafner. 1993. Cospeciation of pocket gophers *(Geomys)* and their chewing lice *(Geomydoecus)*. *Journal of Mammalogy* 74:521–30.

†Douzery, E., and E. Randi. 1997. The mitochondrial control region of Cervidae: Evolutionary patterns and phylogenetic content. *Molecular Biology and Evolution* 14:1154–66.

Farris, J. S., V. A. Albert, M. Källersjö, D. Lipscomb, and A. G. Kluge. 1996. Parsimony jackknifing outperforms neighbor-joining. *Cladistics* 12:99–124.

Gatesy, J., G. Amato, E. Vrba, G. Schaller, and R. DeSalle. 1997. A cladistic analysis of mitochondrial ribosomal DNA from the Bovidae. *Molecular Phylogenetics and Evolution* 7:303–19.

†George, M. J., and O. A. Ryder. 1986. Mitochondrial DNA evolution in the genus *Equus*. *Molecular Biology and Evolution* 3:535–46.

†Source tree for supertrees for Equidae or Cervidae.

Hafner, M. S., and S. A. Nadler. 1988. Phylogenetic trees support the coevolution of parasites and their hosts. *Nature* 332:258–59.

Hafner, M. S., and R. D. M. Page. 1995. Molecular phylogenies and host-parasite cospeciation: Gophers and lice as a model system. *Philosophical Transactions of the Royal Society of London,* ser. B, 349:77–83.

Hafner, M. S., P. D. Sudman, F. X. Villablanca, T. A. Spradling, J. W. Demastes, and S. A. Nadler. 1994. Disparate rates of molecular evolution in cospeciating hosts and parasites. *Science* 265:1087–90.

Herniou, E., J. Martin, K. Miller, J. Cook, M. Wilkinson, and M. Tristem. 1998. Retroviral diversity and distribution in vertebrates. *Journal of Virology* 72:5955–66.

Hugot, J.-P. 1999. Primates and their pinworms parasites: The Cameron hypothesis revisited. *Systematic Biology* 48:523–46.

†Ishida, N. T., S. Oyunsuren, H. Mashima, and N. Saitou. 1995. Mitochondrial DNA sequences of various species of the genus *Equus* with special reference to the phylogenetic relationship between Prezewalskii's wild horse and domestic horse. *Journal of Molecular Evolution* 41:180–88.

Kethley, J. B., and D. E. Johnston. 1975. Resource tracking patterns in bird and mammal ectoparasites. *Miscellaneous Publications of the Entomological Society of America* 9:231–6.

Kumar, S., and S. B. Hedges. 1998. A molecular timescale for vertebrate evolution. *Nature* 392:917–20.

Lyal, C. H. C. 1983. Taxonomy, phylogeny and host relationships of the Trichodectidae (Phthiraptera: Ischnocera). Ph.D. thesis, University of London.

———. 1985. A cladistic analysis and classification of trichodectid mammal lice (Phthiraptera: Ischnocera). *Bulletin of the British Museum (Natural History), Entomology* 51:187–346.

———. 1986. Coevolutionary relationships of lice and their hosts: A test of Fahrenholz's Rule. In *Coevolution and systematics,* edited by A. R. Stone and D. L. Hawksworth, 77–91. Oxford: Clarendon Press.

———. 1987. Co-evolution of trichodectid lice (Insecta: Phthiraptera) and their mammalian hosts. *Journal of Natural History* 21:1–28.

Maddison, W. P., and D. R. Maddison. 1992. MacClade: Analysis of phylogeny and character evolution, version 3. Sunderland, Mass.: Sinauer Associates.

Moran, N. A., C. D. van Dohlen, and P. Baumann. 1995. Faster evolutionary rates in endosymbiotic bacteria than in cospeciating insect hosts. *Journal of Molecular Evolution* 41:727–31.

Novacek, M. J. 1992. Mammalian phylogeny: Shaking the tree. *Nature* 356:121–25.

Page, R. D. M. 1994a. Maps between trees and cladistic analysis of historical associations among genes, organisms, and areas. *Systematic Biology* 43:58–77.

———. 1994b. Parallel phylogenies: Reconstructing the history of host-parasite assemblages. *Cladistics* 10:155–73.

———. 1996. Temporal congruence revisited: Comparison of mitochondrial DNA sequence divergence in cospeciating pocket gophers and their chewing lice. *Systematic Biology* 45:151–67.

Page, R. D. M., and E. C. Holmes. 1998. *Molecular evolution: A phylogenetic approach.* Oxford: Blackwell Science.

Page, R. D. M., P. L. M. Lee, S. A. Becher, R. Griffiths, and D. H. Clayton. 1998. A different tempo of mitochondrial DNA evolution in birds and their parasitic lice. *Molecular Phylogenetics and Evolution* 9:276–93.

Page, R. D. M., R. D. Price, and R. A. Hellenthal. 1995. Phylogeny of *Geomydoecus* and *Thomomydoecus* pocket gopher lice (Phthiraptera: Trichodectidae) inferred from cladistic analysis of adult and first instar morphology. *Systematic Entomology* 20:129–43.

Paterson, A. M., R. D. Gray, and G. P. Wallis. 1993. Parasites, petrels and penguins: Does louse presence reflect seabird phylogeny? *International Journal for Parasitology* 23:515–26.

Peek, A. S., R. A. Feldman, R. A. Lutz, and R. C. Vrijenhoek. 1998. Cospeciation of chemotrophic bacteria and deep sea clams. *Proceedings of the National Academy of Science of the USA* 95:9962–66.

Quicke, D. L. J., J. Taylor, and A. Purvis. 2001. Changing the landscape: A new strategy for the estimation of large phylogenies. *Systematic Biology* 50:60–66.

Sanderson, M. J., A. Purvis, and C. Henze. 1998. Phylogenetic supertrees: Assembling the trees of life. *Trends in Ecology and Evolution* 13:105–9.

Seymour, R. 1997. The subspecies in taxonomy and conservation. M.Sc. thesis, University of London.

Taylor, J. 2000. Diversification of chewing lice and cospeciation with their mammalian hosts. Ph.D. thesis, University of London.

11

COEVOLUTIONARY HISTORY OF ECOLOGICAL REPLICATES: COMPARING PHYLOGENIES OF WING AND BODY LICE TO COLUMBIFORM HOSTS

Kevin P. Johnson and Dale H. Clayton

Phylogenies depict the history of speciation for groups of organisms. Comparing the phylogenies of interacting groups can reveal instances of tandem speciation, or "cospeciation" (Brooks and McLennan, 1991; Hoberg et al., 1997; Paterson and Gray, 1997). Understanding the conditions under which cospeciation takes place is a challenging task. In the case of hosts and their parasites, cospeciation occurs when isolation of host populations also isolates the parasites on those hosts. Patterns of cospeciation can break down owing to dispersal of parasites among host populations, sympatric speciation of parasites on a single host population, or extinction of parasites on a host population (Page and Charleston, 1998). All else being equal, ecologically similar parasites living on the same host should respond to isolation of host populations in the same way, yielding similar coevolutionary histories. In this chapter we compare cospeciation events in two such "replicate" groups of lice living on the same hosts. If forces promoting speciation, such as host speciation, act on these parasites in similar ways, then we would expect cospeciation events to be correlated between these parasite groups. On the other hand, if the parasites respond to isolation differently, then cospeciation events should be independent in the two groups.

We focus on two groups of Ischnoceran feather lice (Insecta: Phthiraptera), both of which are found on pigeons and doves (Aves: Columbiformes). Feather lice are permanent parasites that are restricted to the body of the host by appendages specialized for locomotion on feathers (Clayton, 1991). They complete their entire life cycle on the body of the host, where they feed on feathers and dermal debris. Transmission among hosts usually occurs through physical contact between the feathers of different individual birds, such as that between mated individuals or between parents and their offspring in the nest (Marshall, 1981).

Columbiform feather lice are of two distinct morphological types: wing lice and body lice. Wing lice are long and slender (fig. 11.1, top) and lay their eggs on the wing and tail feathers of their host. Their shape is an adaptation for inserting between feather barbs, which helps them (1) adhere to the host during flight, and (2) avoid being removed by the host when it preens (Clayton, 1991). Body lice, which have a more rounded shape (fig. 11.1, bottom), live primarily on the host's abdominal feathers, where they escape from preening by burrowing in the downy portions of these feathers (Clayton, 1991). Despite their differences in body form and mechanisms of escape, Columbiform wing and body lice are ecologically very similar. Both feed on abdominal contour feathers (Nelson and Murray, 1971) and have similar effects on host fitness (Booth et al., 1993; Clayton and Tompkins, 1995; Clayton et al., 1999). Both are directly transmitted to nestlings (Clayton and Tompkins, 1994); however, both have also been recorded "hitchhiking" phoretically on hippoboscid flies (Couch, 1962).

A historical comparison of ecological replicates would be compromised if their phylogenies were intertwined. In addition to being ecologically similar, Columbiform wing and body lice are phylogenetically independent. A phylogeny based on DNA sequences for a number of Ischnoceran genera indicates that Columbiform wing and body lice are each monophyletic and are distantly related within the Ischnocera (Cruickshank et al., 2001). Body lice are sister to several genera of Galliform lice. The body lice on Columbiformes and Galliformes together form the family Goniodidae (Smith, 2000). Wing lice, on the other hand, appear to be a basal lineage of the Ischnocera, but the sister taxon to Columbiform wing lice currently cannot be identified with certainty (Cruickshank et al., 2001). Most species of pigeons and doves have both wing and body lice (Hopkins and Clay, 1952; Price, unpub. checklist). Based on morphology, wing lice are classified in two genera: *Columbicola* (67 species) and *Turturicola* (8 species). Body lice are classified in five genera containing 141 described species: *Auricotes* (45 species), *Campanulotes* (12 species), *Coloceras* (58 species), *Kodocephalon* (3 species), and *Physconelloides* (23 species) (Hopkins and Clay, 1952; Price, unpub. checklist).

In this chapter we compare phylogenies of wing lice (*Columbicola* only) and body lice (*Auricotes, Campanulotes, Coloceras*, and *Physconelloides*) with the phylogeny of their Columbiform hosts. These phylogenies are based on both nuclear and mitochondrial DNA sequences. We infer nodes in the tree that show apparent cospeciation between dove hosts and louse parasites. A novel aspect of our study is that we examine two parasite groups living on the same hosts. If multiple groups of parasites respond

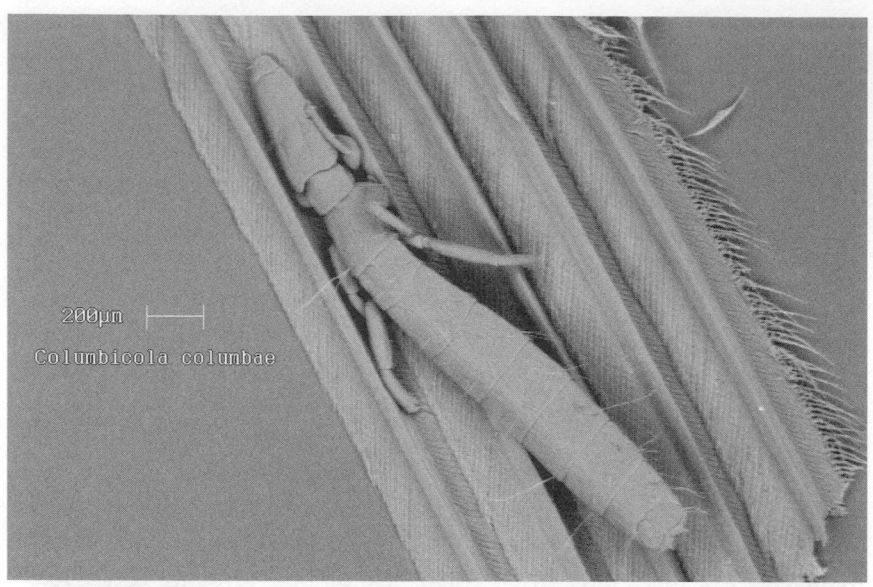

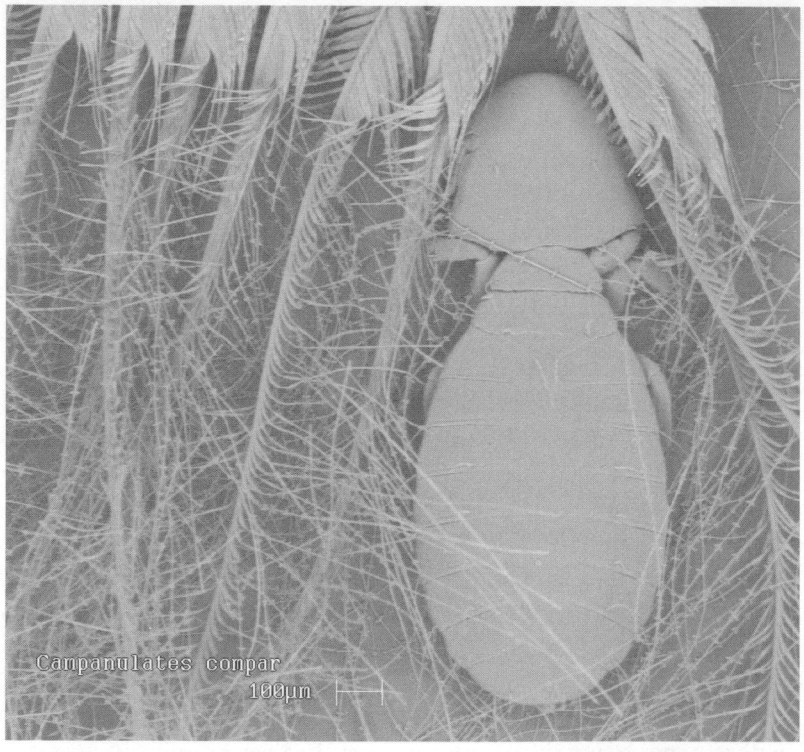

to host speciation in similar ways, we would expect congruence between parasite phylogenies, as well as between host and parasite phylogenies. Although wing and body lice are similar ecologically, detailed differences between the two may influence cospeciation patterns. If one group of parasites can disperse more easily than the other, one might expect that speciation events in the host would be less of a barrier to gene flow among the populations of that group. Differences in survival ability on multiple host species could also create differences in the pattern of cospeciation. Moreover, chance events might play a role in breaking down congruence between the phylogenies of two groups of parasites on the same group of hosts. For example, stochastic extinction of parasites might cause the phylogenies of two groups of parasites on the same group of hosts to show little similarity. In our study, we evaluate the extent to which cospeciation events are common to both wing and body lice, and whether the coincidence of such events is more frequent than expected by chance.

Methods
Samples and DNA Sequencing

We extracted DNA from frozen samples of host tissue using the protocol described by Johnson and Clayton (2000). We included 19 species of Columbiformes in this study with representatives of two divergent subspecies of one of the species: *Leptotila verreauxi* (table 11.1). Lice were sampled from wild hosts using techniques described in Clayton and Walther (1997). Lice were either frozen at $-70°C$ or stored at $-20°C$ in 95% ethanol. For each louse we carefully removed the head from the body and extracted DNA from both using a Qiagen tissue kit. After the DNA extraction procedure, the head and body of the louse were reassembled as a voucher specimen mounted in balsam on a microslide, which was used for identification. PCR and sequencing was done as described by Johnson and Clayton (2000).

For hosts, we sequenced 2,589 base pairs, including portions of the mitochondrial genes cytochrome *b* (cyt *b*) and cytochrome oxidase I (COI) and the nuclear gene β-fibrinogen intron 7 (FIB7) gene. See Johnson and Clayton (2000) for cyt *b* and FIB7 primers, and Hafner et al. (1994) for COI primers (Genbank accession numbers AF182649, AF182650, AF182653, AF182658, AF182661, AF182663, AF182668, AF182670, AF182673, AF182682, AF182686, AF182686, AF182691, AF182697, AF182701,

---

FIGURE 11.1. (*Facing page*) SEMs of feather lice from the Rock Dove (*Columba livia*). Top, *Columbicola columbae,* a wing louse; bottom, *Campanulotes* (*bidentatus*) *compar,* a body louse.

TABLE 11.1 Host and parasite taxa included in study

Host	Wing Louse	Body Louse
Columbina inca	*Columbicola passerinae* 1	*Physconelloides eurysema* 1
Columbina passerina	*Columbicola passerinae* 1	*Physconelloides eurysema* 2
		Physconelloides eurysema 3
Claravis pretiosa	*Columbicola passerinae* 2	*Physconelloides eurysema* 2
Phapitreron amethystina	*Columbicola exilicornis*	*Coloceras clypeatum*
Phapitreron leucotis	*Columbicola veigasimoni*	*Coloceras* n. sp. 1
Ptilinopus occipitalis	*Columbicola xavieri*	*Auricotes rotundus*
Columba speciosa	*Columbicola adamsi*	*Physconelloides spenceri*
Columba plumbea	*Columbicola adamsi*	*Physconelloides anolaimae*
Streptopelia senegalensis	*Columbicola theresae*	*Coloceras* n. sp. 2
Streptopelila capicola	*Columbicola theresae*	*Coloceras* n. sp. 2
Columba livia	*Columbicola columbae* 1	*Campanulotes compar*
Columba guinea	*Columbicola columbae* 2	*Coloceras savoi*
Zenaida asiatica	*Columbicola macrourae* 2	*Physconelloides wisemani*
Zenaida macroura	*Columbicola macrourae* 3	*Physconelloides zenaidurae*
	Columbicola baculoides	
Zenaida galapagoensis	*Columbicola macrourae* 4	*Physconelloides galapagensis*
Geotrygon montana	*Columbicola macrourae* 1	*Physconelloides cubanus*
Leptotila plumbeiceps	*Columbicola macrourae* 1	*Physconelloides ceratoceps* 1
	Columbicola gracilicapitis	*Physconelloides ceratoceps* 2
		Physconelloides ceratoceps 3
Leptotila jamaicensis	*Columbicola gracilicapitis*	*Physconelloides ceratoceps* 1
Leptotila verreauxi angelica	*Columbicola macrourae* 2	*Physconelloides ceratoceps* 3
	Columbicola macrourae 1	
Leptotila verreauxi fulviventris	*Columbicola macrourae* 1	*Physconelloides ceratoceps* 3
	Columbicola gracilicapitis	

AF182703, AF182706, AF279704-AF279743). For wing lice *(Columbicola)*, we sequenced 1,107 base pairs, including portions of the mitochondrial COI and 12S ribosomal genes, as well as the nuclear elongation factor 1-α gene (EF1α). We used the primers L6625 and H7005 (Hafner, et al., 1994) for COI, 12Sai and 12Sbi (Simon et al., 1994) for 12S, and EF1-For3 and EF1-Cho10 (Danforth and Ji, 1998) for EF1α (Genbank accession numbers AF190409, AF190411, AF190412, AF190416, AF190418, AF190420, AF190423, AF190424, AF190426, AF278608-AF278643). For body lice, we sequenced 737 base pairs, including portions of the mitochondrial COI gene and nuclear EF1α gene using the primers listed above (Genbank accession numbers AF278644-AF278679).

We sequenced several individual lice of each species for COI. The COI sequences revealed divergent monophyletic lineages within several described morphological species of lice (Johnson et al., 2002). The divergence between lineages ranged from 3% to 18% uncorrected sequence divergence. However, within each of these lineages, COI sequences were identical or differed at only a few base positions (generally less than 1% sequence divergence). Thus, we sequenced only one representative individual from each of these lineages for EF1α and 12S. We do not give the divergent lineages unique names here, but designate them using arbitrary numbers within each morphological species (e.g., *Columbicola macrourae* 1). These numbered lineages were used as the terminals for cospeciation analyses.

Phylogeny Construction and Comparison

For all three sets of taxa (hosts, wing lice, and body lice), we constructed phylogenies using several different methods in the program PAUP* (Swofford, 1999). In all analyses, we combined gene regions for each taxon. We first reconstructed unordered parsimony trees. Next we used these trees to estimate the best fit maximum likelihood model using the general procedure of likelihood ratio tests as described by Huelsenbeck and Crandall (1997). In each case, the best fit maximum likelihood model incorporated six substitution categories (general time reversible), empirically estimated base frequencies, and rate heterogeneity under the gamma distribution (we partitioned the gamma distribution into eight rate categories). We also used this model with two substitution categories to estimate the transition:transversion ratio under maximum likelihood. We used this estimate (rounded to the nearest whole number) as a weight on transversions in parsimony searches. For both parsimony analyses, we constructed 100 bootstrap replicates (Felsenstein, 1985) to evaluate the relative support for branches in the trees.

To reconstruct trees under maximum likelihood, we used quartet puzzling (Strimmer and von Haeseler, 1996) as a shortcut to heuristic maximum likelihood searches. We used the model derived above in each case for the quartet puzzling replicates, employing a setting of 10,000 puzzling steps because of the relatively large number of taxa in each data set. We also used the reliability values as an indication of relative support for branches in the maximum likelihood analysis. As a third phylogeny reconstruction technique, we used neighbor joining with Kimura two-parameter (Kimura, 1980) distances.

For each type of analysis (unweighted parsimony, transversion weighted parsimony, maximum likelihood quartet puzzling, and neighbor joining) we compared host and parasite trees. Reconciliation analysis (Page, 1990, 1994a) as implemented in the computer program TREEMAP (Page, 1994b) was used to determine the number and position of cospeciation events. We used a randomization of parasite trees to test whether there were significantly more cospeciation events than expected by chance (Page, 1990b, 1994b). These methods assume that both host and parasite phylogenies are known with certainty. However, all the trees had nodes with relatively low support (bootstrap or reliability scores). Thus, to take into account incongruence between phylogenies that might be attributable to weakly supported conflicting nodes (Huelsenbeck et al., 1997), we used the partition homogeneity test (Farris et al., 1994, 1995) and a taxon deletion method (Johnson, 1997; Johnson et al., 2001). Using this method, we assessed which associations resulted in significant incongruence between host and parasite trees. For example, some parasites are associated with multiple, sometimes unrelated, hosts. These associations are likely to cause significant incongruence between host and parasite phylogenies. We removed these incongruent host-parasite associations (sometimes entire taxa) and constructed a combined evidence tree for hosts and the relevant louse taxon. We then constrained this tree and added back in the removed taxa. We conducted parsimony searches with either the host or parasite data set under this constraint. Finally, we used these complete constrained trees in the TREEMAP analyses as indicated above.

To assess whether cospeciation events were correlated between wing and body lice, we tallied host nodes as having (1) no cospeciation events, (2) cospeciation with wing lice only, (3) cospeciation with body lice only, or (4) cospeciation with both wing and body lice. We used these values in a Fisher's exact test for independence. We conducted this test using all five comparisons of host and parasite phylogenies (unordered parsimony, transversion weighted parsimony, maximum likelihood, neighbor joining, and partition homogeneity test/taxon deletion method).

Results
Host Phylogeny

The phylogeny for pigeons and doves resulting from combined analysis of genes is completely resolved and well supported. The maximum likelihood quartet puzzling and unweighted parsimony trees were identical, and this tree was consistent with an analysis of a larger set of taxa (Johnson and Clayton, 2000). In the unweighted parsimony tree, 15 of 17 nodes were supported in greater than 50% of bootstrap replicates. In the quartet puzzling tree (fig. 11.2), all nodes had a reliability index greater than 50%. The transversion weighted parsimony and neighbor joining trees were similar

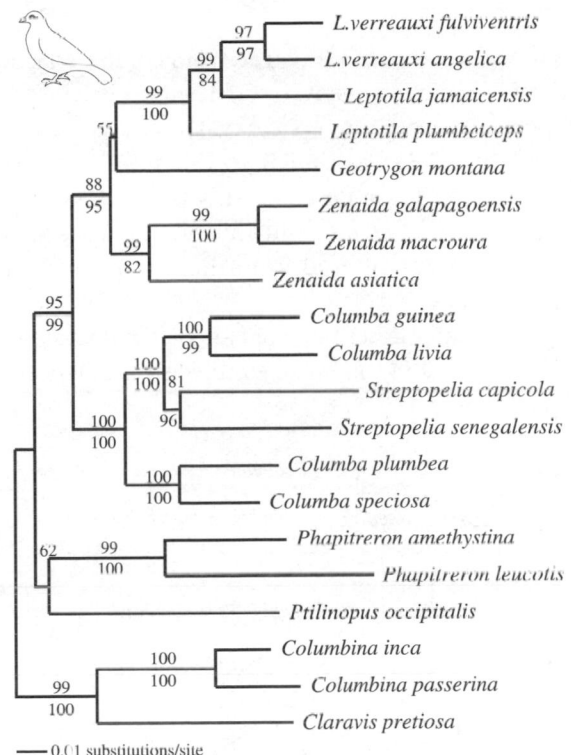

FIGURE 11.2. Phylogeny of Columbiformes derived from maximum likelihood quartet puzzling of COI, cyt *b*, and FIB7 sequences. Model parameters: empirical base frequencies with rate heterogeneity, gamma shape parameter = 0.224, eight rate categories, general time reversible model with transformation parameters 1.39 (A-C), 7.30 (A-G), 0.84 (A-T), 0.71 (C-G), 9.88 (C-T), 1.0 (G-T). Branch lengths are proportional to lengths estimated under the maximum likelihood model (scale indicated). Numbers below branches indicate bootstrap support from 100 replicates using unordered parsimony. Branches unlabeled below had bootstrap support less than 50%. Numbers above the branches indicate reliability indices from 10,000 puzzling replicates.

to the maximum likelihood tree, with most of the differences involving re-arrangements of weakly supported nodes. With the exception of *Columba*, in all trees, Columbiform genera are monophyletic. Old World *Columba* are sister to *Streptopelia*, and New World *Columba* are sister to Old World *Columba* + *Streptopelia*. The minimum and maximum pairwise COI sequence divergences between species of Columbiformes were 3.3% and 15.4%, respectively.

Wing Louse Phylogeny

The single unweighted parsimony tree from combined gene regions for *Columbicola* was completely resolved. Of 12 nodes in the tree, 9 received bootstrap support in over 50% of replicates. The maximum likelihood quartet puzzling tree (fig. 11.3) was similar to this tree, with rearrangements involving weakly supported nodes, and this tree was generally well supported by the quartet puzzling reliability index. The transversion-weighted and neighbor-joining trees differed from these trees in the placement of weakly supported nodes.

Most morphologically described species of *Columbicola* were monophyletic; however, *Columbicola macrourae* was paraphyletic with respect

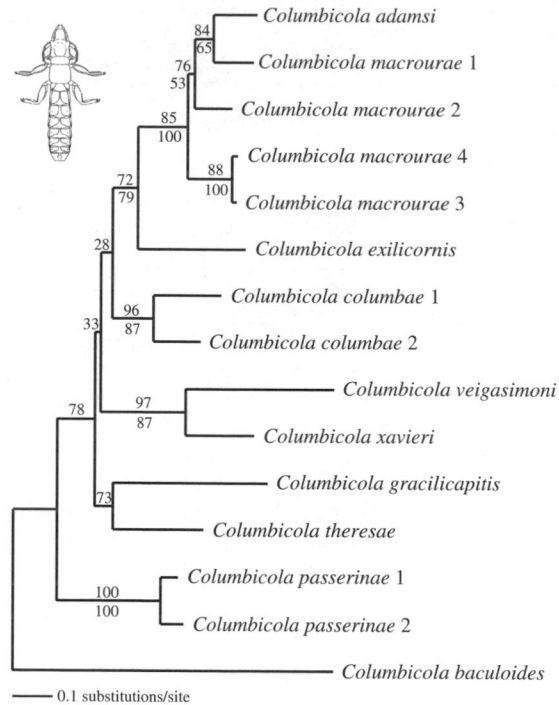

FIGURE 11.3. Phylogeny of wing lice *(Columbicola)* derived from maximum likelihood quartet puzzling of COI, 12S, and EF1α sequences. Model parameters: empirical base frequencies with rate heterogeneity, gamma shape parameter = 0.193, eight rate categories, general time reversible model with transformation parameters 0.53 (A-C), 6.46 (A-G), 1.86 (A-T), 1.39 (C-G), 10.92 (C-T), 1.0 (G-T). Conventions as in figure 11.2 (branch length scale indicated). One branch with 50% bootstrap support in the unordered parsimony analysis is not present in this tree.

to *Columbicola adamsi* in the unweighted parsimony, maximum likelihood, and neighbor joining trees (fig. 11.3). In the transversion weighted parsimony tree, *Columbicola adamsi* fell just outside *Columbicola macrourae*. The EF1α sequences for these two species were identical, so they are undoubtedly closely related. The minimum and maximum pairwise COI sequence divergences between lineages of *Columbicola* were 3.1% and 29.8%, respectively.

Body Louse Phylogeny

Unweighted parsimony analysis of body louse sequences produced a single completely resolved tree. However, several nodes of this tree were not well supported; only 6 of 15 nodes had over 50% support from bootstrap replicates. Furthermore, several relationships changed across the analyses. The maximum likelihood quartet puzzling analysis produced a tree (fig. 11.4) with 11 of 15 nodes receiving a reliability score greater than 50%. In a comparison of this tree with that from unordered parsimony, 8 of 16 nodes were identical between maximum likelihood and parsimony analyses.

Even though the body louse tree was not well supported overall, several relationships were consistent across analyses. The genus *Coloceras* was monophyletic in all analyses. Monophyly of *Coloceras* had strong support by the reliability index (98%), although not by bootstrapping (<50%). All four *Physconelloides* species groups (Price et al., 1999) represented in the maximum likelihood tree (fig. 11.4) were monophyletic. In all analyses, *P. ceratoceps* was paraphyletic, with *P. cubanus* falling within the three divergent *P. ceratoceps* lineages. *Physconelloides* was paraphyletic in all analyses, such that the other genera of body lice (*Auricotes, Campanulotes,* and *Coloceras*) were derived from within *Physconelloides*. The minimum and maximum pairwise COI sequence divergences between lineages of body lice were 3.6% and 19.3%, respectively.

Comparison of Host-Parasite Phylogenies

Comparing host and parasite phylogenies often reveals instances of cospeciation (Hafner and Nadler, 1988; Hafner et al., 1994; Moran and Baumann, 1994; Paterson and Gray, 1997; Page et al., 1998). Host-parasite cospeciation results from concurrent isolation of host and parasite populations, resulting in congruent phylogenies. Incongruence between host and parasite phylogenies can arise from several processes that are difficult to distinguish, making the interpretation of incongruence relatively difficult.

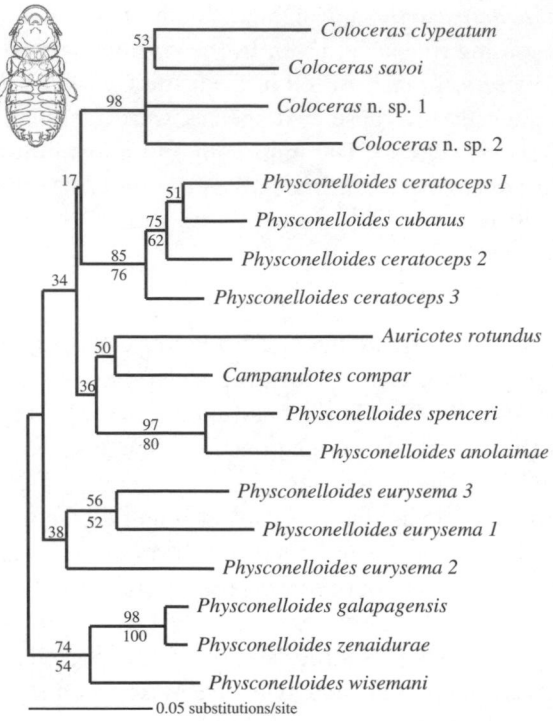

Figure 11.4. Phylogeny of body lice *(Auricotes, Campanulotes, Coloceras, Physconelloides)* derived from maximum likelihood quartet puzzling of COI and EF1α sequences. Model parameters: empirical base frequencies with rate heterogeneity, gamma shape parameter = 0.104, eight rate categories, general time reversible model with transformation parameters 0.02 (A-C), 13.05 (A-G), 3.87 (A-T), 0.50 (C-G), 5.48 (C-T), 1.0 (G-T). Conventions as in figure 11.2 (branch length scale indicated; note difference from fig. 11.2). One branch with 50% bootstrap support in the unordered parsimony analysis is not present in this tree.

Events other than cospeciation can often be difficult to infer with certainty, and often several possible reconstructions exist for any given host and parasite trees. For this reason, in our comparisons of the phylogenies of Columbiformes and their lice, we have chosen to focus on cospeciation events. Reconciliation analysis (Page, 1990a, 1994a) is a straightforward method for recovering cospeciation events. Although it does not allow for host switching, reconciliation analysis is sufficient for the goal of this study, which was to compare the coevolutionary histories of ecological replicates.

We compared trees of hosts and wing lice resulting from each of our five types of phylogenetic analysis (see Methods). In each comparison, we recovered eight cospeciation events (e.g., fig. 11.5A). In each case, there were more cospeciation events between Columbiformes and *Columbicola* than expected by chance ($p < 0.01$ for all five analyses). The two most basal nodes in the host phylogeny showed cospeciation in all five analyses. The *Columbicola* node cospeciating with the basal host node was not the most basal *Columbicola* node, but was higher up in the tree. The most basal *Columbicola* node never showed cospeciation. Three host speciation events always showed cospeciation: (1) *Zenaida macroura—Z. galapagoensis*, (2) *Columba livia—C. guinea*, and (3) *Claravis—Columbina*. Although the lice in these three cases of cospeciation are conspecific on morphological grounds (Clayton and Price, 1999), their DNA sequences are highly divergent (fig. 11.3).

For body lice, eight cospeciation events (fig. 11.5B) were recovered by the four methods of analysis that did not exclude weakly supported nodes. In contrast, 10 cospeciation events were recovered when weakly supported nodes were taken into account using the partition homogeneity test/taxon deletion method (see Methods). Eight cospeciation events were more than expected by chance ($p = .05$), or nearly so ($p = .07$), depending on the type of phylogenetic analysis. Ten cospeciation events were considerably more than expected by chance ($p = .003$). As for the wing lice, body lice showed cospeciation with the two most basal nodes in the host tree, regardless of analytical method. Other cospeciation events consistent across analyses included one event involving *Columbina*, one involving *Columba*, and two events involving *Zenaida*. In the case of the four straight tree comparisons, the most basal node in the parasite tree did not show cospeciation. However, the fifth method (partition homogeneity test/taxon deletion) recovered basal cospeciation of body lice.

Testing Independence of Cospeciation Events

We tested for the independence of wing and body louse cospeciation events by evaluating the host nodes that showed no cospeciation, cospeciation in one taxon only, and cospeciation in both wing and body lice (see table 11.2 for an example using the maximum likelihood trees). For all five analyses, the two-tailed p-value (Fisher's exact test) was 1.0, indicating that speciation events in wing and body lice are independent. Out of 19 host nodes, only 3 or 4, depending on the analysis, exhibited cospeciation with both wing and body lice.

If cospeciation events in the two parasite groups were correlated, we would expect the parasite phylogenies themselves to be somewhat congruent. However, the parasite phylogenies are largely incongruent. For

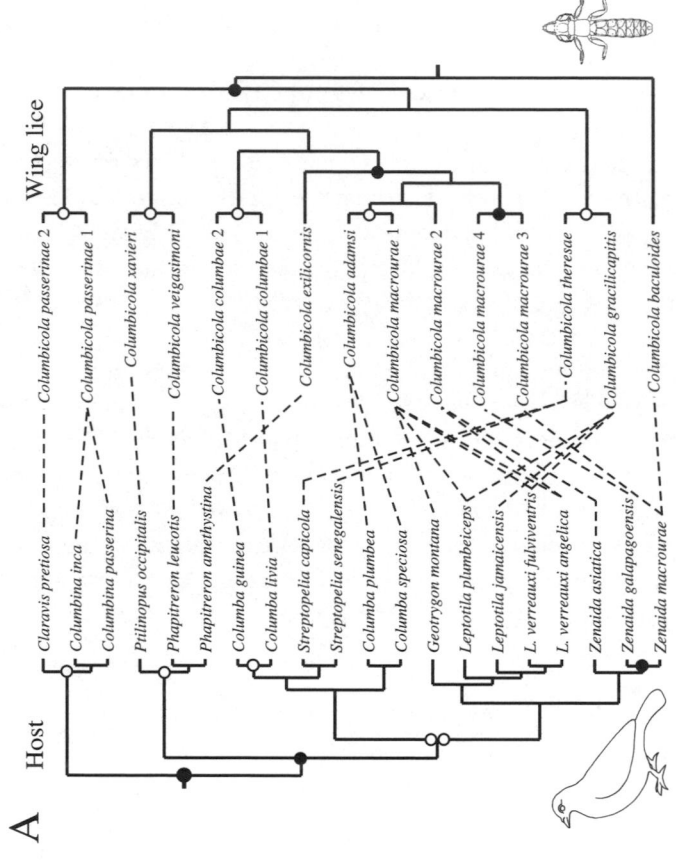

A

Host   Wing lice

Claravis pretiosa — — — — — — — *Columbicola passerinae 2*
Columbina inca — — — — — *Columbicola passerinae 1*
Columbina passerina — — — — — *Columbicola xavieri*
Ptilinopus occipitalis — — — — — *Columbicola veigasimoni*
Phapitreron leucotis — — — — — *Columbicola columbae 2*
Phapitreron amethystina — — — — — *Columbicola columbae 1*
Columba guinea — — — — — *Columbicola exilicornis*
Columba livia — — — — — *Columbicola adamsi*
Streptopelia capicola — — — — — *Columbicola macrourae 1*
Streptopelia senegalensis — — — — — *Columbicola macrourae 2*
Columba plumbea — — — — — *Columbicola macrourae 4*
Columba speciosa — — — — — *Columbicola macrourae 3*
Geotrygon montana — — — — — *Columbicola theresae*
Leptotila plumbeiceps — — — — — *Columbicola gracilicapitis*
Leptotila jamaicensis — — — — — *Columbicola baculoides*
L. verreauxi fulviventris
L. verreauxi angelica
Zenaida asiatica
Zenaida galapagoensis
Zenaida macrourae

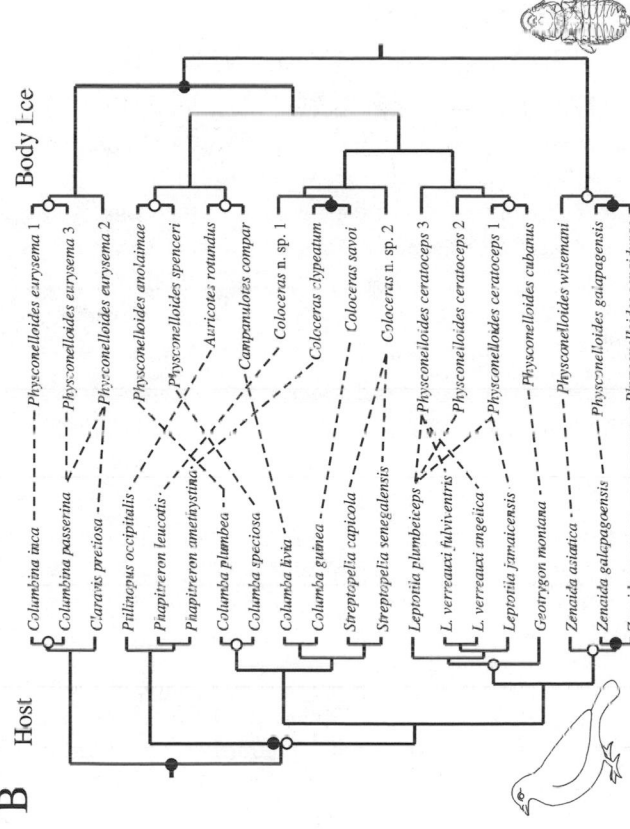

B Host

Body Lice

Columbina inca — — — — — — — Physconelloides eurysema 1
Columbina passerina — — — — — Physconelloides eurysema 3
Claravis pretiosa — — — — — — Physconelloides eurysema 2
Ptilinopus occipitalis — — — — — Physconelloides anolaimae
Paupireron leucotis — — — — — Physconelloides spenceri
Paupireron amethystina — — — — Auricotes rotundus
Columba plumbea — — — — — — Campanulotes compar
Columba speciosa — — — — — — Coloceras n. sp. 1
Columba livia — — — — — — — Coloceras clypeatum
Columba guinea — — — — — — Coloceras savoi
Streptopelia capicola — — — — — Coloceras n. sp. 2
Streptopelia senegalensis — — — Physconelloides ceratoceps 3
Leptotila plumbeiceps — — — — Physconelloides ceratoceps 2
L. verreauxi fulviventris — — — — Physconelloides ceratoceps 1
L. verreauxi angelica — — — — — Physconelloides cubanus
Leptotila jamaicensis — — — — — Physconelloides wisemani
Geotrygon montana — — — — — — Physconelloides galapagensis
Zenaida asiatica — — — — — — Physconelloides zenaidurae
Zenaida galapagoensis
Zenaida macroura

FIGURE 11.5. Comparison of quartet puzzling maximum likelihood trees for Columbiformes and (A) wing lice and (B) body lice. Lines connecting taxa indicate host-parasite associations. Circles represent nodes inferred by reconciliation analysis to have cospeciated. Closed circles are nodes that are cospeciation events shared by wing and body lice. Open circles are nodes that are not shared cospeciation events.

TABLE 11.2 Comparison of nodes with and without
cospeciation in wing vs. body lice

		Body Lice	
		Cospeciation	No Cospeciation
Wing lice	Cospeciation	3	4
	No cospeciation	4	8

Fisher's exact test $p = 1.0$. Quartet puzzling maximum likelihood trees
used in the analysis.

example, the wing lice on the host sister species *Columba plumbea* and *C. speciosa* are the same species, having failed to speciate, while the body lice on these hosts are sister taxa. Another example is that the wing lice on the host sister species *Columba livia* and *C. guinea* are also sister species, whereas the body lice on these same hosts are in different genera. In summary, concordance between wing and body louse phylogenies is minimal.

Discussion

Some portions of the host-parasite trees we reconstructed showed evidence of cospeciation, whereas other portions were incongruent. Cospeciation events in wing and body lice were not significantly correlated, suggesting that factors promoting cospeciation in wing lice may be independent of those promoting cospeciation in body lice. A number of issues relevant to comparisons of host and parasite phylogenies are evidenced by this study. We explore each of these issues below.

Estimating the Frequency of Events in Host-Parasite Histories

Reconciliation analysis (Page, 1990a, 1994a) identifies three types of events when comparing host and parasite phylogenies: cospeciation, parasite duplication, and sorting events (such as parasite extinction). A more refined analysis (TREEMAP: Page, 1994b) allows for the possibility of host switching. In our analysis, we observed a fifth phenomenon not explicitly incorporated into existing tree comparison algorithms: failure to speciate. We uncovered three instances of wing lice failing to cospeciate with speciating hosts. We also found two instances of failure to speciate in body lice. Analyses comparing host and parasite trees make assumptions about the relative frequency of historical events when arriving at an optimal reconstruction. However, it is largely unknown how common each of the five types of events listed above are in nature.

A conservative way to evaluate the frequency of cophylogenetic events is to examine terminal sister taxa. Terminal taxa comparisons circumvent many of the difficulties of phylogenetic inference for deeper nodes because these comparisons are independent of one another, as well as of other nodes in the tree. We can evaluate the relative frequency of cospeciation, parasite duplication, sorting, and failure to speciate by examining relationships of the lice on terminal host sister taxa. The frequency of host switching cannot be evaluated using this approach, since by definition, host switching involves nonsister species of hosts. We examined seven pairs of terminal host sister taxa (*Columbina, Phapitreron,* New World *Columba, Streptopelia,* Old World *Columba, Leptotila verreauxi,* and *Zenaida*) and recorded whether their associated lice showed (1) cospeciation, (2) failure to speciate, or (3) other incongruence events with multiple possible explanations. We also evaluated the relative frequency of parasite duplication (speciation in the parasite not accompanied by host speciation) by examining each host species and determining whether a speciation event had occurred between its associated parasites (table 11.3). We were not able to evaluate the relative frequency of sorting events for the two parasite groups because, in our study, we intentionally included species of hosts from which we had samples of both wing and body lice. Nearly all species of Columbiformes that have been thoroughly sampled are known to have both wing and body lice, so recent sorting events appear to be rare. However, no wing lice have been found on one well-sampled species, the New Zealand Pigeon *(Hemiphaga novaeseelandiae)* (Paterson et al., 1999; R. Palma, pers. comm.). To our knowledge, this is the only evidence suggesting a possible extinction of feather lice on an extant Columbiform host. Another interesting case is the extinct Passenger Pigeon *(Ectopistes migratorius),* from which body lice have never been recovered, despite concerted efforts to find them on museum skins from which many wing lice have been collected (Price et al., 2000).

In the case of wing lice from terminal host sister taxa, we observed two cospeciation events, three failure to speciate events, and two other incongruence events (involving deeper combinations of duplications, sorting events, and/or host switches: table 11.3). We found no evidence of wing

TABLE 11.3 Numbers of cophylogenetic events for host sister taxa comparisons

Event	Wing Lice	Body Lice
Cospeciation	2	3
Failure to speciate	3	2
Other incongruence event(s)	2	2
Duplications	0	0

Note: Duplications based on examination of 20 terminal host taxa. All other events based on seven congeneric sister taxa comparisons.

louse duplication on any of the 20 extant species of hosts. For the seven comparisons of body lice on terminal host sister taxa (table 11.3), we found three cases of cospeciation, two failure to speciate events, and two other incongruence events. Again, we found no evidence of body louse duplication on any of the 20 extant host species.

Our evaluation of events in closely related taxa suggests that failure to speciate may be a seriously overlooked event in reconciling host and parasite phylogenies. Conversely, it appears that the importance of parasite duplication may be overemphasized when seeking explanations to reconcile host and parasite phylogenies. It may be that by inferring failure to speciate rather than parasite duplication, host and parasite phylogenies could be more easily reconciled (in terms of the number of events needed to explain the differences). For example, consider the hypothetical biogeographic scenario depicted in figure 11.6. The once contiguous host species A is fragmented by a geographic barrier, and the host speciates into A and B. However, there could still be sufficient gene flow between parasite populations to prevent speciation in the parasite X (a "failure to speciate" event). This scenario is not unreasonable; for example, Dybdahl and Lively (1996) found much higher levels of gene flow between populations in a trematode parasite than in its snail host. (In lice, failure to speciate might occur by phoresis of lice on hippoboscid flies between diverging host populations.) If host species A then colonizes a new isolated area by dispersal, one could imagine a new speciation event in the host (producing host species C). Coincident with host speciation is a speciation event in the parasite (producing parasite species Y) because of a complete lack of further gene flow between the more completely isolated parasite populations. This scenario would produce the host-parasite phylogenies shown in figure 11.6. Using TREEMAP for host-parasite history reconstruction, and invoking cospeciation, duplication, sorting, and host switching, we recovered four events needed to explain the pattern: one duplication and three sorting events. However, allowing the parasite to fail to speciate, only two events are needed: a failure to speciate event and a cospeciation event. We suggest that future work on methods of host-parasite phylogeny reconciliation explicitly take into account failure of the parasite to speciate as a possible event in the history of the host-parasite association.

Host Specificity and the Significance of Cospeciation

For all comparisons of host and parasite trees we found the same number or more cospeciation events in body lice than in wing lice. However, in most analyses body lice showed only a marginally significant amount of

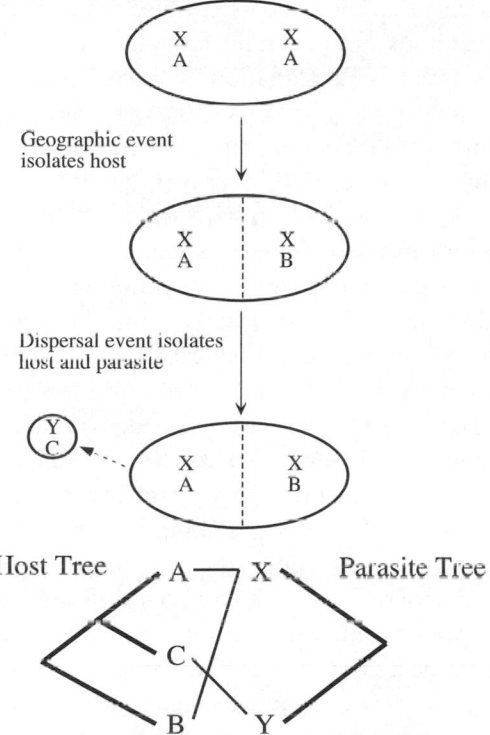

FIGURE 11.6. Hypothetical scenario for speciation in host and parasites and resulting phylogenies. Contiguous host species A fragments by a geographic barrier into host species A and host species B. Parasite species X is not affected by this barrier and fails to speciate. Subsequently, host species A colonizes an isolated area and this results in speciation in both the host and parasite, producing host C and parasite Y.

cospeciation, while the *p*-value for wing lice was always low ($< .01$). How can these differences be explained?

The first possibility is that we have the wrong body louse phylogeny. Many of the nodes in the body louse tree are poorly supported compared with the trees for the hosts and wing lice. This poor support is most likely a result of the fact that we have the least amount of total sequence for body lice. A further indication that an incorrect louse phylogeny may be a contributing factor is the marginal significance of cospeciation when the body louse phylogeny is taken to be correct, compared with the strongly significant cospeciation observed when we used the partition homogeneity test/taxon deletion method. This method explicitly takes into account differences between host and parasite trees owing to weak support (Johnson et al., 2001).

However, the fact that the same number of cospeciation events (eight) between wing and body lice occurs in most of the comparisons suggests there might be an additional explanation for the differences in p-values. The basic technique for evaluating whether more cospeciation is observed than expected by chance is to randomize the parasite phylogeny and count the number of resulting "cospeciation" events (Page, 1990b, 1994b). This procedure produces a null distribution with which the observed number of cospeciation events can be compared. An example of these distributions is shown in figure 11.7A for the maximum likelihood trees for both wing and body lice. The distribution for body lice is shifted to the right, compared with the distribution for wing lice. One possible explanation for this shift in the null distribution is that body lice are more host specific than wing lice. High host specificity may tend to make recovering a large number of cospeciation events more likely by chance. To examine the impact of host specificity on the null distribution, we arbitrarily pruned host associations from the body louse and host trees, making each body louse species perfectly host specific. When this is done, the distribution shifts even further to the right; the number of randomizations with a high number of cospeciation events increases, while the number of randomizations with a low number of cospeciation events decreases (fig. 11.7B). This effect, when combined with the fact that there are more body louse species (18) than wing louse species (15) on the same hosts, may explain why wing lice showed highly significant cospeciation, while body lice generally showed marginal p-values. While the shift in the randomized distribution due to increased host specificity is not dramatic, it has the potential to alter the significance level assigned to the amount of cospeciation recovered. These observations indicate that caution should be used when comparing the results of randomization tests for hosts and parasites across parasite groups that differ in host specificity.

Relative Ages of Host-Parasite Associations

A striking difference between the wing and body louse sequence data sets is that wing lice were much more divergent than body lice (note branch lengths and difference in scales between figs. 11.3 and 11.4). Pairwise uncorrected sequence divergences between wing louse species generally range between 18% and 30% for COI, and between 0% and 11% for EF1α. In contrast, pairwise uncorrected sequence divergences between body louse species generally range between 8% and 20% for COI, and between 0% and 3% for EF1α. One possible explanation for this difference is that wing lice are evolving faster at the molecular level than body lice. However, examination of a few correlated recent cospeciation events indicates that, if anything, body lice are evolving faster than wing lice.

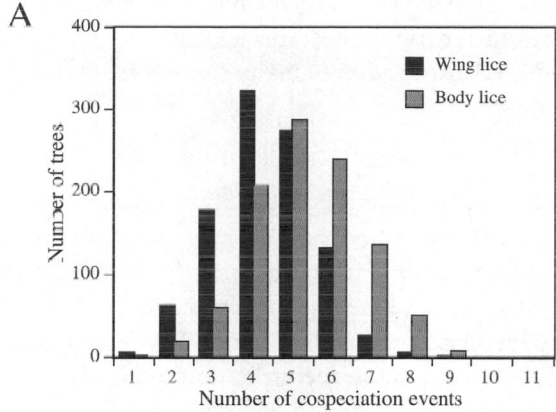

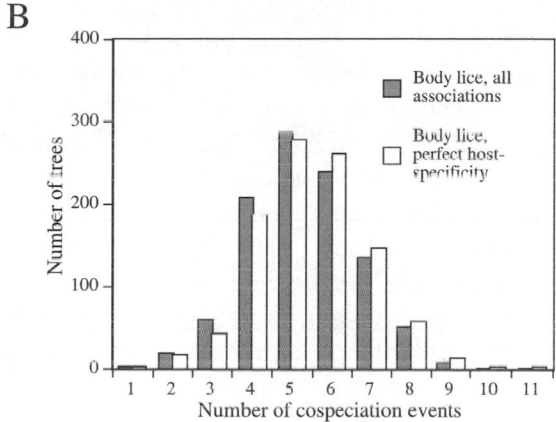

FIGURE 11.7. Comparison of distributions of number of cospeciation events for 1,000 random louse trees: *A*, wing lice and body lice; *B*, body lice with all host associations maintained and body lice with host associations removed such that each species of louse is found on only one host.

A second possibility is that Columbiform wing lice are an older radiation than Columbiform body lice. In the partition homogeneity test/taxon deletion analysis, which has the highest probability of inferring cospeciation events, the first speciation event in wing lice is not a cospeciation event, while it is a cospeciation event in body lice. If the oldest node in the Columbiform phylogeny is a cospeciation event, but the oldest node in the *Columbicola* phylogeny is not one, then *Columbicola* must be older than modern Columbiform hosts. If modern *Columbicola* lineages did evolve before modern Columbiform lineages, this suggests there may be lineages of hosts previously parasitized by these old *Columbicola* lineages, but which are now extinct. These old *Columbicola* lineages may have survived by colonizing more recently evolving host species, and this possibility is consistent with the broad host distribution of many *Columbicola* species.

Independence of Cospeciation Events

We found that cospeciation events between wing and body lice were largely independent. Two of the three cospeciation events common to both wing and body lice were the two most basal nodes of the host tree. While this may reflect actual history, it also seems probable that this may be an artifact of the way reconciliation methods work. When host and parasite trees are not completely congruent, reconciliation methods tend to map shallow parasite speciation events onto deep host speciation events. This has a tendency to "push" cospeciation events back in the host tree, which may explain why basal nodes in the host tree often showed cospeciation in our analyses. If this is the case, one should be cautious when using deep cospeciation events to compare rates of evolution between hosts and parasites.

The third cospeciation event common to both wing and body lice involved the node between the Mourning Dove *(Zenaida macroura)* and Galapagos Dove *(Z. galapagoensis)*. Since the Galapagos Dove is the only species of dove on the Galapagos Islands, it probably brought its lice with it upon colonization. This biogeographic event, which was common to both wing and body lice, would have caused speciation in both of them. When biogeographic isolating events are responsible for speciation in the host, and this isolation is extreme as in the colonization of an island, we would expect speciation events in replicate parasite groups to be correlated. However, when hosts and parasites are more broadly distributed, isolating events for one parasite group may not affect the other, even when the parasites share similar ecologies.

Understanding reasons for the similarities (few as they are) between the two parasite trees is easier than understanding their differences. One possibility is that the differences arise from differences in chance events (e.g., chance parasite extinction). In such a case, we would expect to see little pattern to the differences. On the other hand, there may be predictable patterns underlying the differences between parasite phylogenies. We suspect that wing lice are less likely to speciate than body lice in response to host speciation, which gives rise to the differences in phylogenies. Although body lice showed only one less failure to speciate event than wing lice (table 11.3), population level genetic data indicated significant differentiation in one of the cases (*Physconelloides ceratoceps* 3 on *Leptotila verreauxi fulviventris* and *L. v. angelica:* Johnson et al., 2002). Perhaps *P. ceratoceps* 3 is cospeciating with its hosts, but is lagging behind in the process. Unlike body lice, no genetic differentiation can be detected in populations of *Columbicola* on multiple species of host, suggesting that *Columbicola* is capable of dispersing to multiple hosts (Johnson et al., 2002). Wing lice can survive for a longer period of time off the host than body lice (unpub. data), which is consistent with a higher probability of dispersal by wing lice. Ecological studies comparing dispersal of wing and body lice among host species are needed to confirm this differential dispersal hypothesis. The lower dispersal ability of body lice could cause them to speciate in response to host speciation more often than wing lice. Uncovering this possible link between ecology and macroevolutionary pattern is an exciting future prospect, which we discuss in chapter 13.

Acknowledgments

We are extremely grateful to the following individuals and institutions who provided critical assistance, without which the fieldwork would not have been possible: D. Drown, L. Heaney, N. Ingle, R. Moyle, A. Navarro, A. T. Peterson, M. Robbins, J. Weckstein, and Texas Parks and Wildlife. R. J. Adams and R. D. Price prepared slide mounts and provided identifications for the voucher specimens. We are grateful to J. Burtt and J. Ichida for the SEMs of lice, V. Smith for line drawings of lice, and S. Al-Tamimi for line drawings of Columbiformes. We thank R. J. Adams, D. Drown, and P. Ewald, R. Page, and A. Paterson for valuable discussion or comments on the manuscript. Several host tissue samples were provided by the Kansas Museum of Natural History, the Field Museum of Natural History, the Louisiana State University Museum of Natural Science, and the Tracy Aviary. We thank the DNA Sequencing Facility at the University of Utah, supported in part by NCI grant #5p30CA42014. This work was supported by NSF-CAREER award DEB-9703003 to D.H.C.

References

Booth, T. D., D. H. Clayton, and B. A. Block. 1993. Experimental demonstration of the energetic cost of parasitism in free-ranging hosts. *Proceedings of the Royal Society of London,* ser. B, 253:125–29.

Brooks, D. R., and D. A. McLennan. 1991. *Phylogeny, ecology, and behavior.* Chicago.: University of Chicago Press.

Clayton, D. H. 1991. Coevolution of avian grooming and ectoparasite avoidance. In *Bird-parasite interactions: Ecology, evolution, and behaviour,* edited by J. E. Love and M. Zuk. Oxford: Oxford University Press.

Clayton, D. H., P. L. M. Lee, D. M. Tompkins, and E. D. Brodie III. 1999. Reciprocal natural selection on host-parasite phenotypes. *American Naturalist* 154:261–70.

Clayton, D. H., and R. D. Price. 1999. Taxonomy of New World *Columbicola* (Phthiraptera: Philopteridae) from the Columbiformes (Aves), with descriptions of five new species. *Annals of the Entomological Society of America* 92:675–85.

Clayton, D. H., and D. M. Tompkins. 1994. Ectoparasite virulence is linked to mode of transmission. *Proceedings of the Royal Society of London,* ser. B, 256:211–17.

———. 1995. Comparative effects of lice and mites on the reproductive success of rock doves *(Columba livia). Parasitology* 110:195–206.

Clayton, D. H., and B. A. Walther. 1997. Collection and quantification of arthropod parasites of birds. In *Host-parasite evolution: General principles and avian models,* edited by D. H. Clayton and J. Moore, 419–40. Oxford: Oxford University Press.

Couch, A. B. J. 1962. Phoretic mallophaga from hiposboscids of mourning doves *Zenaidura macroura. Journal of Parasitology* 48:497.

Cruickshank, R. H., K. P. Johnson, V. S. Smith, R. J. Adams, D. H. Clayton, and R. D. M. Page. 2001. Phylogenetic analysis of partial sequences of elongation factor 1 alpha identifies major groups of lice (Insecta: Phthiraptera). *Molecular Phylogenetics and Evolution* 19:202–15.

Danforth, B. N., and S. Ji. 1998. Elongation factor-1α occurs as two copies in bees: Implications for phylogenetic analysis of EF-1α sequences in insects. *Molecular Biology and Evolution* 15:225–35.

Dybdahl, M. F., and C. M. Lively. 1996. The geography of coevolution: Comparative population structures for a snail and its trematode parasite. *Evolution* 50:2264–75.

Farris, J. S., M. Källersjö, A. G. Kluge, and C. Bult. 1994. Testing significance of incongruence. *Cladistics* 10:315–19.

———. 1995. Constructing a test for incongruence. *Systematic Biology* 44:570–72.

Felsenstein, J. 1985. Confidence limits on phylogenies: An approach using the bootstrap. *Evolution* 39:783–91.

Hafner, M. S., and S. A. Nadler. 1988. Phylogenetic trees support the coevolution of parasites and their hosts. *Nature* 332:258–59.

Hafner, M. S., P. D. Sudman, F. X. Villablanca, T. A. Spradling, J. W. Demastes, and S. A. Nadler. 1994. Disparate rates of molecular evolution in cospeciating hosts and parasites. *Science* 265:1087–90.

Hoberg, E. P., D. R. Brooks, and D. Seigel-Causey. 1997. Host-parasite cospeciation: History, principles, and prospects. In *Host-parasite evolution: General principles and avian models,* edited by D. H. Clayton and J. Moore, 212–35. Oxford: Oxford University Press.

Hopkins, G. H. E., and T. Clay. 1952. A checklist of the genera and species of Mallophaga. London: British Museum of Natural History.

Huelsenbeck, J. P., and K. A. Crandall. 1997. Phylogeny estimation and hypothesis testing using maximum likelihood. *Annual Review of Ecology and Systematics* 28:437–66.

Huelsenbeck, J. P., B. Rannala, and Z. Yang. 1997. Statistical tests of host-parasite cospeciation. *Evolution* 51:410–19.

Johnson, K. P. 1997. The evolution of behavior in the dabbling ducks (Anatini): A phylogenetic approach. Ph.D. diss., University of Minnesota.

Johnson, K. P., and D. H. Clayton. 2000. Nuclear and mitochondrial genes contain similar phylogenetic signal for pigeons and doves (Aves: Columbiformes). *Molecular Phylogenetics and Evolution* 14:141–51.

Johnson, K. P., D. M. Drown, and D. H. Clayton. 2001. A data based parsimony method of cophylogenetic analysis. *Zoologica Scripta* 30:79–87.

Johnson, K. P., B. L. Williams, D. M. Drown, R. J. Adams, and D. H. Clayton. 2002. The population genetics of host specificity: Genetic differentiation in dove lice (Insecta: Phthiraptera). *Molecular Ecology* 11:25–38.

Kimura, M. 1980. A simple model for estimating evolutionary rates of base substitutions through comparative studies of nucleotide sequences. *Journal of Molecular Evolution* 16:111–20.

Marshall, A. G. 1981. *The ecology of ectoparasitic insects.* London: Academic Press.

Moran, N., and P. Baumann. 1994. Phylogenetics of cytoplasmically inherited microorganisms of arthropods. *Trends in Ecology and Evolution* 9:15–20.

Nelson, B. C., and M. D. Murray. 1971. The distribution of Mallophaga on the domestic pigeon *(Columba livia). International Journal for Parasitology* 1:21–29.

Page, R. D. M. 1990a. Component analysis: A valiant failure? *Cladistics* 6:119–36.

———. 1990b. Temporal congruence and cladistic analysis of biogeography and cospeciation. *Systematic Zoology* 39:205–26.

———. 1994a. Maps between trees and cladistic analysis of historical associations among genes, organisms, and areas. *Systematic Biology* 43:58–77.

———. 1994b. Parallel phylogenies: Reconstructing the history of host-parasite assemblages. *Cladistics* 10:155–73.

Page, R. D. M., and M. A. Charleston. 1998. Trees within trees: Phylogeny and historical associations. *Trends in Ecology and Evolution* 13:356–59.

Page, R. D. M., P. L. M. Lee, S. A. Becher, R. Griffiths, and D. H. Clayton. 1998. A different tempo of mitochondrial DNA evolution in birds and their parasitic lice. *Molecular Phylogenetics and Evolution* 9:276–93.

Paterson, A. M., and R. D. Gray. 1997. Host-parasite cospeciation, host switching, and missing the boat. In *Host-parasite evolution: General principles and avian models,* edited by D. H. Clayton and J. Moore, 236–50. Oxford: Oxford University Press.

Paterson, A. M., R. L. Palma, and R. D. Gray. 1999. How frequently do avian lice miss the boat? *Systematic Biology* 48:214–23.

Price, R. D., R. J. Adams. and D. H. Clayton. 2000. Pigeon lice down under: Taxonomy of the Australian *Campanulotes* (Phthiraptera: Philopteridae), with a description of *C. durdeni* n. sp. *Journal of Parasitology* 86:948–50.

Price, R. D., D. H. Clayton, and R. A. Hellenthal. 1999. Taxonomic review of *Physconelloides* (Phthiraptera: Philopteridae) from the Columbiformes (Aves), including descriptions of three new species. *Journal of Medical Entomology* 36:195–206.

Simon, C., F. Frati, A. Beckenbach, B. Crespi, H. Liu, and P. Flook. 1994. Evolution, weighting, and phylogenetic utility of mitochondrial gene sequences and a compilation of conserved polymerase chain reaction primers. *Annals of the Entomological Society of America* 87:651–704.

Smith, V. S. 2000. Basal ischnoceran louse phylogeny (Phthiraptera: Ischnocera: Gonioididae and Heptapsogasteridae). *Systematic Entomology* 25:73–94.

Strimmer, K., and A. von Haeseler. 1996. Quartet puzzling: A quartet maximum-likelihood method for reconstructing tree topologies. *Molecular Biology and Evolution* 13:964–69.

Swofford, D. L. 1999. *PAUP*. Phylogenetic Analysis Using Parsimony (*and other methods).* Sunderland, Mass.: Sinauer Associates.

12

DROWNING ON ARRIVAL, MISSING
THE BOAT, AND X-EVENTS: HOW LIKELY
ARE SORTING EVENTS?

*Adrian M. Paterson, Ricardo L. Palma,
and Russell D. Gray*

Introduction

One of the major goals of coevolution studies is to infer the chronicle of events that has determined the present distribution of parasites on their hosts. This is a difficult task that involves unraveling the four types of coevolutionary events that interact to produce parasite distributions: cospeciation, sorting, host switching, and intrahost speciation (duplication) events (Clay, 1949; Paterson and Gray, 1997; Paterson et al., 1999) (see fig. 12.1a–d). Cospeciation events occur when host and parasite species codiverge (fig. 12.1a). For example, the isolation of a host population will often result in the isolation of the parasite population and its subsequent speciation. Sorting events occur when parasite species are entirely removed from host species (fig. 12.1b). Host switching events occur when a parasite species colonizes a host species other than its current host (fig. 12.1c). Intrahost speciation occurs when a parasite lineage diverges without the stimulus of host speciation and results in multiple closely related species on the descendant host lineage (fig. 12.1d). Here we will demonstrate that sorting events are very common in some avian host-parasite systems and that this is not an artifact of small sample sizes (or *x-events;* see fig. 12.1b). We will argue that this has major implications both for the methods used to analyze host-parasite coevolution, and for the types of questions we might ask about comparative patterns of such systems.

Coevolutionary Events

Several methods have been developed to make inferences about coevolutionary history that find various combinations of coevolutionary events (see Paterson and Gray, 1997, for a review). Early methods simply compared host and parasite trees or classifications by eye to assess their

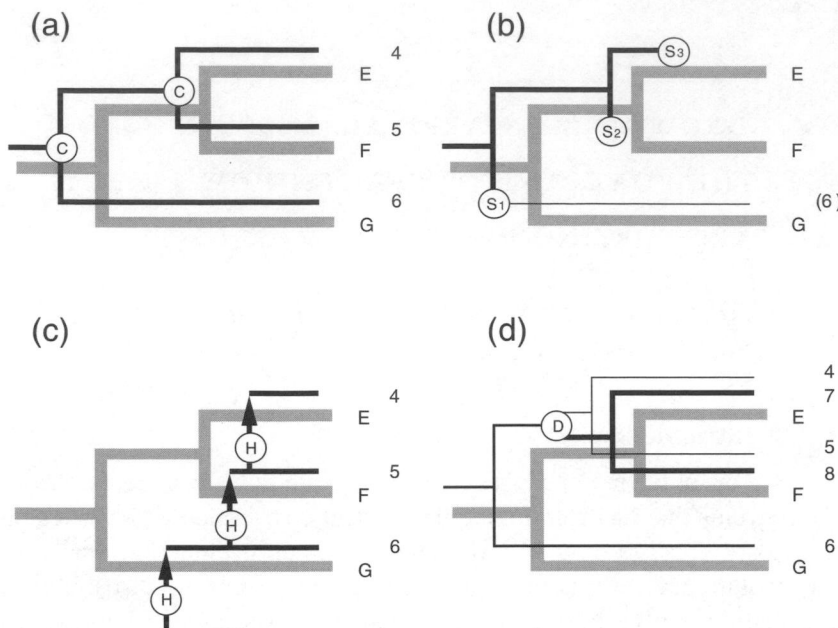

FIGURE 12.1. The four different types of coevolutionary events inferred in
reconciliation analysis. Each figure shows a phylogeny of three host taxa (shaded line)
with a parasite lineage (line) mapped upon it. *(a)* Cospeciation events (C): the parasite
taxa speciate or codiverge at the same point as their host. *(b)* Sorting events (S):
parasite taxa are lost from their host lineage. There are three types of sorting events:
S1, the parasite taxa are present but have not been detected (*x*-events); S2, the parasite
taxa were not present on the founding host population (MTB—"missing the boat");
and S3, the parasite taxa have gone extinct on a host lineage (DOA—"drowning on
arrival"). *(c)* Host switching events (H): a parasite taxon has colonized the host taxon
from a different host lineage and then successively colonized the host's close relatives.
(d) Duplication event (D): the parasite taxon has speciated on a host without an
accompanying host speciation event and has produced multiple parasite lineages on
the host's descendants.

congruence, and thus infer cospeciation, which Brooks (1981) termed a nar-
rative approach. More recently, the extent of cospeciation in host-parasite
systems has been examined quantitatively using Brooks Parsimony Analysis (Brooks, 1981) and reconciliation analysis. Page (1990a, 1990b, 1993,
1994, 1996) has developed a method that reconstructs cospeciation, intrahost speciation, sorting, and suspected host switching events from phylogenetic trees of parasites and their hosts. This method by Page, termed
reconciliation analysis, postulates the minimal numbers of intrahost speciation and sorting events needed to reconcile host and parasite trees without

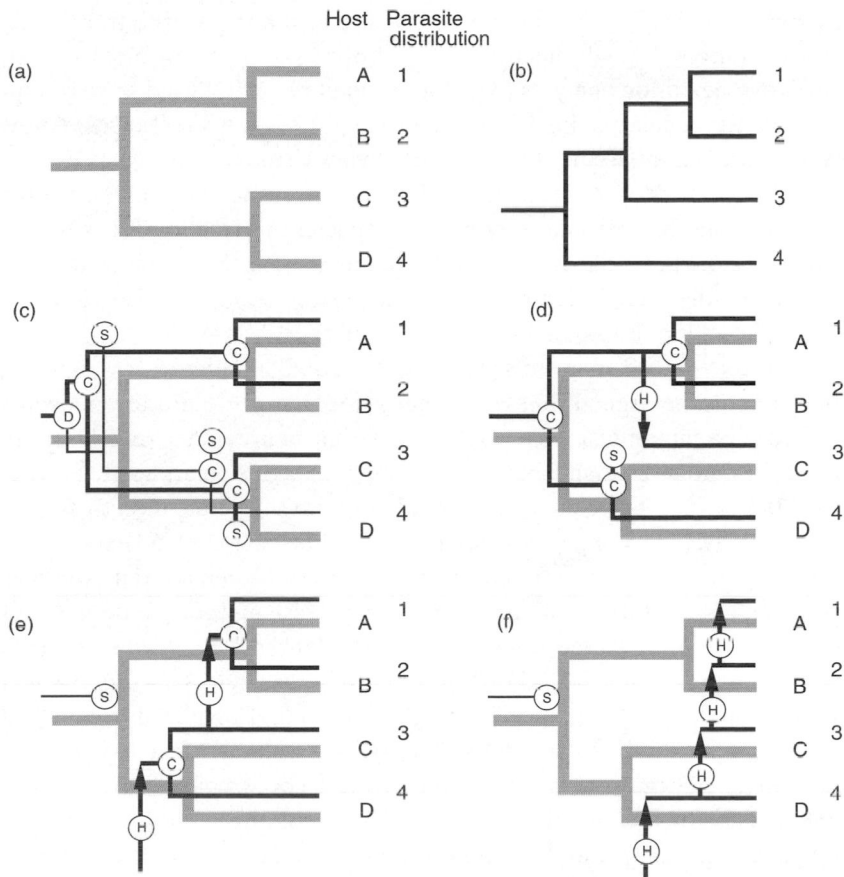

Host Parasite
      distribution

FIGURE 12.2. *(a)* The phylogeny for a group of hosts A–D and the distribution of the parasites 1–4 over these hosts. *(b)* The phylogeny for the parasites 1–4. Reconciliation analysis provides many different scenarios for reconciling these trees. *(c)* This reconciliation assumes no host switching. The lines of different widths refer to the two different lineages descending from the duplication event. *(d)* This reconciliation allows one host switching event. Note that the host switching event makes a prediction that the host clade (C, D) is older than the host clade (A, B). *(e)* This reconciliation allows two host switching events. *(f)* This reconciliation only allows host switching events.

postulating host switching. The program TREEMAP (Page, 1994) implements an important refinement on reconciliation analysis that allows host switching to be addressed in a systematic manner.

TREEMAP uses a host tree (fig. 12.2a) and a usually incongruent parasite tree (fig. 12.2b) together with the parasite distribution (fig. 12.2a) to make specific predictions about how these groups have coevolved. There are many scenarios that would reconcile the host and parasite trees given various combinations of the different coevolving or cophylogenetic events. Early reconciliation analyses (e.g., Paterson et al., 1993) used the assumption that no host switching had occurred (fig. 12.2c), while TreeMap now examines all possible combinations with a view to maximizing cospeciation events (fig. 12.2d–e). A further model of host-parasite evolution, resource tracking, generally assumes that parasite species are free to host switch to available host taxa (fig. 12.2f). The forthcoming TreeMap 2 program will implement the "jungles" algorithm (Charleston, 1998; see also chap. 3), which allows the differential weighting of all these events.

If all coevolutionary events are accorded equal likelihood or weighting, then it could be argued that those scenarios that postulate fewer events are more parsimonious. This would imply that figures 12.2e and 12.2f are more parsimonious than figures 12.2c and 12.2d. It is clear, however, that the different events are not equally likely but, rather, relate directly to parasite ecology. For example, fleas are highly mobile, so we would expect host switching to occur more often than cospeciation. Molecular data that are evolving in a clocklike fashion are the only precise method to use in order to discriminate among the various scenarios (Hafner et al., 1994; Page et al., 1998; Paterson et al., 2000). For example, the louse genus *Saemundssonia* is found on the gulls (Charadriformes) and petrels (Procellariformes) bird orders but not on the closely related penguins (Sphenisciformes) and cormorants (Pelecaniformes). There are at least two scenarios to test. The first scenario postulates that *Saemundssonia* colonized the petrels from the gulls (fig. 12.3a), and the second scenario postulates that the penguins and the cormorants have lost this genus (fig. 12.3b). These two scenarios imply different levels of genetic divergence between the two *Saemundssonia* lineages. If the genus has host switched from the gulls to the petrels, then the genetic divergence between them will be less than between that of their hosts. Alternatively, if *Saemundssonia* has been passed down through the host lineages, then the level of divergence between the louse lineages will be at least as great as their hosts.

A differential rate of molecular evolution could explain the difference in genetic divergence between host and parasite and not provide any information about the relative depth of divergence. To avoid this, Paterson et al. (2000) used only cospeciating nodes, where both seabird and louse have speciated contemporaneously, to correct for the greater rate of molecular evolution in lice. Paterson et al. (2000) showed that the level of 12S rRNA genetic divergence between the lice (corrected distance $d = 0.45$) was considerably greater than that of their hosts ($d = 0.22$), thereby supporting the scenario of descent. The same arguments are made for testing the reality of

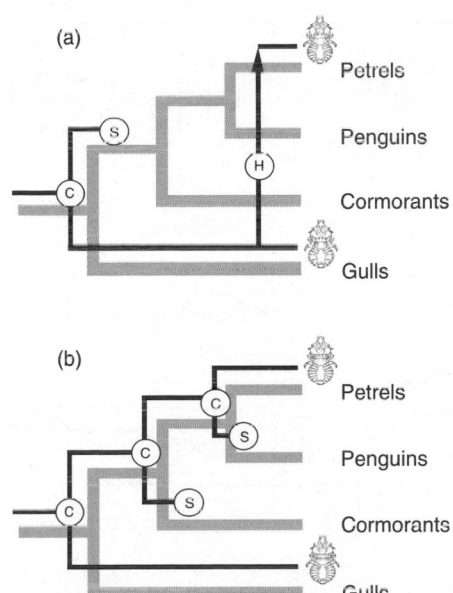

FIGURE 12.3 Two possible scenarios (or reconciliations) for explaining the presence of *Saemundssonia* lice on gulls and petrels and its absence from penguins and cormorants. *(a) Saemundssonia* has colonized the petrels from the gulls. *(b) Saemundssonia* has passed down the lineage from a common ancestor of gulls and petrels. The hypothesized louse phylogeny (dark line) is mapped onto the hypothesized host phylogeny (shaded line).

inferred cospeciation events (host and parasite relative genetic distances should be similar) and duplication events (parasite relative genetic distances should be at least as large as their hosts).

Sorting Events

Sorting events, however, cannot be directly tested. How can one test for a lineage that is not there? There is nothing that can be sequenced—a doubly unfortunate circumstance, as there are often high numbers of sorting events predicted relative to other coevolutionary events in reconciliation analyses. For example, Paterson et al. (2000), in their analysis of seabird and chewing louse cospeciation, found 11 sorting events (fig. 12.4; relative to seven cospeciation events), and Paterson and Poulin, (1999), in their analysis of parasitic copepods and teleost fish, found 9 sorting events (fig. 12.5; relative to three cospeciation events).

A further complication is that there are three processes that produce the pattern of absence of parasites from their hosts (i.e., three types of sorting events: fig. 12.1b). First, parasites may occur in low numbers on the

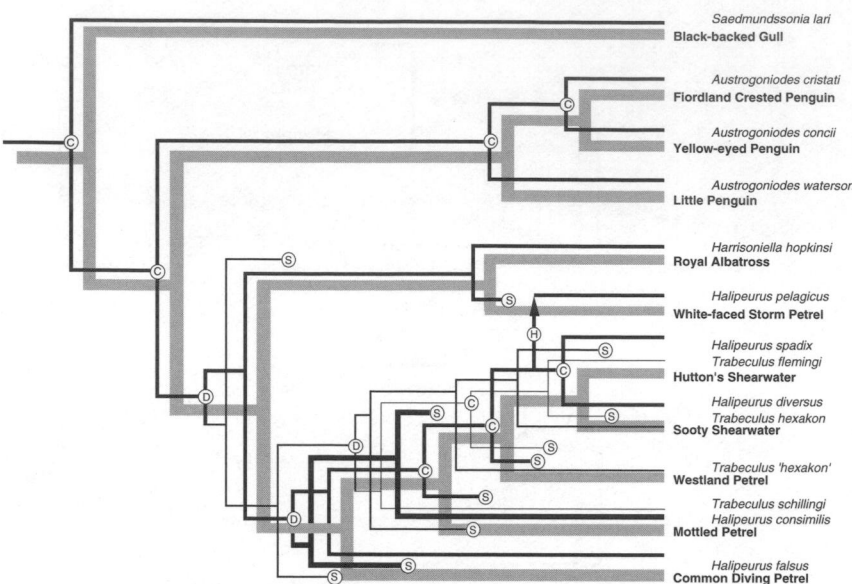

FIGURE 12.4. A reconciliation tree for the coevolution of seabirds and chewing lice (after Paterson et al., 2000, fig. 5b). The louse phylogeny (*italicized* taxon names and dark lines—the varying width of lines reflects lineages derived from duplication events) are mapped onto the host phylogeny (**boldface** names and shaded line). This reconciliation hypothesizes 11 sorting, 9 cospeciation, 1 host switching, and 3 duplication events.

extant host population but have escaped detection by sampling error (S1 in fig. 12.1b), which we term an "*x*-event." Note that because sampling error generates false absence of parasites from hosts it is only a pseudosorting event. Second, parasites may have been absent from the host founder population at a speciation event because of the patchy distribution of a parasite throughout the range of the host or some other stochastic event (S2 in fig. 12.1b). We have previously (Paterson et al., 1997; 1999) referred to this as "missing the boat" (MTB). Third, parasites may have gone extinct from a host lineage after a host speciation event (S3 in fig. 12.1b), which we term "drowning on arrival" (DOA).

To test the idea that sorting events may be common in parasite-host coevolution, we analyzed Australasian bird-louse distributions in which we could clearly identify a parent-daughter relationship between bird taxa (i.e., a taxon that was clearly derived from another taxon). We added 17 Australian examples (appendix 12.1) to the 48 New Zealand examples from a previous study (Paterson et al., 1999). Nine of these examples were

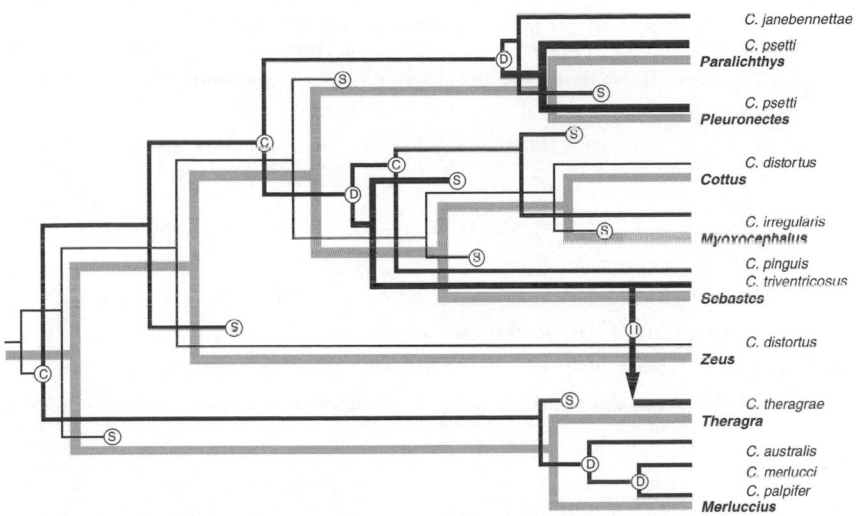

FIGURE 12.5. A reconciliation tree for the coevolution of teleost fish and
Chondracanthus parasitic copepods (after Paterson and Poulin, 1999, fig. 2f). The
copepod phylogeny (dark lines—the varying width of lines reflects lineages derived
from duplication events) are mapped onto the known host phylogeny (**boldface** taxon
names and shaded line). This reconciliation hypothesizes 9 sorting, 3 cospeciation,
1 host switching, and 3 duplication events.

bird species introduced into both New Zealand and Australia. We divided
the records into three groups: human-introduced bird taxa (introduced),
bird taxa with a history of repeatedly colonizing an island (cosmopolitan),
and closely related taxa (such as regional subspecies). Australasia is well
placed for such a study, as it is relatively isolated, has an unfortunate history
of successful human introductions of bird species, and has many offshore
islands, which have led to host differentiation. We assumed that these were
ideal conditions for sorting events to have occurred in and predicted that
there would be a reduction in the louse species present on the daughter
taxon relative to the parent taxon. A summary of the data is given in
table 12.1. This study also extends that of Paterson et al. (1999) by examining the likelihood of x-events.

Sixty-five bird taxa were identified as having a parent-daughter connection; of these, 47 showed a reduction in louse species number. This reduction was significant (one-tailed binomial test, binomial probability = 0.5:
$p = .0002$). When partitioned into the three categories, both the introduced
and closely related taxa showed a significant reduction in louse species
number ($p < .0001$ and $p = .0178$, respectively), whereas the cosmopolitan

TABLE 12.1 Australasian examples of parent-daughter host taxa and the status of their louse species. Same = no change in louse species composition; reduced = fewer louse species on the daughter bird taxon relative to its parent taxon.

	Human Introduced	Cosmopolitan	Closely Related
Same	3	7	8
Reduced	24	3	20

species showed no significant decrease ($p = .945$). These results are re-flected in the mean decrease in louse species number from parent to daugh-ter taxa (overall decrease: 1.5 ± 1.7 species; introduced decrease: 2.5 ± 2.0; closely related taxa decrease: 0.9 ± 0.9; cosmopolitan decrease: 0.3 ± 0.5).

X-Events

A potential problem with the interpretation of these data is the degree to which apparent sorting events are really a result of poor or unequal sampling effort (x-events). X-events will be of particular importance if the daughter taxon is usually sampled less well relative to the parent taxon. Might we expect poorer sampling in the daughter taxa of our Australasian data set? Many of the daughter taxa in this study are found on small oceanic islands that are relatively inaccessible—for example, Kermadec Storm Petrel *(Pelagadroma marina albiclunis),* Bounty Island Shag *(Leucocarbo campbelli),* and Antipodes Pipit *(Anthus novaeseelandiae steindachneri)*—while their parent taxa are found on the mainland—for example, White-Faced Storm Petrel *(Pelagadroma marina maoriana),* Stewart Island Shag *(Leucocarbo carunculatus),* and New Zealand Pipit *(Anthus novaeseelandiae novaeseelandiae),* respectively. Sampling of remote species is of-ten a matter of brief collecting trips or fortuitous beach-cast specimens. Introduced taxa may be insufficiently sampled for a quite different reason, as their ubiquity and low priority for research result in their being little studied. An additional problem is that our data only represent positive hosts, hosts that actually have lice when sampled, rather than how the lice are distributed. If 100 individuals from a species were sampled and only one of that 100 has lice, there would be a positive host sample size of just one, but it would be represented as a positive host sample in this data set.

In order to determine whether x-events were a problem in our data, we examined louse species collected from 136 New Zealand bird taxa that are held in the Museum of New Zealand Te Papa Tongarewa collec-tion. We collected data on the number of positive individuals sampled for each host taxon and the number of louse species found on that host taxon

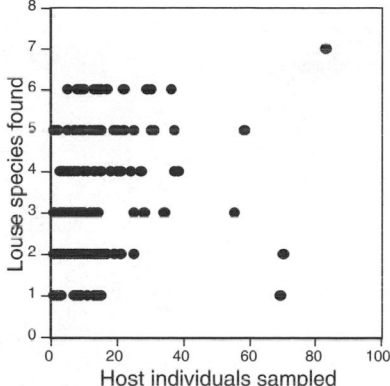

FIGURE 12.6. The relationship ($p = .001, r^2 = .078$) between host sample size and numbers of louse species found for 136 New Zealand bird taxa (appendix 12.2).

(appendix 12.2). If x-events are common in our data, we would expect to see a strong positive relationship between louse species number and host sample size—that is, the more hosts searched, the greater the chance of finding louse species that are present on a host taxon. A linear regression of the data showed that there was a significant relationship ($p = .001$), but that it explained very little of the variation observed ($r^2 = .078$: fig. 12.6). It is likely that much of this variation is generated by those host taxa from which only a few individuals had been sampled. This idea was assessed by cumulatively excluding hosts by sample size and repeating the regression, in other words excluding those hosts that had only been sampled once, then those that had only been sampled once or twice, and so on. This analysis (fig. 12.7) showed that hosts that had more than seven individuals sampled, and sometimes hosts that had more than five samples, no longer showed a significant relationship between sample effort and finding louse species. It appears, therefore, that there is a threshold below which sample size is important in determining presence of louse species. What does this say about x-events in our data? Most (75%) of our hosts have greater than four samples, and even for those that may have had less, there is only a small chance that a louse species may have been missed. It therefore seems unlikely that x-events are responsible for many of the observed sorting events.

In order to examine the generality of our finding, we analyzed a data set of louse presence on host species collected by one of us (R.L.P.) in the Galapagos Islands in 1992. Forty-seven of the island group's 58 bird species were examined and numbers of hosts sampled; positive hosts and louse species found were recorded (appendix 12.3). A linear regression of

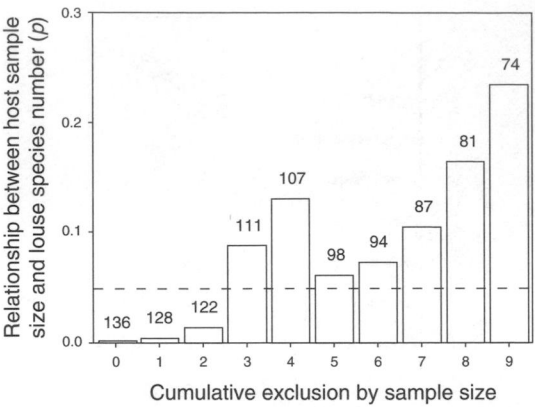

FIGURE 12.7. The effect of cumulatively excluding host taxa by sample size on the significance of the relationship between host sample size and louse species number for 136 New Zealand bird taxa. There is a significant positive relationship only for host species that have been sampled for lice from as few as 1, 2, or 3 individuals. There is no significant relationship after hosts from which more than three individual samples are included. The host species sample size remaining is shown over each bar.

positive host versus louse species number showed no significant relationship ($p = .915$, $r^2 < .001$: fig. 12.8). Sampling more hosts did not equate to finding more louse species. This result agrees with the New Zealand data by showing that x-events are unlikely to be common explanations for sorting events. It appears that, in general, our identified sorting events are likely to be real (e.g., MTB or DOA events). The common nature of these events is not a surprise given that the distribution of parasites is patchy (Rékási et al., 1997), and the size of host populations in speciation events is small (e.g., a small founder population).

Missing the Boat or Drowning on Arrival?

Given that the majority of sorting events identified in our data are real, is it possible to determine which event type, MTB or DOA, plays a greater role in this data set? This differentiation should be possible because the two types of sorting events predict very different effects of founding events on parasite diversity. Time since a founding event is the most important factor for DOA in reducing parasite diversity (e.g., by the extinction of louse species), as the likelihood of a species going extinct will increase with time. In contrast, the most important factor for MTB in reducing parasite diversity is the founding event itself. For MTB the likelihood of a species going extinct will not increase with time. An examination of the

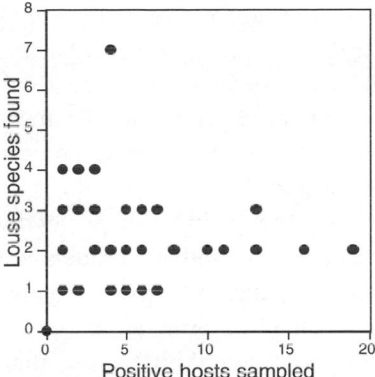

FIGURE 12.8. The nonsignificant relationship ($p = .915, r^2 < .001$) between host sample size and numbers of louse species found for 47 Galapagos Islands bird taxa (appendix 12.3).

human-introduced data should provide some answers. Human-introduced species into Australasia have been through the greatest bottleneck event (most populations were established from only a few to 100 individuals) but have had only about 100 years since founding. Given the above argument, a DOA scenario would predict that there will have been relatively few sorting events in the short time since the introduction of the human-introduced hosts, whereas an MTB scenario would predict relatively large numbers of sorting events because of the small size of the founder populations. In our data set human-introduced hosts lost 2.4 louse species per event, whereas closely related host species, which probably better reflect the more usual situation (and thus represent an appropriate comparison), lost about one louse species per founding event. This seems to support the idea that MTB events are more important than DOA events in determining louse species diversity in Australasian daughter host taxa.

Generalizing

Is it possible to generalize from the louse and bird species of Australasia and the Galapagos to louse and birds (or even parasites and hosts) in general? The geographical area encompassed by these regions is very large and, more important, contains examples of most habitat types. Also, the majority of bird species, and hence speciation events, are (or were, prior to human-caused extinction) island based (Steadman, 1997). Sorting events may be much more frequent in these types of areas compared with other continental areas due to high levels of isolation between islands. These conditions make the opportunities for continuing contact between founding and parent populations rare. In a continental area it may be more difficult to achieve these levels of isolation. Individuals may periodically continue to arrive in the founding population, sometimes carrying louse species that were lost after the initial founding. The cosmopolitan species that we see in the Australasian data set may illustrate this point. These species showed virtually no reduction in louse species diversity. Cosmopolitan bird species appear to have the ability to repeatedly colonize isolated island groups: individuals periodically arrive carrying louse species that have been lost from the original founding population. A future test of this idea would be to examine a continental bird-louse fauna in the same detail as we have done here. We would predict lower levels of sorting events, as bird populations are less likely to become isolated or to produce founding populations.

Is there a bias in our sorting event data due to an overrepresentation of species from particular orders? The most represented order is that of Passeriformes, with 15 species pairs, of which 13 show a loss of louse species; then Galliformes, with 11 species, of which 8 show a loss. Most of the other orders are represented by 4 to 6 species. We reanalyzed the data at the order level and found the same general result. Ten of the 13 orders contained a majority of taxa that lost species, which was significant (one-tailed binomial test, binomial probability $= 0.5$: $p = .046$). We also performed a concentrated changes test (Maddison, 1990) by mapping the pattern of loss of louse species onto a phylogeny of the parent taxa, obtained from Sibley and Ahlquist (1990). This test examines whether sorting events covary phylogenetically (i.e., certain host clades are more likely to lose lice by sorting) or whether the events are randomly distributed with respect to phylogeny such that the likelihood of sorting is not dependent on the host clade. The sorting events were found to be randomly distributed on the tree ($p = 0.999$), suggesting louse taxa were lost independent of host phylogeny.

A further impediment to generalizing further to other parasite taxa may be that lice are ideally suited to experiencing sorting events. The distribution of lice is naturally patchy. Fowler and Price (1987) found that the distribution of the louse *Philoceanus robertsi* over a population of Wilson's Storm Petrel *(Oceanites oceanicus)* best fit a negative binomial curve. This result agrees with our own and other observations about the patchiness of lice on birds. Given a patchy distribution of lice on their host species, then, a large founding group of hosts would be required before it was likely that all louse species found in the parent host population were present in the new population. The numbers of hosts needed to qualify as a "large" founding group will vary with such host traits as body mass (larger body masses carry more lice: Rózsa, 1997) and coloniality (territorial species show greater variation in louse distributions: Rózsa et al., 1996). It may be that other parasite species with different life histories would be less susceptible to sorting events. The ecology of individual hosts and parasites will largely determine the frequency of sorting events.

Sorting Out Sorting Events

Reconciliation approaches to measuring host-parasite coevolution typically infer many sorting events. We believe that, far from representing a problem with the reconciliation approach, this reflects a more realistic view of the relative probabilities of the processes involved in host-parasite coevolution. Sorting events have the potential to be highly informative about the historical ecological processes that have occurred in a particular host-parasite relationship. Most important, these quantitative studies will go beyond the mere chronicling of coevolutionary patterns to address the testing of hypotheses about the origin of these patterns. There are several directions in which future studies of sorting events might move. First, it might be argued that the number of louse species inferred to have been on ancestral bird hosts is excessive. For example, figure 12.4 shows that the ancestor of the Mottled Petrel had four louse lineages, of which two were lost to sorting events. For these particular louse genera, all extant hosts have just two species. However, host species such as the Mottled Petrel *(Pterodroma inexpectata)* currently play host for up to nine louse species, and 24% of 423 bird species in the Australasian area have four or more louse species. By looking at the distributions of louse species within orders, we will estimate whether inferred numbers of louse lineages on ancestors are realistic. Second, future studies will examine whether parasite population structure underlies the likelihood of sorting events: for example, is population distribution more uniform on hosts experiencing fewer sorting events? Third, further studies will test whether ecological and life-history parameters of both host and parasite reflect the likelihood of sorting events occurring: for example, are sorting events related to the number of niches present on the host? Fourth, and most definitely not finally, studies will address the hypothesis that hosts are islands: for example, can sorting events be explained by island biogeography theory? We look forward to sorting out these questions.

Acknowledgments

We thank Jonathan Banks, Fiona Jordan and Richard Duncan, Martyn Kennedy, and Kevin Johnson for useful comments that improved the manuscript. This work was funded by a University of Auckland Research Grant (R.D.G.) and a Lincoln University New Developments Grant (A.M.P.).

REFERENCES

Brooks, D. R. 1981. Hennig's parasitological method: A proposed solution. *Systematic Zoology* 30:229–49.

Charleston, M. A. 1998. Jungles: A new solution to the host/parasite phylogeny reconciliation problem. *Mathematical Biosciences* 149:191–223.

Clay, T. 1949. Some problems in the evolution of a group of ectoparasites. *Evolution* 3:279–99.

Fowler, J. A., and R. A. Price. 1987. A comparative study of the ischnoceran Mallophaga of Wilson's Petrel *Oceanites oceanicus* and British Storm Petrel *Hydrobates pelagicus. Seabird* 10:43–49.

Hafner, M. S., P. D. Sudman, F. X. Villablanca, T. A. Spradling, J. W. Demastes, and S. A. Nadler. 1994. Disparate rates of molecular evolution in cospeciating hosts and parasites. *Science* 265:1087–90.

Maddison, W. P. 1990. A method for testing the correlated evolution of two binary characters: Are gains or losses concentrated on certain branches of a phylogenetic tree? *Evolution* 44: 539–57.

Page, R. D. M. 1990a. Component analysis: A valiant failure? *Cladistics* 6:119–36.

———. 1990b. Temporal congruence and cladistic analysis of biogeography and cospeciation. *Systematic Zoology* 39:205–26.

———. 1993. Genes, organisms, and areas: The problem of multiple lineages. *Systematic Biology* 42:77–84.

———. 1994. Parallel phylogenies: Reconstructing the history of host-parasite assemblages. *Cladistics* 10:155–73.

———. 1996. Temporal congruence revisited: Comparison of mitochondrial DNA sequence divergence in cospeciating pocket gophers and their chewing lice. *Systematic Biology* 45:151–67.

Page, R. D. M., P. L. M. Lee, S. A. Becher, R. Griffiths, and D. H. Clayton. 1998. A different tempo of mitochondrial DNA evolution in birds and their parasitic lice. *Molecular Phylogenetics and Evolution* 9:276–93.

Palma, R. L., and S. C. Barker. 1996. Phthiraptera. In *Zoological Catalogue of Australia,* edited by A. Wells, vol. 26, pp. 81–247, 33–361 (app. 1–4), 73–96 (index). Melbourne: CSIRO Publishing.

Paterson, A. M., and R. D. Gray. 1997. Host-parasite cospeciation, host switching, and missing the boat. In *Host-parasite evolution: General principles and avian models,* edited by D. H. Clayton and J. Moore, 236–50. Oxford: Oxford University Press.

Paterson, A. M., R. D. Gray, and G. P. Wallis. 1993. Parasites, petrels and penguins: Does louse presence reflect seabird phylogeny? *International Journal for Parasitology* 23:515–26.

Paterson, A. M., R. L. Palma, and R. D. Gray. 1999. How frequently do avian lice miss the boat? *Systematic Biology* 48:214–23.

Paterson, A. M., and R. Poulin. 1999. Have chondracanthid copepods co-speciated with their teleost hosts? *Systematic Parasitology* 44:79–85.

Paterson, A. M., G. P. Wallis, and R. D. Gray. 2000. Seabird and louse coevolution: Complex histories revealed by sequence data and reconciliation analyses. *Systematic Biology* 49:383–99.

Rékási, J., L. Rózsa, and B. J. Kiss. 1997. Patterns in the distribution of avian lice (Phthiraptera: Amblycera, Ischnocera). *Journal of Avian Biology* 28:150–56.

Rózsa, L. 1997. Patterns in the abundance of avian lice (Phthiraptera: Amblycera: Ischnocera). *Journal of Avian Biology* 28:249–54.

Rózsa, L., J. Rékási, and J. Reiczigel. 1996. Relationship of host coloniality to the population ecology of avian lice (Insecta: Phthiraptera). *Journal of Animal Ecology* 65:242–48.

Sibley, C. G., and J. E. Ahlquist. 1990. *Phylogeny and classification of birds: A study in molecular evolution.* New Haven, Conn.: Yale University Press.

Steadman, D. W. 1997. Human-caused extinction of birds. In *Biodiversity II: Understanding and protecting our biological resources,* edited by M. L. Reaka-Kudla, D. E. Wilson, and E. O. Wilson, 139–61. Washington, D.C.: Joseph Henry Press.

Appendix 12.1

Louse records for 17 Australian bird parent-daughter taxa (after Palma and Barker, 1996)

Host taxa *Louse taxa*	Present on	
	Parent Taxon	Daughter Taxon
Human introduced		
Mute Swan (*Cygnus olor* Gmelin, 1789)		
Anatoecus icterodes oloris Zlotorzycka, 1970	•	•
Anatoecus dentatus magnicornutus Zlotorzycka, 1970		•
Anatoecus penicillatus Keler, 1960	•	
Ciconiphilus cygni Price & Beer, 1965	•	
Ornithobius bucephalus (Giebel, 1874)	•	•
Trinoton anserinum cygni Eichler, 1943	•	
Mallard Duck (*Anas platyrhynchos platyrhynchos* Linnaeus, 1758)		
Anaticola crassicornis (Scopoli, 1763)	•	•
Anatoecus dentatus (Scopoli, 1763)	•	•
Anatoecus icterodes (Nitzsch, 1818)	•	•
Holomenopon leucoxanthum (Burmeister, 1838)	•	•
Holomenopon maxbeieri Eichler, 1954	•	
Trinoton querquedulae (Linnaeus, 1758)	•	•

APPENDIX 12.1 *continued*

| Host taxa | Present on | |
Louse taxa	Parent Taxon	Daughter Taxon
Ring-Necked Pheasant (*Phasianus colchicus* Linnaeus, 1758)		
Amyrsidea perdicis (Denny, 1842)	•	
Goniocotes chrysocephalus Giebel, 1874	•	
Goniodes colchici Denny, 1842	•	•
Lagopoecus colchicus Emerson, 1949	•	
Lipeurus maculosus maculosus Clay, 1938	•	
Menacanthus phasiani (Modrzejewska & Zlotorzycka, 1977)	•	
Oxylipeurus mesopelios colchicus Clay, 1938	•	
Lipeurus caponis (Linnaeus, 1758)		•
Peafowl (*Pavo cristatus* Linnaeus, 1758)		
Amyrsidea minuta Emerson, 1961	•	•
Amyrsidea phaeostoma (Nitzsch [in Giebel], 1866)	•	
Colpocephalum tausi (Ansari, 1951)	•	
Goniocotes parviceps (Piaget, 1880)	•	
Goniocotes rectangulatus Nitzsch (in Giebel), 1866	•	
Goniocotes mayuri Lakshminarayana & Emerson, 1971		•
Goniodes meinertzhageni Clay, 1940	•	
Goniodes pavonis (Linnaeus, 1778)	•	•
Lipeurus pavo Clay, 1938	•	
Wild Turkey (*Meleagris gallopavo* Gray, 1843)		
Chelopistes meleagridis (Linnaeus, 1758)	•	•
Menacanthus stramineus (Nitzsch, 1818)	•	•
Oxylipeurus corpulentus Clay, 1938	•	
Oxylipeurus polytrapezius polytrapezius (Burmeister, 1838)	•	•
Rock Pigeon (*Columba livia* Gmelin, 1789)		
Bonomiella columbae Emerson, 1957	•	•
Campanulotes bidentatus compar (Burmeister, 1838)	•	•
Coloceras aegypticum (Kellogg & Paine, 1911)	•	
Coloceras damicorne (Nitzsch, 1866)		•
Colpocephalum turbinatum Denny, 1842	•	•
Columbicola columbae columbae (Linnaeus, 1758)	•	•
Hohorstiella lata (Piaget, 1880)	•	•
Song Thrush (*Turdus philomelos* Brehm, 1831)		
Brueelia merulensis (Denny, 1842)	•	
Brueelia turdinulae Ansari, 1956	•	
Menacanthus eurysternus (Burmeister, 1838)	•	
Myrsidea iliaci Eichler, 1951	•	
Philopterus turdi (Denny, 1842)	•	•

APPENDIX 12.1 *continued*

Host taxa / *Louse taxa*	Present on Parent Taxon	Daughter Taxon
Ricinus elongatus (Olfers, 1816)	•	
Sturnidoecus melodicus Eichler, 1951	•	
Common Blackbird (*Turdus merula* Linnaeus, 1758)		
Brueelia amsel (Eichler, 1951)	•	
Brueelia merulensis (Denny, 1842)	•	
Brueelia oudhensis Ansari, 1956	•	
Menacanthus eurysternus (Burmeister, 1838)	•	•
Myrsidea thoracica (Giebel, 1874)	•	
Philopterus turdi (Denny, 1842)	•	•
Ricinus elongatus (Olfers, 1816)	•	
European Starling (*Sturnus vulgaris* [Linnaeus, 1758])		
Brueelia nebulosa (Burmeister, 1838)	•	•
Menacanthus eurysternus (Burmeister, 1838)	•	•
Myrsidea cucullaris (Nitzsch, 1818)	•	
Sturnidoecus sturni (Schrank, 1776)	•	•
Closely related		
Little Penguin (*Eudyptula minor minor* [Forster, 1781])[1]		
Austrogoniodes waterstoni (Cummings, 1914)	•	•
Brown Quail (*Synoicus ypsilophorus ypsilophorus* Bosc, 1792)[2]		
Cuclotogaster synoicus (Clay, 1938)	•	•
Goniodes retractus Le Souef, 1902	•	•
Australian Bush-Turkey (*Alectura lathami lathami* Gray, 1831)[3]		
Colpocephalum alecturae Price and Beer, 1964	•	•
Colpocephalum lathami Price and Beer, 1964	•	•
Goniodes fissus (Rudow, 1869)	•	•
Oxylipeurus ischnocephalus (Taschenberg, 1882)	•	•
Scrub Fowl (*Megapodius reinwardt tumulus* Gould, 1842)[4]		
Goniodes biordinatus Clay, 1940	•	
Goniodes minor (Piaget, 1880)	•	•
Lipeurus sinuatus Taschenberg, 1882	•	•
Brown Goshawk (*Accipiter fasciatus fasciatus* (Vigors & Horsfield, 1827))[5,6]		
Degeeriella fulva (Giebel, 1874)	•	
Degeeriella fusca (Denny, 1842)	•	•
Purple Swamp Hen (*Porphyrio porphyrio melanotus* Temminck, 1820)[7]		
Pseudomenopon concretum (Piaget, 1880)	•	•
Rallicola lugens (Giebel, 1874)	•	•

APPENDIX 12.1 *continued*

Host taxa *Louse taxa*	Present on	
	Parent Taxon	Daughter Taxon
Common bronzewing (*Phaps chalcoptera* [Latham, 1790])[8]		
Campanulotes flavus flavus (Rudow, 1863)	•	
Coloceras grande Tendeiro, 1973	•	
Columbicola angustus (Rudow, 1869)	•	
Columbicola tasmaniensis Tendeiro, 1967		•
Physconelloides strangeri Tendeiro, 1980	•	
Physconelloides australiensis Tendeiro, 1969		•

Daughter taxon: [1]Fairy Penguin (*Eudyptula minor novaehollandiae* [Stephens, 1826]), [2]Brown Quail (*Synoicus ypsilophorus australis* [Latham, 1801]), [3]*Alectura lathami purpureicollis* Le Souf, 1898, [4]*Megapodius reinwardt yorki* Mathews, 1929, [5]*Accipiter fasciatus didimus* (Mathews, 1912), [6]*Accipiter fasciatus natilis* (Lister, 1889), [7]*Porphyrio porphyrio bellus* Gould, 1840, [8]Tasmanian Common Bronzewing (*Phaps chalcoptera* [Latham, 1790])

APPENDIX 12.2

New Zealand bird taxa sampled for louse species (see fig. 12.6)

Host Taxa	Host Individuals Sampled	Louse Species Found
Sphenisciformes		
Eudyptes chrysocome chrysocome	13	3
Eudyptes chrysocome filholi	5	3
Eudyptes pachyrhynchus	9	2
Eudyptes robustus	19	2
Eudyptes sclateri	7	3
Eudyptula minor	69	1
Megadyptes antipodes	6	2
Podicipediformes		
Podiceps cristatus	2	1
Podiceps rufopectus	1	1
Procellariiformes		
Daption capense	38	4
Diomedea bulleri	30	6
Diomedea cauta cauta	22	6
Diomedea cauta eremita	9	5
Diomedea cauta salvani	14	6
Diomedea chrysostoma	10	6
Diomedea epomophora epomophora	29	6
Diomedea epomophora sandfordi	13	6
Diomedea exulans	31	5

APPENDIX 12.2 *continued*

Host Taxa	Host Individuals Sampled	Louse Species Found
Diomedea melanophris	9	6
Fregretta tropica	5	4
Fulmaris glacialoides	15	4
Garrodia nereis	12	2
Halobaena caerulea	14	5
Lugensa brevirostris	19	5
Macronectes giganteus	30	6
Macronectes halli	10	5
Oceanites oceanicus	1	1
Pachyptila belcheri	20	5
Pachyptila crassirostris	15	4
Pachyptila desolata	13	5
Pachyptila salvini	22	5
Pachyptila turtur	58	5
Pachyptila vittata	37	5
Pagodroma nivea	3	2
Pelagodroma marina	24	4
Pelecanoides urinatrix	55	3
Phoebetria palpebrata	7	5
Procellaria aequinoctialis	10	4
Procellaria cinerea	5	4
Procellaria parkinsoni	13	4
Procellaria westlandica	27	4
Pterodroma auxillaris	3	2
Pterodroma cookii	15	6
Pterodroma externa	10	3
Pterodroma inexpectata	36	6
Pterodroma leucoptera	4	4
Pterodroma longirostris	5	5
Pterodroma macroptera	22	6
Pterodroma magentae	17	6
Pterodroma mollis	12	5
Pterodroma neglecta	6	4
Pterodroma nigripennis	18	4
Pterodroma pycrofti	6	4
Puffinus assimilis elegans	6	3
Puffinus assimilis haurakiensis	3	4
Puffinus assimilis kermadecensis	11	4
Puffinus bulleri	10	4
Puffinus carneipes	5	6
Puffinus gavia	10	3
Puffinus griseus	37	4
Puffinus huttoni	25	5

Host Taxa	Host Individuals Sampled	Louse Species Found
Puffinus pacificus	8	6
Puffinus tenuirostris	8	4
Thalassoica antarctica	7	4
Pelecaniformes		
Leucocarbo campbelli campbelli	3	1
Leucocarbo campbelli ranfurlyi	2	1
Leucocarbo carunculatus chalconotus	5	2
Leucocarbo carunculatus onslowi	3	1
Pelecanus conspicillatus	1	2
Phalacrocorax melanoleucos brevirostris	13	2
Phalacrocorax sulcirostris	4	2
Phalacrocorax carbo novaehollandiae	10	2
Phalacrocorax varius varius	8	2
Stictocarbo punctatus	25	2
Sula bassana serrator	70	2
Sula dactylatra personata	4	2
Ciconiiformes		
Ardea novaehollandiae	11	2
Botaurus stellari	15	2
Egretta alba	11	2
Anseriformes		
Anas platyrhynchus	30	5
Anas rhynchotis	3	4
Anas superciliosa	21	4
Branta canadensis	20	5
Cygnus atratus	83	7
Cygnus olor	3	2
Tadorna variegata	12	5
Falconiformes		
Circus approximans	34	3
Falco cenchroides	1	3
Falco novaeseelandiae	7	2
Galliformes		
Alectoris chukar	8	3
Meleagris gallopavo	4	3
Pavo cristatus	5	3
Perdix perdix	1	2
Phasianus colchicus	5	6
Synoicus ypsilophorus	2	2
Gruiformes		
Gallirallus australis australis	16	2
Gallirallus australis scotti	11	1

APPENDIX 12.2 *continued*

Host Taxa	Host Individuals Sampled	Louse Species Found
Porphyrio mantelli	13	1
Porphyrio porphyrio	15	2
Charadriiformes		
Anarhynchus frontalis	8	1
Calidris canutus	5	5
Charadrius bicinctus bicinctus	2	2
Charadrius bicinctus exilis	2	2
Charadrius obscurus	28	3
Coenocorypha aucklandica	9	2
Haematopus chathamensis	14	2
Haematopus ostralegus	20	4
Haematopus unicolor	5	4
Himantopus himantopus leucocephalus	7	4
Himantopus novaezealandiae	3	2
Limosa lapponica	13	5
Limosa limosa	1	5
Pluvialis fulva	3	3
Vanellus miles novaehollandiae	9	3
Columbiformes		
Columba livia	15	5
Hemiphaga novaeseelandiae	21	2
Psittaciformes		
Cyanoramphus auriceps	1	1
Cyanoramphus malherbi	1	1
Cyanoramphus novaeseelandiae	15	1
Cyanoramphus unicolor	3	2
Nestor meridionalis	9	3
Nestor notabilis	18	4
Strigops habroptilus	3	1
Strigiformes		
Athene noctua	7	1
Ninox novaeseelandiae	17	2
Coraciiformes		
Halycon sancta vagans	14	1
Passeriformes		
Acridotheres tristis	25	2
Anthornis melanura melanura	7	3
Corvus frugilegus	8	1
Gymnorhina tibicen	25	3
Passer domesticus	15	2
Prunella modularis	9	1

APPENDIX 12.2 *continued*

Host Taxa	Host Individuals Sampled	Louse Species Found
Sturnus vulgaris	12	3
Turdus merula	2	5
Turdus philomelos	14	3
Zosterops lateralis	14	1

APPENDIX 12.3

Galapagos Islands bird taxa sampled for louse species (see fig. 12.8)

Host Taxa	Total Hosts Sampled	Hosts with Lice	Louse Species Found
Sphenisciformes			
Sphenicus mendiculus	5	5	1
Procellariiformes			
Diomedea irrorata	4	4	7
Oceanodroma castro	6	6	3
Oceanites gracilis galapagoensis	4	4	1
Pterodroma phaeopygia	2	2	4
Puffinus subalaris	5	5	3
Pelecaniformes			
Fregata magnificans	3	3	3
Fregata minor	3	3	3
Nannopterum harrisi	7	7	1
Pelecanus occidentalis	4	4	2
Phaethon aethereus	3	3	2
Sula dactylatra	5	5	2
Sula nebouxii	10	10	2
Sula sula	5	5	1
Charadriiformes			
Anous stolidus	3	3	4
Arenaria interpres	1	1	2
Creagus furcatus	7	7	3
Haematopus palliatus	3	3	4
Larus fuliginosus	9	5	1
Larus pipixcan	2	2	3
Tringa incana	1	1	3
Ciconiiformes			
Butorides sundevalli	4	2	1
Nyctanassa violacea	5	4	1
Phoenicopterus ruber	3	3	2

APPENDIX 12.3 *continued*

Host Taxa	Total Hosts Sampled	Hosts with Lice	Louse Species Found
Anseriformes			
Anas bahamensis	2	1	4
Falconiformes			
Buteo galapagoensis	3	3	3
Strigiformes			
Asio flammeus	1	1	2
Gruiformes			
Laterallus spilonotus	4	2	3
Columbiformes			
Zenaida galapagoensis	19	19	2
Passeriformes			
Camarhynchus pallidus	1	1	1
Camarhynchus pauper	6	1	1
Camarhynchus parvulus	1	0	0
Camarhynchus psittacula	12	5	2
Certhidea olivacea	6	0	0
Dendroica petechia	8	6	2
Geospiza conirostris	42	8	2
Geospiza difficilis	47	6	2
Geospiza fortis	46	6	2
Geospiza fuliginosus	86	13	3
Geospiza magnirostris	7	3	2
Geospiza scandens	6	0	0
Myarchus magnirostris	65	16	2
Nesiomimus parvulus	12	11	2
Nesomimus trifasciatus	6	6	1
Nesomimus macdonaldi	18	13	2
Platyspiza crassirostris	5	1	1
Pyrocephalus rubinus	6	2	1

13

THE ECOLOGICAL BASIS
OF COEVOLUTIONARY HISTORY

Dale H. Clayton, Sarah Al-Tamimi,
and Kevin P. Johnson

Macroevolutionary patterns are difficult to interpret because they are the product of a time scale so vast that deterministic and chance events are hard to distinguish. Although the macroevolutionary history of a group can be reconstructed from extant species, determining the ecological context in which that group evolved is a tall order. Ecology involves interactions between organisms and both the living and nonliving components of their environments. These interactions are important because they influence selection, dispersal, drift, and other microevolutionary processes that govern macroevolution. Short of inventing time travel, the best bet for obtaining data on ecological history has traditionally been to focus on groups that have an unusually good fossil record. A more recent approach, however, is to focus on groups having a history of prolonged coevolution that yields congruent phylogenies.

Congruent phylogenies are produced by repeated bouts of parallel speciation in unrelated lineages. If every speciation event in one group is accompanied by a "cospeciation" event in the other group, and if no species are lost from their original associations, then phylogenies will be completely congruent (although branch lengths may differ). In reality, however, phylogenies seldom show absolute congruence. Generally speaking, the degree of congruence is correlated with the ecological intimacy of the groups, whose interactions vary from obligate association to opportunistic encounters. One end of the spectrum is represented by mitochondria, chloroplasts, and other eukaryotic organelles evolved from free-living prokaryotic ancestors. The other end of the spectrum consists of far less intimate interactions, such as those between generalist herbivores and their host plants. Most interactions lie between these two extremes.

Inferences about the ecological history of interacting groups are perhaps easiest when the environment of one species is delineated completely by the members of another species, as in the case of "permanent" parasites. For such parasites, which carry out their entire life cycle on the body of the host, the branching pattern of the host phylogeny provides a detailed record of vicariance events that may influence the parasites. The host phylogeny also can be used to draw inferences about the habitat parameters of ancestral parasites. For example, estimating the body size of an ancestral host is tantamount to knowing the size of the resource base available to its parasites. Unfortunately, it is not usually possible, even in cases of extensive congruence, to reconstruct specific ecological processes, such as demographic fluctuations, or competitive interactions. On the other hand, processes generating congruence can be illuminated to some extent by extrapolating backwards from data on modern ecological parameters. In this chapter we adopt such a reverse engineering approach, using information about the ecology of extant species to explain differences in the degree of phylogenetic congruence among related host-parasite systems. Although we have chosen to focus on host-parasite interactions, we make an effort to address issues that are applicable to coevolving systems in general.

The ecological basis of coevolutionary history can be explored by comparing interactions that vary in their degree of phylogenetic congruence. Unfortunately, adequate comparative ecological data are not available for many of the systems that have been subjected to cophylogenetic analysis. A fortunate exception is provided by parasitic lice (Insecta: Phthiraptera), which occur on birds and mammals. For a few genera of lice, enough data are now available to begin exploring the relationship between ecology and congruence. Making comparisons among taxa of lice helps ensure against spurious conclusions drawn from comparisons of distantly related taxa that may have evolved in entirely different environmental contexts.

Phylogenetic congruence is governed by several kinds of macroevolutionary events, which we review below. We then consider the impact of various ecological factors on the relative frequency of these macroevolutionary events. We conclude the chapter by comparing the ecology of four genera of lice that have histories ranging from extensive phylogenetic congruence with their hosts, to a complete lack of congruence.

Macroevolutionary Events governing Phylogenetic Congruence

Phylogenetic congruence is a historical pattern produced by repeated bouts of *cospeciation*. Cospeciation is a process in which speciation in one lineage is accompanied by speciation in an associated, but unrelated lineage (fig. 13.1a). All else being equal, phylogenies containing a high proportion

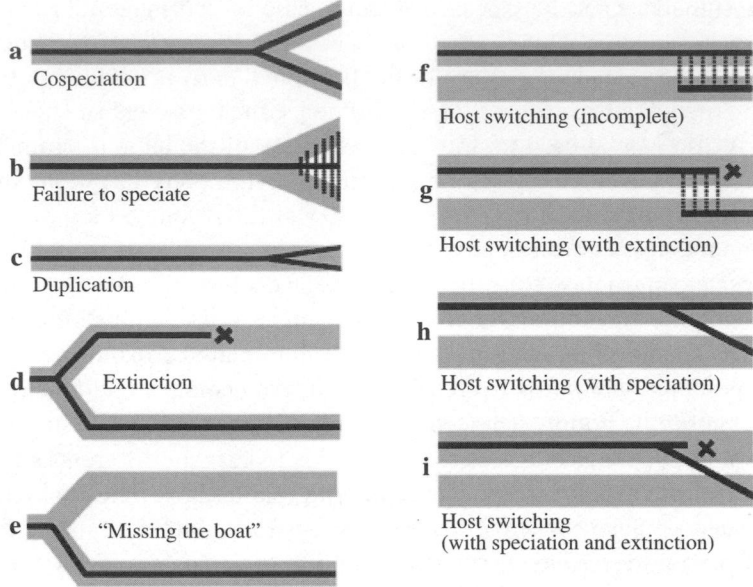

FIGURE 13.1. Macroevolutionary events that influence phylogenetic congruence. Gray lines represent hosts; black lines represent parasites. Black stippling represents gene flow between parasite populations. See text for discussion.

of cospeciated nodes will show more congruence than those containing a low proportion. However, the amount of cospeciation cannot be inferred directly from the amount of congruence between two phylogenies because it is possible for "false" congruence to be generated by processes other than cospeciation, such as extinction. Moreover, incongruence between phylogenies does not necessarily mean that there has been no cospeciation. As outlined below, other macroevolutionary events can reduce congruence between lineages that have undergone a considerable amount of cospeciation.

Aside from cospeciation (fig. 13.1a), all the events depicted in figure 13.1b–i effectively reduce congruence. For example, if a reproductive barrier affects a host lineage, but not its parasite, the parasite will *fail to speciate,* thereby reducing congruence (fig. 13.1b). Conversely, reproductive barriers that affect only the parasite lineage promote parasite *duplication* (fig. 13.1c), which also reduces congruence. Congruence is further reduced when parasites are lost from a host lineage, either through parasite *extinction* (fig. 13.1d), or by *"missing the boat,"* which occurs when parasites fail to disperse onto one of two new host clades (fig. 13.1e).

The final macroevolutionary event governing congruence is *host switching*, in which parasites colonize a "foreign" host species on which they did not previously occur. Host switching involves an initial expansion of the parasite's host range (fig. 13.1f). This expansion is not a host switch, in and of itself, because the parasite persists on the original host. The host switch is completed by extinction of the parasite on the original host (fig. 13.1g), speciation of the parasite on the new host (fig. 13.1h), or by concurrent extinction and speciation (fig. 13.1i). It is not possible, using phylogenies reconstructed from extant species, to distinguish switching with extinction (fig. 13.1g) from switching with speciation and extinction (fig. 13.1i). However, information on the ecology of the descendant species can suggest the relative likelihood of the two types of host switches. A parasite with poor dispersal ability is more likely to speciate after colonizing a new host, since there will be little or no parasite gene flow between the original and new host species. In such cases, the host switch is completed by the speciation event, which may later be followed by extinction (fig. 13.1i). In contrast, when parasite dispersal is common, incomplete host switching may be a frequent event (fig. 13.1f).

Relationship of Ecological Factors to Macroevolutionary Events

Ecological factors can have a fundamental impact on the probability of host switching and other macroevolutionary events governing the congruence of interacting clades (fig. 13.1). Ecological factors, which, by definition, affect the *distribution* and *abundance* of organisms (Begon et al., 1990), influence congruence through their impact on the host, the parasite, or both. For example, any factor that causes a parasite to be patchily distributed over the range of its host may increase the probability of parasite duplication (fig. 13.1c). Ecological factors that affect the abundance of the host and/or its parasite can also have an important influence. For example, stochastic extinction is far more likely in the case of a parasite that is typically found only in small numbers on host individuals.

Generally speaking, the distributions of host taxa have been reasonably well documented (e.g., Sibley and Monroe, 1990; Nelson, 1994; Nowak, 1999). In contrast, the distributions of most parasite taxa remain poorly known (Brooks and McLennan, 1993; Clayton and Moore, 1997). Parasites with indirect life cycles are especially problematic in this regard, since they involve one or more intermediate host species, in addition to free living stages and the final host. Distributions of parasites with direct life cycles are much simpler to characterize, being tied largely to the distribution of a single host species. Permanent parasites, which complete their entire life cycle on the body of the host, have distributions that are particularly easy to characterize.

Given their close association with the host, the abundance of permanent parasites also can be measured accurately. This is particularly true for permanent *ecto*parasites, such as lice, which can even be observed and counted on live hosts (Clayton and Drown, 2001). These advantages make it possible to track the distribution and abundance of such parasites over the course of longitudinal studies. It is also possible to add or remove the parasites in controlled experiments designed to test the relative importance of the ecological factors thought to influence phylogenetic congruence (see below).

It is difficult to overemphasize the need for rigorous data on parasite abundance. Sampling errors can create the false impression of parasite-free host individuals, populations, or species, leading to erroneous conclusions about extinction or missing the boat (Paterson and Gray, 1997). Parasite ecologists normally measure two main components of parasite abundance: *prevalence* and *intensity* (Bush et al., 1997). Prevalence is the percent of individuals in a host population that actually have parasites. Intensity is the number of parasites on a parasitized individual; mean intensity is the average number of parasites across all parasitized individuals. Accurate measures of parasite prevalence and intensity require sampling methods that have demonstrated efficacy. Such methods have been tested thoroughly for lice (Clayton and Drown, 2001).

Ecological factors relevant to phylogenetic congruence vary in both time and space. Data from long-term studies are often required to document how ecological factors vary over time. In contrast, spatial variation can be relatively easy to document. However, is important to recognize that spatial variation exists on a variety of scales, ranging from variation among the microhabitats on a single host individual, to variation among host individuals, populations, and species (fig. 13.2). In the next section we consider the relevance of each of these four scales to the macroevolutionary events that govern phylogenetic congruence.

Variation among Microhabitats within a Host Individual

Different species of parasites appear to partition microhabitats on individual hosts (Poulin, 1998). For example, species of helminth worms tend to be concentrated in different regions of the host intestinal tract (Stock and Holmes, 1988). This pattern is consistent with competitive displacement owing to interspecific competition for limited resources, such as food. If competitive displacement is pervasive, it could influence the composition of parasite communities, leading to an absence of congeneric species on

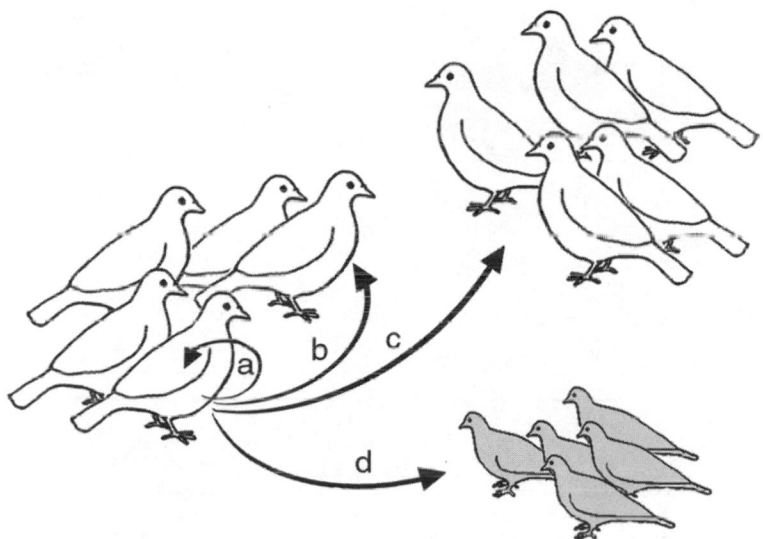

FIGURE 13.2. Ecological factors that influence host-parasite congruence vary among *(a)* microhabitats within a host individual (e.g., body versus wing); *(b)* host individuals; *(c)* host populations; and *(d)* host species.

a single species of host (e.g., Barker and Close, 1990; see also chap. 8). Competitive displacement may also operate between higher taxa. For example, since dispersing to dogs from its original host (a wallaby), the louse *Heterodoxus spiniger* appears to have displaced the louse *Trichodectes canis* from much of its former range (Barker, 1994).

Although intriguing, such patterns do not constitute robust evidence for competition (Simberloff, 1990; Page et al., 1996). Rigorous tests of competition require an experimental approach in which the population response of one or more species to the removal of a potential competitor(s) is carefully monitored (Poulin, 1998). Figure 13.3 summarizes the results of such a test that we recently conducted with wing lice *(Columbicola columbae)* and body lice *(Campanulotes bidentatus compar)* on Feral Pigeons *(Columba livia)*. Wing lice spend most of their time on the host's wing and tail feathers, whereas body lice reside primarily on the abdominal feathers (see chap. 11). Despite these microhabitat differences, both species depend on abdominal contour feathers for food (Nelson and Murray, 1971). Figure 13.3 shows that body lice have a negative impact on the population growth of wing lice. Although the reason for this negative impact is unknown, it may have to do with better foraging ability on the part of body lice, assuming abdominal feathers are a limiting resource.

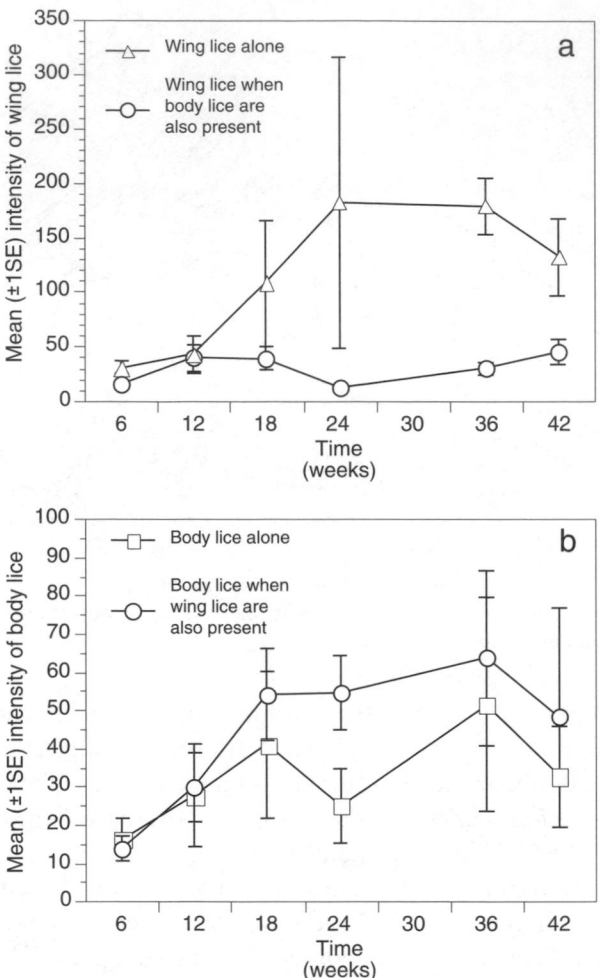

FIGURE 13.3. Competitive interaction between wing lice and body lice on Feral Pigeons. Population growth of wing lice *(a)* was slower on birds also parasitized by body lice, than on birds only parasitized by wing lice (repeated-measures ANOVA, $p = .02$). In contrast, population growth of body lice *(b)* was similar with and without wing lice present ($p = .38$). At the start of the experiment birds were "seeded" with identical numbers of lice, as follows: 16 pigeons were captured in Salt Lake City, Utah, then housed at low relative humidity (<30%) for 10 weeks, to exterminate their natural populations of lice and eggs (Moyer et al., 2002). The birds were then isolated in 16 cages and assigned randomly to three "seeding" treatments: 100 wing lice, 100 body lice, or 50 wing and 50 body lice. The data for mixed species of lice (circles in panels *a* and *b*) are thus from the same individual birds ($n = 6$). The preening behavior of all birds was impaired with harmless plastic bits to facilitate increases in their lice (Clayton et al., 1999). Louse populations were monitored for approximately 10 months

If interspecific competition influences the structure of parasite communities in general, then it is reasonable to predict that it could influence host switching in particular (Barker, 1994). For example, successful host switches should be more likely in the case of parasites dispersing to hosts that have depauperate parasite communities, because such hosts will represent greater ecological opportunity, that is, more untapped resources. This prediction could be tested by comparing the survival and reproductive success of parasites experimentally transferred to foreign hosts with and without other species of parasites already present. If interspecific competition reduces the fitness of parasites transferred to foreign hosts, then it clearly has the potential to influence host switching.

The *direction* of interspecific competition could also influence host switching. Asymmetric competitive effects have been documented in a number of parasite communities (Poulin, 1998). Figure 13.3 provides an example of asymmetric competition in which wing lice do worse in the presence of body lice, while body lice are unaffected by wing lice. These results predict that it would be easier for body lice to switch to foreign hosts already parasitized by wing lice, than for wing lice to switch to hosts with body lice. Body lice are significantly more host specific than wing lice (Johnson et al., 2002), further suggesting that host specific parasites may have a competitive edge over generalists. Data on the underlying causes of asymmetric competition are needed to begin exploring this intriguing possibility.

Microhabitat diversity and resource availability may also influence parasite duplication. A possible example involves the human body louse *(Pediculus humanus)*, which is the sister species of the human head louse *(Pediculus capitis)* (Amevigbe et al., 2000). The two species are so similar that they are often classified as subspecies rather than full species (e.g., Durden and Musser, 1994). [Crab lice, which are confined to pubic hair, are members of another genus *(Pthirus pubis)*]. Body lice attach their eggs to clothing, while head lice attach their eggs to hair. Body lice probably diverged from head lice concurrent with the loss of hair and adoption of clothing by hominids (Busvine, 1978). Clothing may therefore have provided a novel microhabitat that facilitated a duplication event, in which the

---

FIGURE 13.3 *(continued)* by periodically counting all lice on the underside of one wing, the tail, and lice observed during timed intervals on three additional body regions: back and keel (30 seconds each), and rump (60 seconds). The sum of lice on these five regions significantly predicts the total number of *C. columbae* ($R^2 = .66$) and *C. b. compar* ($R^2 = .79$) on feral pigeons ($p < .0001$ for both species; Clayton and Drown, 2001).

common ancestor of head and body lice diverged through speciation on a single species of host. Alternatively, head and body lice may have speciated allopatrically and subsequently come into secondary contact (Busvine, 1978). Under either scenario of parasite speciation, the probability of duplication may increase with the diversity of microhabitats available on a given host. This hypothesis could be tested by comparing the number of potential microhabitats among different hosts to the number of congeneric parasites they support.

Microhabitat specialization could conceivably also increase the probability of parasite extinction, assuming the abundance of microhabitat specialists is lower than that of microhabitat generalists. For example, chimpanzees *(Pan troglodytes)* have a single species of louse *(Pediculus schaeffi)* (Durden and Musser, 1994) that presumably roams all of the host's body, given its continuous distribution of hair. If so, then one can predict that the mean abundance of this louse will be greater than that of human lice, all three species of which are more restricted in their microhabitat distributions (see above). The chimp louse may therefore be less prone to extinction if it really is more abundant. Of course, this ignores differences in host body size, population size, and other factors that are already known to influence parasite abundance (Poulin, 1998). The impact of microhabitat specialization on risk of extinction could be explored with comparative analyses that control for these and other factors.

Variation among Individuals within a Host Population

Parasite abundance varies within host populations, ranging from heavily parasitized individuals to individuals having no parasites. Low parasite prevalence increases the probability that a parasite will miss the boat, since a dispersing founder population of hosts is more likely to be comprised solely of individuals having no parasites (Paterson et al., 1999). Variation in parasite intensity—the number of parasite individuals on a host (see above)—may also influence the probability that parasites will miss the boat or go extinct (Rózsa, 1993). At the very least, parasites that are rare are more likely to miss the boat or go extinct than those that are abundant. In contrast, parasites that are abundant might increase the probability of host switching, assuming that dispersal is density dependent, that is, attempts at host switching will presumably be higher when parasites are dispersing at a higher rate.

Parasite abundance is usually quite variable even in populations that have a high mean parasite intensity. This is particularly the case for "macroparasites," such as helminths and arthropods (Anderson and May,

1979). Macroparasites tend to concentrate on a minority of heavily infested individuals, forming an aggregated distribution with the property that the variance in parasite intensity exceeds the mean. Proximal reasons for aggregation are unclear, although variation in host susceptibility and the contagious nature of parasite transmission are probably important factors (Hudson and Dobson, 1995).

Aggregation should increase the probability of parasites missing the boat. A founder population dispersing from a parent population containing aggregated parasites is more likely to be parasite free than one dispersing from a parent population containing an even distribution of parasites. Aggregation could also increase the probability of parasite extinction if heavily parasitized individuals die before their parasites are transmitted to new hosts. This is probably not an overly likely scenario, since parasite virulence is generally correlated with the ease of transmission (Ewald, 1994). Parasites that are transmitted vertically from parent to offspring, such as feather lice, are usually fairly benign, since their reproductive fitness is linked closely with that of the host (Clayton and Tompkins, 1994). More virulent parasites tend to be transmitted horizontally, which prevents them from being marooned on a dead or dying host (Herre, 1993).

Variation among Host Populations

Parasites can be rare or absent from some host populations, while abundant on others. For example, feather lice have a patchy distribution in which they are abundant on doves in many areas of the world, yet virtually absent from the same species of doves in arid regions (Moyer et al., 2002). Patchiness can also be generated when host population size falls below the minimum threshold required to support parasites (Rózsa, 1993). Thus, an area of low host density can be a dispersal barrier for parasites, or a sink that limits gene flow between parasite populations in different parts of the host's geographic range.

Patchiness has the potential to influence macroevolutionary events. Nested within their host limits, parasites often show geographic limits to their distribution (see Clay, 1964, 1972). These geographic limits increase the likelihood of parasites missing the boat, since parasites may be absent from hosts involved in founder events (Brown and Wilson, 1975). For example, 18 bird species were historically introduced to New Zealand by humans. The species richness of lice on these birds in New Zealand is significantly lower than that on the same birds in their native environments (Paterson et al., 1999). The founding populations of most of these introduced birds were small—a few individuals to as many as 100 individuals. Although stochastic loss of ectoparasite species has undoubtedly contributed to the lower species richness of lice on these birds, it is conceivable that some of the founder hosts came from native populations that were relatively free from parasites in the first place. Gaps in parasite distribution may also facilitate parasite duplication if little or no parasite gene flow occurs across the gaps (Clay, 1949). In summary, variation in the distribution of parasites among host populations is another potential factor with direct relevance to the macroevolutionary events that govern phylogenetic congruence.

Variation among Host Species

The distribution of parasites among host species is, of course, directly relevant to phylogenetic congruence. Host specificity is essentially an index to a parasite's distribution among host species. Parasites range from highly specific, being restricted to a single species or subspecies of host, to generalists found on a variety of host taxa. Although host specificity is a necessary condition for phylogenetic congruence, it is by no means a sufficient condition. Just because a parasite is currently host specific does not mean that its ancestors were host specific, much less that they underwent cospeciation with their hosts (Hoberg, 1992; Hoberg et al., 1997). Specificity describes a pattern of current association that may or may not reflect macroevolutionary history or even current adaptation to the host. Some parasites may be specific simply because they are incapable of dispersing among host taxa (Tompkins and Clayton, 1999). Other parasites may indeed be adapted to a particular host. In such cases, specificity can be viewed as a phenotypic trait of the parasite that, assuming a heritable component, has the potential to evolve just like any other trait (Secord and Kareiva, 1996).

Host specificity is integral to three macroevolutionary events in addition to cospeciation: failure to speciate, extinction, and host switching. All three of these events can reduce host-parasite congruence. Failure to speciate occurs when gene flow is maintained between populations of parasites on hosts that have already undergone speciation (Johnson et al., in review.b.).

Host specificity is also relevant to extinction, particularly in the case of the coextinction of parasites on hosts that have themselves gone extinct (Brooks and Hoberg, 2000). For example, the louse *Columbicola extinctus* was considered extinct because it was assumed to be a host specific parasite of the extinct Passenger Pigeon *(Ectopistes migratorius)* (Stork and Lyal, 1993). However, Clayton and Price (1999) recently showed that this species is synonymous with extant wing lice on the Band-Tailed Pigeon *(Columbicola fasciata)*, thus resurrecting the species from extinction, so to speak. True coextinction of host specific parasites with their hosts can alter phylogenetic congruence, depending on how deletion of one (or both) of the "missing" clades alters reconstruction of the rest of the phylogenetic tree(s).

Host specificity is, of course, directly relevant to host switching, in which a parasite colonizes a foreign host on which it did not previously occur. Successful colonization of a foreign host requires that the parasite *disperse* to that host, and then *establish* a viable breeding population on it. Inability to disperse, establish, or both will prevent a host switch from taking place. In cases of incomplete host switching (fig. 13.1f), the parasite merely incorporates a new host species, thereby reducing its host specificity. If the parasite switches to a new host while going extinct on the original host (13.1g), host specificity effectively remains the same. Likewise, if the parasite speciates after switching (13.1h), the specificity of each sister species is the same as that of the common ancestor, assuming each daughter species is unable to colonize additional hosts. Finally, if the parasite speciates after switching, then goes extinct on the original host (13.1i), specificity again remains the same as that on the ancestral species.

The distribution of parasites among host species is often documented in a qualitative sense, that is, whether or not a parasite occurs on a given species of host. Qualitative thinking leads to the erroneous assumption that parasites found on foreign hosts are "stragglers" that are of little evolutionary significance. As Rózsa (1993) has argued, however, stragglers may represent the initial (dispersal) stage in host switching. Although most stragglers are presumably doomed, it may take but a single breeding pair— or single inseminated (or asexual) female—to establish a viable population of parasites on a foreign host. Thus straggling may be of considerable significance, particularly given the expanse of evolutionary time over which repeated dispersal events can eventually yield a successful host switch.

Understanding straggling and host switching requires a more quantitative concept of host specificity that is based on large scale sampling of parasites from many host individuals. Tompkins and Clayton (1999) recently compared the host specificity of six species of *Dennyus* lice on four species of Bornean Cave Swiftlets (Apodiformes: Collocalliini). For each species of host they carefully removed all the lice from hundreds of host individuals. The study documented at least three instances of lice that would normally be labeled as stragglers, since they were present on foreign hosts at very low frequencies (< 3%). The study also revealed species of lice present on foreign hosts at frequencies of 5.3%, 8.6%, and 8.9%. Should these also be deemed stragglers? What really matters is the frequency with which host specific parasites disperse to foreign hosts, and the establishment ability of the parasites following dispersal. Dispersal and establishment are influenced by a number of variables, which we consider below.

Variables influencing Dispersal

Dispersal is constrained by the morphology, physiology, ecology, and behavior of the parasite. Parasites that have limited powers of dispersal, such as lice, can disperse only between host species that are both sympatric and syntopic (sharing habitat). However, this does not necessarily imply that individuals of different host species must be in direct physical contact. Clay (1949) and Timm (1983) postulated four ways in which bird lice can move between species of hosts: (1) by dispersal on detached feathers, (2) via shared dust baths, (3) via shared nest holes, and (4) by phoresis on hippoboscid flies.

Lice dislodged during dust bathing could conceivably move onto the next species of bird to use the same dusting arena (Clay, 1949). For example, Hoyle (1938) provided anecdotal evidence suggesting that lice dislodged from dust bathing chickens could end up on house sparrows that subsequently dust bathe in the same spots. However, the hypothesis that dusting facilitates dispersal of lice has not been rigorously tested. Likewise, the hypothesis that lice disperse on feathers has not been tested, although anecdotes of lice on molted waterfowl feathers do exist (Eichler, 1963). Clayton (1990) provided evidence concerning owl lice that is pertinent to the shared nest hole hypothesis. Species of *Strigiphilus* owl lice found on more than one species of host invariably occur on species with overlapping ranges, habitats, and nest habits (Clayton, 1990). Nest holes are a limiting resource that, if used in rapid succession by different species of birds, may well provide an ecological opportunity for lice to disperse between species.

The final means of dispersal concerns the ability of lice to hitch rides on other, more mobile species such as hippoboscid flies. This process, known as phoresis, is suprisingly common. Several hundred records of Ischnoceran lice riding on hippoboscid flies have been published (Keirans, 1975). The phenomenon can also be common at a local level. For example, Corbet (1956) documented lice attached to 43.5% of the hippoboscid flies removed from a large sample of freshly netted European Starlings *(Sturnus vulgaris)*. Since hippoboscids are not as host specific as many lice, they may provide a means of dispersal between host species. Like straggling, phoresis has often been considered a red herring of little evolutionary significance. We suspect, however, that it actually plays an important role in the ecology and evolution of lice.

Variables influencing Establishment

As merely the first step in successful colonization of a new host species, dispersal is of little consequence if it is not followed by successful establishment of a breeding population on the new host. Like dispersal, establishment may be influenced by many ecological variables, such as the ability of the parasite to remain attached to the host (tenacity), the nutritive value of the host, severity of host defense, and intensity of competition from other parasites already living on the host.

One powerful approach for assessing establishment ability is to transfer parasites to a "foreign" host species, then compare the survival and reproductive success of those parasites to that of parasites transferred to new individuals of the usual host. Transfer experiments have recently been published for gopher lice (Reed and Hafner, 1997) and swiftlet lice (Tompkins and Clayton, 1999). The results of these experiments show that lice are able to survive and breed on reasonably closely related foreign hosts, both in captivity and in the field, at least over the course of short term experiments. Tompkins and Clayton (1999) showed that relative fitness on a foreign host is highly correlated with feather barb size. Transfers of lice to swiftlet hosts that had feathers >2 microns different from the original host resulted in greatly reduced survival. When transferred to hosts that differed only slightly in size, swiftlet lice shifted their microhabitat to prefer feathers closer in size to those of their original host. These results suggest that lice on feathers too dissimilar in size may simply have trouble hanging on to the host.

Another factor that could conceivably block host switching is competition from resident lice. No test of this hypothesis has been carried out, although we do know that pigeon wing and body lice compete (fig. 13.3). To our knowledge, these are the only rigorous data relevant to interspecific competition in lice. Furthermore, we are unaware of tests in any other host-parasite system designed to measure the impact of competition on host switching or other macroevolutionary events relevant to phylogenetic congruence.

A final variable that appears to contribute substantially to host specificity is host defense (see below). Wing and body lice transferred among species of pigeons and doves have much higher fitnesses on foreign hosts having impaired preening ability, the main defense against lice (unpub. data). Experiments we are currently conducting will shed additional light on the importance of host defense and interspecific competition in host switching and other macroevolutionary events.

Case Studies

As discussed earlier, lice are an excellent group for exploring the relationship of ecology to coevolutionary history. Different genera of lice vary considerably in their degree of congruence with their hosts, ranging from extensive congruence (Hafner et al., 1994) to a lack of congruence (Johnson et al., in review.a.). In addition, different genera of lice differ in ecological characteristics relevant to processes affecting the degree of congruence between host and parasite phylogenies. Below we review four case studies involving genera of lice with recently published phylogenies based on DNA sequences. Relevant ecological information is also available for all four genera (table 13.1).

Pocket Gophers and *Geomydoecus*

Pocket gophers (Rodentia: Geomyidae) and their lice are a textbook example of cophylogenetic congruence. Species in the genus *Geomydoecus* (Ischnocera: Trichodectidae) are extremely host specific, and often different subspecies of gophers harbor different species of lice. Phylogenetic

TABLE 13.1 Ecological factors promoting phylogenetic *incongruence* between lice and their hosts. Entries are relative assessments among the four genera, with no absolute meaning. Categories are not mutually exclusive; see text for discussion. "Syntopy" refers to populations that are in close physical proximity because they share the same habitat (Lincoln et al., 1982).

	Geomydoecus	*Dennyus*	*Columbicola*	*Brueelia*
Distributional factors				
Sympatry of host species[1,2]	-	-	+	+
Syntopy of host species[1,2]	-	+	+	+
Host populations without lice[3,4]				
(patchiness)	-	-	+	+?
Dispersal to foreign hosts[1,2]	-	-	+	+
Abundance factors				
Prevalence usually low[4,5]	-	-	-	+
Mean intensity usually low[5]	-	+	-	+
Establishment on foreign				
hosts[1,2]				
Survival	+	+	+	+
Reproduction	+	?	+	?

Increase in factor (+) promotes
[1]Failure to speciate
[2]Host switching
[3]Duplication
[4]Missing the boat
[5]Extinction

trees, using many types of data, have been produced for many species of gophers and for their respective parasitic lice in the genus *Geomydoecus*. In addition, a great deal is known about the ecological details of the interaction between these hosts and parasites (reviewed by Hafner et al., chap. 8).

Phylogenies based on mitochondrial cytochrome oxidase I (COI) sequences for 15 taxa of gophers and 15 species of *Geomydoecus* show considerable congruence (Hafner et al., 1994). More detailed comparisons of these phylogenies reveal 8 out of 12 nodes (67%) in the in group with potential cospeciation events (Page and Hafner, 1996). This amount of cospeciation is considerably more than expected by chance alone ($p < .01$). These comparisons indicate that cospeciation between gophers and lice is widespread.

Several ecological parameters have undoubtedly contributed to the extensive history of cospeciation between gophers and lice (table 13.1). First, the distribution of the hosts themselves plays a major role. Most individual gophers build extensive tunnel systems from which they exclude other individuals over most of their life cycle. Sympatry of gopher species is rare and syntopy is even rarer (table 13.1). In addition, gophers have some of the lowest dispersal distances known for mammals. Individuals rarely travel far from their natal homes, and populations are very patchily distributed. Together these factors provide little opportunity for dispersal of lice between individuals of the same host species, and even less opportunity for dispersal between different host species. Thus, opportunities for host switching are few and far between, promoting congruence of host and parasite phylogenies.

Second, the intrinsic ability of *Geomydoecus* lice to move between hosts is low. These lice are specialized for climbing on the hairs of the host, but they are not very mobile off the body of the host. In addition, dispersal routes other than vertical transmission are not known for gopher lice (although apparently they do exist: Demastes et al., 1998; Hafner et al., 1998). Hippoboscid flies do not occur on gophers, meaning that gopher lice cannot disperse phoretically. The low dispersal ability of *Geomydoecus* thus makes host switching and failure to speciate unlikely events.

In contrast with low dispersal ability, species of *Geomydoecus* do seem to be able to establish themselves on foreign host species, at least in transfer experiments in which no competitors were present (Reed and Hafner, 1997). However, lice have difficulty surviving on hosts that are only distantly related to their own. Establishment ability may be related to host defense. Individuals of *Geomydoecus* hang on to host hairs using a rostral groove, in addition to their legs and mouth parts. Studies of the size of this groove indicate a close match between louse groove size and host hair diameter (Morand et al., 2000; Reed et al., 2000). This close match may help gopher lice avoid being removed by the host during grooming. Thus, the ability of lice to establish on foreign hosts may be dictated by the match between the size of the louse and the size of the host. Although establishment may well be possible when the match is close enough, the inability of gopher lice to disperse is probably a major factor preventing widespread establishment and switching to new hosts.

The probability of duplication and sorting events also appears to be low in *Geomydoecus* because of several underlying ecological factors. Virtually all populations of gophers are infested with lice. Thus gene flow in gophers is likely to correspond to gene flow in lice, reducing the possibility for parasite duplication. Similarly, nearly all gophers in a population seem to have lice (high prevalence), thereby reducing the risk of extinction or missing the boat. Finally, the mean intensity of lice on gophers is quite high, often numbering several hundred individuals, which greatly reduces the risk of extinction.

Taken together, these ecological factors appear to promote a history of cospeciation between gophers and *Geomydoecus*. However, this degree of cospeciation and phylogenetic congruence is not the norm, as will become evident in the additional case studies below.

Apodidae and *Dennyus*

Another system that has received scrutiny at both the ecological and phylogenetic scales are members of the amblyceran louse genus *Dennyus*, which are parasites of swifts and swiftlets (Aves: Apodidae). Species of *Dennyus* are quite host specific, but not to the same degree as *Geomydoecus*. Page et al. (1998) conducted a preliminary analysis of cytochrome *b* (cyt *b*) sequences for both hosts and parasites. Here we present expanded results for more taxa and additional sequences. Our cophylogenetic analysis (fig. 13.4) recovered 13 cospeciation events, which are more than expected by chance ($p < .001$). A total of 12 out of 20 host nodes (60%) were associated with a cospeciation event. This analysis indicates that *Dennyus* cospeciates extensively with its hosts, but a smaller fraction of host nodes showed cospeciation than for the gopher-*Geomydoecus* system.

Several ecological factors can be identified that contribute to this intermediate degree of congruence between host and parasite phylogenies (table 13.1). Many species of swiftlets are endemic to isolated oceanic islands, a situation which obviously provides their lice with little prospect

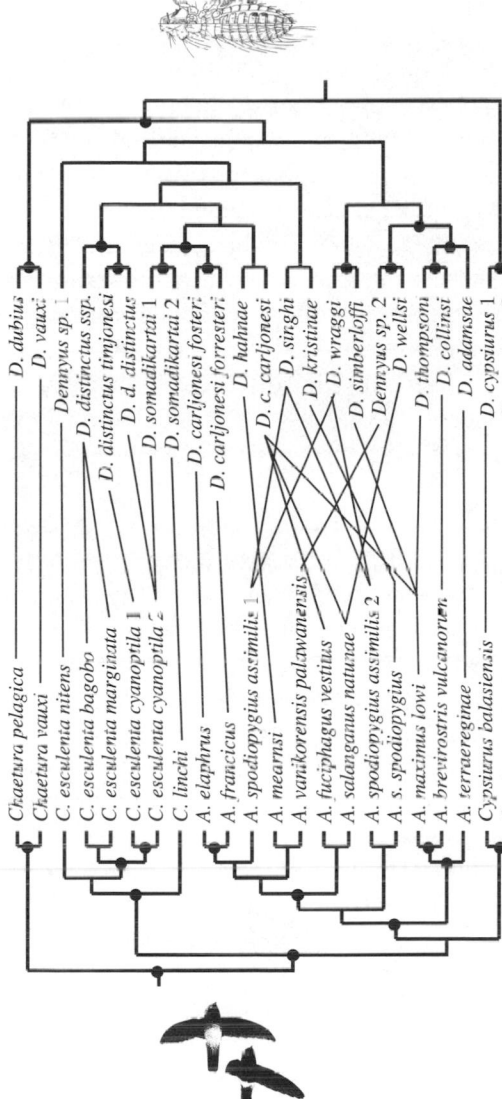

FIGURE 13.4. Comparison of swift(let) (Apodidae) phylogenetic tree to louse tree (*Dennyus* spp). The phylogenies are derived from neighbor joining (Kimura two-parameter distances) of combined mitochondrial cyt *b* and COI sequences (unpub.) for both groups. Bulleted nodes are cospeciation events inferred from reconciliation analysis (Page, 1990) in TREEMAP (Page, 1994). Numbered terminal taxa are those distinguished by molecular differences. Lines connecting birds and lice indicate host-parasite associations. *C. = Collocalia,* *A. = Aerodramus, D = Dennyus.*

for moving between host species. However, in some locations up to four species of swiftlets may be sympatric and syntopic, nesting together in clusters in caves. The lice on these species have ample opportunity for dispersal between hosts, at least at a low rate. Collections of *Dennyus* lice from syntopic species show that some species of lice occur on multiple hosts in the same location, albeit at low intensity (Tompkins and Clayton, 1999).

The intrinsic dispersal ability of species of *Dennyus,* while higher than species of *Geomydoecus,* is still fairly low. *Dennyus,* like other members of the suborder Amblycera, have some locomotory capabilities when off the body of the host. It is probably possible for individual *Dennyus* species to crawl between closely positioned nests in caves where several species of hosts co-occur. However, prospects for long distance dispersal are far more limited. Phoresis on hippoboscid flies has never been observed for *Dennyus,* nor for other species of louse in the suborder Amblycera (Keirans, 1975).

While prospects for dispersal in *Dennyus* are limited, there is evidence for the ability of species to establish on foreign hosts provided they can get there. Transfer experiments by Tompkins and Clayton (1999) showed that, when transferred to a foreign host having feathers that are not too different in size from the usual host, species of *Dennyus* can survive on the foreign host. Since swifts and swiftlets have few defenses against lice, being inefficient preeners, the importance of size probably relates to tenacity, the ability to hang on to the host during flight. Like *Geomydoecus,* species of *Dennyus* tend to match the size of their hosts (fig. 13.5). However, the match isn't perfect, and different subgroups of *Dennyus* appear to have different louse-to-host body size ratios. The ability of species of *Dennyus* to alter their microhabitat preferences when on different sized hosts (Tompkins and Clayton, 1999) may be a factor in their ability to survive on hosts over a relatively wide size range.

Prevalence of species of *Dennyus* is usually high (Lee and Clayton, 1995; unpub. data), which should make extinction and "missing the boat" events relatively infrequent. In general, most populations of swifts and swiftlets appear to have lice. However, there may be some absences of lice from small populations on islands. For example, lice have not been found on *Aerodramus bartshii* (Hawaii) or *A. sawtelli* (Cook Islands), despite concerted sampling of both species (unpub. data). Small body size may also be a factor: no lice have been recovered from the Philippine endemic species *Collocalia troglodytes* (unpub. data), which, although abundant, is the smallest bodied species of swiftlet. One factor promoting extinction of louse populations—low mean intensity—does seem to be the rule in

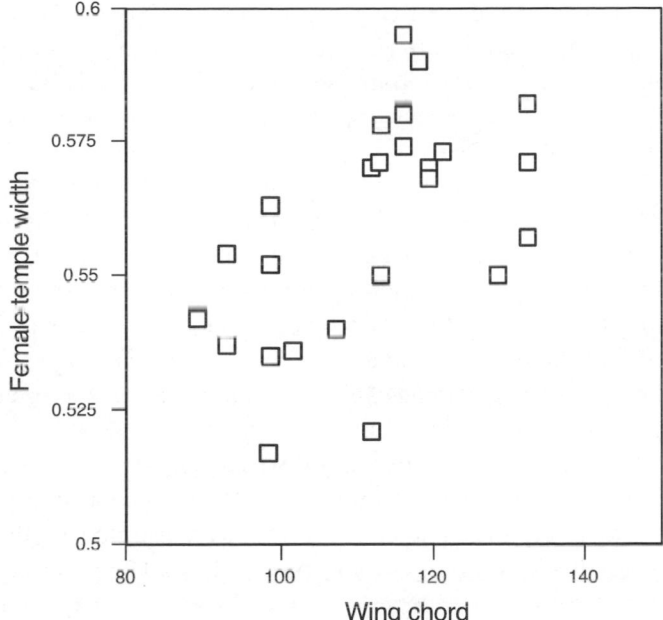

FIGURE 13.5. Plot of average temple width of female *Dennyus* species (in millimeters) against wing chord length of host species (in millimeters). Independent contrasts analysis (Felsenstein, 1985) over the *Dennyus* phylogeny produces a significant association between changes in host body size and louse body size ($p < .01$).

Dennyus. Most parasitized host individuals harbor less than 10 individual lice, owing to the small body size of swifts and swiftlets.

In sum, opportunities and ability to disperse between host species are limited for species of *Dennyus,* but do occur on measurable ecological time scales. Species of *Dennyus* have the ability to establish on a foreign host, provided that host is not too different in size from the usual one. These lice also exhibit a relatively high prevalence and low intensity. Taking these factors together, we would predict some degree of incongruence between host and parasite phylogenies. Moreover, it should be possible to predict from the ecological details which clades are the most likely ones to be incongruent.

Columbiformes and *Columbicola*

A third well studied system consists of species of the ischnoceran louse genus *Columbicola,* which are parasites of pigeons and doves (Aves: Columbiformes). Studies of many aspects of the biology of *Columbicola* have been conducted. Indeed, the species *Columbicola columbae* is considered the "white rat" of bird lice (Eichler, 1963; Nelson and Murray, 1971). Species of *Columbicola* range from those that are completely host specific, to those that are parasitic on many species of hosts over a wide geographic area (Johnson et al., 2002).

A phylogeny for *Columbicola* has been reconstructed on the basis of both mitochondrial (12S and COI) and nuclear (EF1-α) DNA sequences. Comparisons of the *Columbicola* phylogeny to that of its pigeon and dove hosts indicated eight cospeciation events (see chap. 11), which is more cospeciation than expected by chance ($p < .01$). Of 19 nodes in the host phylogeny, 7 of these (37%) had an associated cospeciation event. While cospeciation occurs in *Columbicola,* several factors appear to break down congruence with the host phylogeny in this genus.

First, many species of pigeons and doves are sympatric and syntopic (table 13.1), which provides an opportunity for dispersal between host species. In addition, widespread host species often overlap with the ranges of host species that have more restricted distributions. This pattern provides an opportunity for lice to disperse between allopatric species via dispersal on more widespread host species, which may have contributed to failure of the lice to speciate. For example, the species *Columbicola theresae* occurs on two widespread species of hosts, *Oena capensis* and *Streptopelia senegalensis*. In addition, this louse occurs on two allopatric sister taxa having more restricted distributions: *Streptopelia capicola* (South and East Africa) and *S. vinacea* (sub-Saharan North Africa). Populations of nonspecific lice on widespread host species may keep that louse species from diverging on other allopatric hosts.

In addition to proximity of hosts, species of *Columbicola* appear to be able to take advantage of dispersal opportunities. There are records of phoresis by species of *Columbicola* on hippoboscid flies (Couch, 1962; Keirans, 1975). Studies of the genetics of populations of *Columbicola* on different host species generally indicate a lack of structure, suggesting high continuous capabilities for dispersal between hosts (Johnson et al., 2002).

Although *Columbicola* lice can disperse between host species, they still must be able to establish breeding populations on those hosts. Preliminary transfer experiments to birds whose preening ability is impaired show that pigeon wing lice *(Columbicola columbae)* may be able to establish on foreign hosts. Several hundred *C. columbae* have been recovered from Mourning Doves and Common Ground Doves four months after being "seeded" with 50 lice (unpub. data). These results are noteworthy, particularly given that Common Ground Doves are an order of magnitude smaller

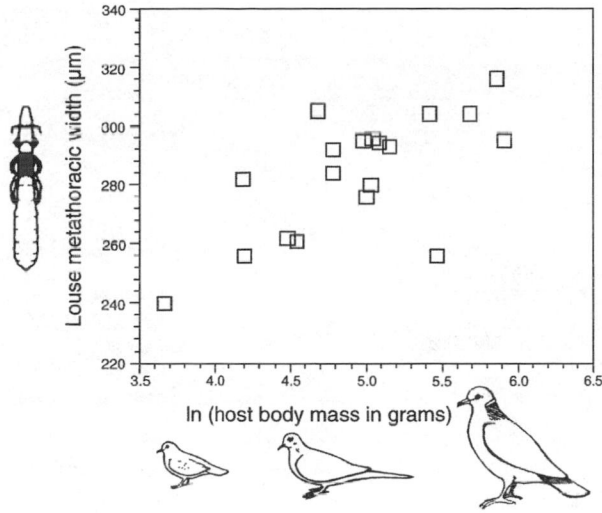

FIGURE 13.6. Plot of average metathoracic width of female *Columbicola* species (in micrometers) against ln body mass (in grams) of Columbiform host species (body mass is significantly correlated with feather barb size; unpub. data). Independent contrasts analysis (Felsenstein, 1985) of these data over the *Columbicola* phylogeny produces a significant association between changes in louse body size and host body size ($p = 0.001$).

than Feral Pigeons. With host defense in place, however, *Columbicola* may not be able to establish over such a broad range of host body size.

Evidence consistent with some limitation to acceptable host body size comes from comparisons of parasite to host body sizes (fig. 13.6), which show a somewhat tighter relationship than in *Dennyus*. This match between parasite and host body size may be related to the mechanism *Columbicola* uses to escape from host preening defense. When exposed to preening, individuals of *Columbicola* insert themselves between the barbs of the wing feathers for protection (fig. 13.7). The size of individual *Columbicola* must therefore approximate the size of the interbarb space in order to prevent being removed by preening. *Columbicola* can probably establish on foreign hosts provided those hosts are not too different in body size from the usual host. This establishment ability, combined with the dispersal ability of *Columbicola*, makes it possible for the members of this genus to move between host species, thereby increasing the probability of host switching and failure to speciate events.

In addition to the ability to move between host species, the population structure of *Columbicola* makes them somewhat prone to duplication

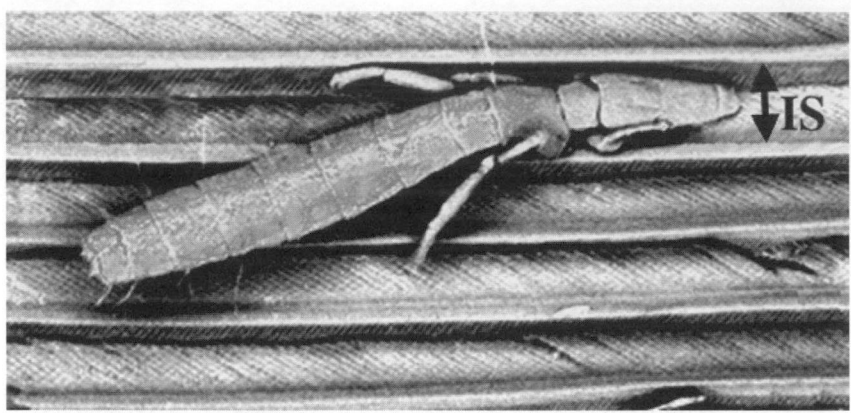

FIGURE 13.7. SEM of *Columbicola* inserting between feather barbs. IS = interbarb space (see text).

and missing the boat events, but makes them unlikely to experience an extinction event. Species of *Columbicola* can often have patchy distributions, possibly resulting from climatic factors. For example, several species of *Columbicola* are almost completely absent from doves in arid Arizona, while these same species on the same hosts are abundant in the humid climate of southern Texas (Moyer et al., 2002). These gaps in the geographic distribution of parasites increase the probability of duplication and missing the boat. On the other hand, species of *Columbicola* can often have a high prevalence (80%) and high intensity (50–100 lice per host), making extinction unlikely. Together these factors appear to generate an intermediate level of congruence between host and parasite phylogenies (chap. 11).

Birds and *Brueelia*

As a final case study, we consider the avian ischnoceran louse genus *Brueelia*, which is known from four orders of birds: Passeriformes (songbirds), Coraciiformes (kingfishers, bee eaters, and rollers), Piciformes (barbets and woodpeckers), and Trogoniformes (trogons). Based on current taxonomy, species of *Brueelia* appear to be quite host specific. While ecological interactions between *Brueelia* and their hosts are not as well studied as in the previous three examples, enough is known to make meaningful comparisons.

The phylogeny of 15 species in the louse genus *Brueelia* was reconstructed by Johnson et al. (in review.a.) on the basis of nuclear EF1-α and mitochondrial COI DNA sequences. The phylogeny of *Brueelia,* when compared with that of their avian hosts (fig. 13.8), indicates only seven cospeciation events, well within the number expected by chance alone ($p = .25$). Only 5 of 24 (20%) nodes in the host tree have a cospeciation event associated with them. Thus, despite fairly high host specificity, there is very little evidence of cospeciation between species of *Brueelia* and their hosts. Several aspects of the ecology of *Brueelia* suggest a basis for the lack of phylogenetic congruence between this genus of lice and their hosts.

First, opportunities for host switching and failure to speciate are high. Many species of passerines (often >100) co-occur in the same geographic region, and many of these co-occur in the same habitat. In addition, many species of passerines and nearly all the nonpasserine hosts of *Brueelia* nest in holes. Competition for holes is high among species of birds, and interspecific takeovers of hole nests often occur (Merilä and Wiggins, 1995). Johnson et al. (unpub.) found that species that nest in holes often share species of *Brueelia*. The possibilities for short-term survival of *Brueelia* off the host are high (Dumbacher. 1999), and takeovers of nests provide an opportunity for dispersal to a new host species.

In addition to the sympatry and syntopy of many hosts of *Brueelia*, the lice themselves seem to be excellent dispersers, at least via phoresis on hippoboscid flies. In fact, about 80% of the nearly 350 records of phoresis summarized by Keirans (1975) involve passerine lice. Of these, the majority are *Brueelia* and the closely related *Sturnidoecus*, with most of the remaining records involving *Philopterus* (the other major genus of passerine ischnoceran louse). While little is known about the establishment ability of *Brueelia*, some species are found on multiple host families (Johnson et al., unpub.). In addition, Clay (1951) describes *Brueelia* as a generalist louse in relation to its habitat on the host's body. Taken together these factors suggest high potential for host switching and failure to speciate, which would break down congruence between host and parasite phylogenies.

In addition to dispersal opportunities and abilities, several factors promoting duplication and sorting events are also evident in *Brueelia*. The single population level study of *Brueelia* to date (Clayton, Price, and Peterson, in prep.) reveals localities where *Brueelia* is absent. Such patchiness could lead to duplication and missing the boat events. While the extent of distributional patchiness in most species of *Brueelia* is unclear, patterns in prevalence and intensity are well documented. Prevalence of species of *Brueelia* tends to be low (< 10%) (Clayton et al., 1992; Hahn et al., 2000). Low prevalence increases the chance of missing the boat and extinction events. In addition to their low prevalence, the intensity of *Brueelia* is also often rather low (< 10 lice) (Clayton et al., 1992; Hahn et al., 2000). These

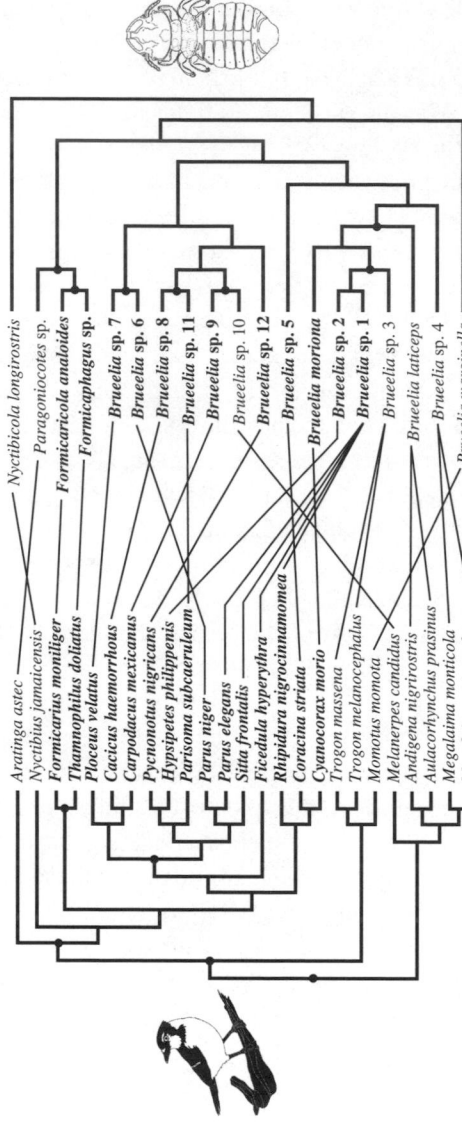

FIGURE 13.8. Comparison of bird phylogenetic tree to louse tree (*Brueelia* spp) (Johnson et al., in review.b). The host phylogeny is derived from Sibley and Ahlquist (1990). Bold taxa indicate Passeriformes and associated louse species. The louse phylogeny is derived from maximum likelihood analysis of mitochondrial COI and nuclear EF1-α sequences. Bulleted nodes are cospeciation events inferred from reconciliation analysis (Page, 1990). Lines between birds and lice show host-parasite associations.

low intensities increase the chance of extinction. In summary, nearly all aspects of the biology of *Brueelia* (table 13.1) appear to favor events that reduce congruence between host and parasite phylogenies.

Conclusions

The ecological basis of coevolutionary history is discernible, as illustrated by the four case studies compared above. Lice having phylogenies that are more congruent with their host phylogeny tend to have fewer factors that promote failure to speciate, host switching, duplication, missing the boat, and/or extinction. As factors responsible for these events become more common, congruence declines. This is not to say, however, that the factors set out in table 13.1 are sufficient to explain all patterns of host-parasite coevolutionary history. For example, rare events, such as dispersal, can be extremely important over long periods of evolutionary time, even though they may be difficult or impossible to measure in ecological time.

A case in point is rock wallabies (Marsupialia: Macropodidae) and their lice (Amblycera: *Heterodoxus*). Several allopatric species of rock wallabies are found along the east coast of Australia. One might expect to find a different species of *Heterodoxus* on each of these species of wallabies. However, this is not the case. A single species of wallaby often harbors more than one species of *Heterodoxus,* even though individual wallabies seldom have more than one species of louse (Barker, 1991). The geographic distribution of a single species of *Heterodoxus* often spans the boundary between host species, although it seldom spans the entire range of both hosts. Not surprisingly, the phylogenies of rock wallabies and their lice show little congruence (Barker, 1991; Barker et al., 1992). These geographic patterns suggest that the lice are capable of dispersing between host species, but that they are not necessarily capable of dispersing among all the populations within a given host species. What's going on? The probable explanation is suggested by the fact that rock wallabies live in small groups on rock outcrops that are often separated by many miles of unsuitable habitat. Thus, wallaby populations are effectively patchier within species than between species. In addition to representing uneven dispersal opportunities for lice, such distributions favor parasite duplication, which further reduces the extent of host-parasite congruence. As additional case studies accumulate, more ecological factors will undoubtedly be found to have an influence on coevolutionary history.

Although caution is advised when drawing conclusions about coevolutionary history from studies of extant species, valid generalizations can still be made. For example, the case studies outlined above suggest that dispersal is a more fundamental barrier to host switching among related hosts than is establishment. Transfer experiments show that host-specific parasites can survive on foreign hosts that are phenotypically similar to the usual host, so long as the parasites can get to these hosts. However, opportunities for dispersal appear quite limited in some systems, such as pocket gophers and their lice. To date, studies of parasite dispersal have been mainly inferential. A better understanding of the role of dispersal will require more direct data on dispersal frequency and distances.

Although limitations to dispersal are clearly important in maintaining the specificity of parasites among *related* hosts, limitations to establishment appear to be important in maintaining the specificity of parasites among *unrelated* hosts. Data regarding the lice on brood parasitic birds are informative in this regard, since they represent a kind of "natural" dispersal experiment. Brood parasites lay their eggs in the nests of other species, which are then tricked into rearing the young of the brood parasites. Recent studies of lice on brood parasitic cuckoos and cowbirds show that lice disperse from foster species to nestling brood parasites. Although these foster species of lice survive on the brood parasites in the short term, there is little evidence for long term establishment (de L. Brooke and Nakamura, 1998; Lindholm et al., 1998; Hahn et al., 2000). Among other things, these results suggest a major shortcoming of transfer experiments— they may not be of sufficient duration to simulate what actually happens under natural conditions. For example, Hahn et al. (2000) recovered nearly as many species of lice from cowbird fledglings ($n = 11$) as they did from all 30 foster species combined at their study site. However, they recovered only 5 species of lice from adult cowbirds, which is about the number expected for a bird of this size and sampling effort (Clayton and Johnson, 2001).

Although Hahn et al. (2000) concluded that limitations to dispersal are the fundamental factor maintaining the host specificity of lice, their results actually indicate the opposite: limitations to establishment have been the most important factor. If dispersal governed specificity, then adult brood parasites should have thriving populations of most of the lice that are found on juvenile brood parasites (the same ones found on the foster species). Like most birds, however, brood parasites appear to have their own host specific lice, despite continually being exposed to lice from other species. In this case, limitations to establishment must be the reason.

Indeed, limitations to establishment are the basis for the phenomenon known as "resource tracking," in which parasites occur on hosts that share some critical resource, such as feather size, rather than being confined to hosts that have descended from a common ancestor, which we will call "phyletic tracking." If the distribution of the resource being tracked is independent of host phylogeny, and if parasites are able to disperse broadly among hosts, then there will be little congruence between host and parasite phylogenies (Kethley and Johnston, 1975). On the other hand, if dispersal opportunities are severely limited, and/or the resource being tracked is correlated with host phylogeny, then congruence is expected. Thus, resource tracking and phyletic tracking make opposing predictions only when the resource being tracked is *not* correlated with the host phylogeny. However, the similarity of many resources on related hosts means that resource tracking may often *increase* congruence. In other words, phyletic tracking and resource tracking are both processes that can contribute to phylogenetic congruence.

To conclude, this chapter has merely scratched the surface of the potential for synergy between ecological and cophylogenetic studies. As additional data sets become available (e.g., Paterson et al., 2000) we will be able to conduct these analyses in a more comprehensive way. The phylogenies and ecological interface for most coevolving species remain largely a mystery. It is our hope that this chapter has served to convince the reader of the potential for improving our understanding of coevolutionary history by continuing to extrapolate backwards from data on the ecology of extant species.

Acknowledgments

Portions of this chapter grew from seeds planted during conversations more than a decade ago with John Kethley, Town Peterson, and Patrick Phillips. We are grateful to J. Burtt and J. Ichida for the louse SEM. This work was supported by NSF-CAREER award DEB-9703003 to D.H.C. We thank Mark Hafner, Bob Minckley, Adrian Paterson, David Reed. and Vince Smith for comments that greatly improved the manuscript.

REFERENCES

Amevigbe, M. D. D., D. A. Ferrer, S. Champorie, N. Monteny, J. Deunff, and D. Richard-Lenoble. 2000. Isoenzymes of human lice: *Pediculus humanus* and *P. capitis*. *Medical and Veterinary Entomology* 14:419–25.

Anderson, R. M., and R. M. May. 1979. Population biology of infectious diseases: Part I. *Nature* 280:361–67.

Barker, S. C. 1991. Evolution of host-parasite associations among species of lice and rock-wallabies: Coevolution? (J. F. A. Sprent Prize Lecture, August 1990). *International Journal for Parasitology* 21:497–501.

————. 1994. Phylogeny and classification, origins, and evolution of host associations of lice. *International Journal for Parasitology* 24:1285–91.

Barker, S. C., D. A. Briscoe, and R. L. Close. 1992. Phylogeny inferred from allozyme in the *Heterodoxus octoseriatus* group of species (Phthiraptera: Boopiidae). *Australian Journal of Zoology* 40:411–22.

Barker, S. C., and R. L. Close. 1990. Zoogeography and host associations of the *Heterodoxus octoseriatus* group and *H. ampullatus* (Phthiraptera: Boopidae) from rock wallabies (Marsupialia: *Petrogale*). *International Journal for Parasitology* 20:1081–87.

Begon, M., J. L. Harper, and C. R. Townsend. 1990. *Ecology*. Oxford: Blackwell Scientific.

Brooks, D. L., and E. P. Hoberg. 2000. Triage for the biosphere: The need and rationale for taxonomic inventories and phylogenetic studies of parasites. *Comparative Parasitology* 67:1–25.

Brooks, D. R., and D. A. McLennan. 1993. *Parascript: Parasites and the language of evolution*. Washington, D.C.: Smithsonian Institution Press.

Brown, N. S., and G. I. Wilson. 1975. A comparison of the ectoparasites of the house sparrow *(Passer domesticus)* from North America and Europe. *American Midland Naturalist* 94:154–65.

Bush, A. O., K. D. Lafferty, J. M. Lotz, and J. M. Shostak. 1997. Parasitology meets ecology on its own terms: Margolis et al. revisited. *Journal of Parasitology* 83:575–83.

Busvine, J. R. 1978. Evidence from double infestations for the specific status of human head lice and body lice (Anoplura). *Systematic Entomology* 3:1–8.

Clay, T. 1949. Some problems in the evolution of a group of ectoparasites. *Evolution* 3:279–99.

————. 1951. An introduction to a classification of the avian Ischnocera (Mallophaga): Part I. *Transactions of the Royal Entomological Society of London* 102:171–94.

————. 1964. Geographical distribution of the avian lice (Phthiraptera: Insecta). *Bulletin of the British Ornithological Club* 84:14–16.

————. 1972. Geographical distribution of the avian lice (Phthiraptera): A review. *Journal of the Bombay Natural History Society* 71:536–47.

Clayton, D. H. 1990. Taxonomy of the *Strigiphilus cursitans* group (Ischnocera: Philopteridae), parasites of owls (Strigiformes). *Annals of the Entomological Society of America* 77:340–63.

Clayton, D. H., and D. M. Drown. 2001. Critical evaluation of five methods for quantifying chewing lice (Insecta: Phthiraptera). *Journal of Parasitology* 87:1291–1300.

Clayton, D. H., R. D. Gregory, and R. D. Price 1992. Comparative ecology of neotropical bird lice. *Journal of Animal Ecology* 61:781–95.

Clayton, D. H., and K. P. Johnson. 2001. What's bugging brood parasites? *Trends in Ecology and Evolution* 16:9–10.

Clayton, D. H., P. L. M. Lee, D. M. Tompkins, and E. D. Brodie III. 1999. Reciprocal natural selection on host-parasite phenotypes. *American Naturalist* 154:261–70.

Clayton, D. H., and R. D. Price. 1999. Taxonomy of New World *Columbicola* (Phthiraptera: Philopteridae) from the Columbiformes (Aves), with descriptions of five new species. *Annals of the Entomological Society of America* 92:675–85.

Clayton, D. H., and D. M. Tompkins. 1994. Ectoparasite virulence is linked to mode of transmission. *Proceedings of the Royal Society of London*, ser. B, 256:211–17.

Clayton, D. H., and J. Moore, eds. 1997. *Host-parasite evolution: General principles and avian models.* Oxford: Oxford University Press.

Corbet, G. B. 1956. The phoresy of Mallophaga on a population of *Ornithomyia fringillina* Curtis (Dipt., Hippoboscidae). *Entomologist's Monthly Magazine* 92:207–11.

Couch, A. B. Jr. 1962. Phoretic mallophaga from hiposboscids of mourning doves *Zenaidura macroura*. *Journal of Parasitology* 48:497.

de L. Brooke, M., and H. Nakamura. 1998. The acquisition of host-specific feather lice by common cuckoos *(Cuculus canorus)*. *Journal of Zoology* 244:167–73.

Demastes, J. W., M. S. Hafner, D. J. Hafner, and T. A. Spradling. 1998. Pocket gophers and chewing lice: A test of the maternal transmission hypothesis. *Molecular Ecology* 7:1065–69.

Dumbacher, J. P. 1999. Evolution of toxicity in pitohuis: I. Effects of Homobactrachotoxin on chewing lice (order Phthiraptera). *The Auk* 116:957–63.

Durden, L. A., and G. G. Musser. 1994. The sucking lice (Insecta, Anoplura) of the world. A taxonomic checklist with records of mammalian hosts and geographic distributions. *Bulletin of the American Museum of Natural History* 218: 90 pp.

Eichler, W. D. 1963. Arthropoda, Insecta, Phthiraptera. 1. Mallophaga. *Bronn's, Klassen und Ordnungen des Tierreichs* (Ent.) 5:1–290.

Ewald, P. W. 1994. *Evolution of infectious disease.* Oxford: Oxford University Press.

Felsenstein, J. 1985. Confidence limits on phylogenies: An approach using the bootstrap. *Evolution* 39:783–91.

Hafner, M. S., J. W. Demastes, D. J. Hafner, T. A. Spradling, P. D. Sudman, and S. A. Nadler. 1998. Age and movement of a hybrid zone: Implications for dispersal distance in pocket gophers and their chewing lice. *Evolution* 52:278–82.

Hafner, M. S., P. D. Sudman, F. X. Villablanca, T. A. Spradling, J. W. Demastes, and S. A. Nadler. 1994. Disparate rates of molecular evolution in cospeciating hosts and parasites. *Science* 265:1087–90.

Hahn, D. C., R. D. Price, and P. C. Osenton. 2000. Use of lice to identify cowbird hosts. *Auk* 117:947–55.

Herre, E. A. 1993. Population structure and the evolution of virulence in nematode parasites of fig wasps. *Science* 259:1442–45.

Hoberg, E. P. 1992. Congruent and synchronic patterns in biogeography and speciation among seabirds, pinnipeds, and cestodes. *Journal of Parasitology* 78:601–15.

Hoberg, E. P., D. R. Brooks, and D. Seigel-Causey. 1997. Host-parasite cospeciation: History, principles, and prospects. In *Host-parasite evolution: General principles and avian models,* edited by D. H. Clayton and J. Moore, 212–35. Oxford: Oxford University Press.

Hoyle, W. L. 1938. Transmission of poultry parasites by birds with special reference to the "English" or House Sparrow and chickens. *Transactions of the Kansas Academy of Science* 41:379–84.

Hudson, P. J., and A. P. Dobson. 1995. Macroparasites: Observed patterns in naturally fluctuating animal populations. In *Ecology of infectious diseases in natural populations,* edited by B. T. Grenfell and A. P. Dobson. Cambridge: Cambridge University Press.

Johnson, K. P., R. J. Adams, and D. H. Clayton. In review.a. The phylogeny of the louse genus *Brueelia* does not reflect host phylogeny.

Johnson, K, P. , R. J. Adams, R. D. M. Page, and D. H. Clayton. In review.b. When do parasites fail to speciate in response to host speciation?

Johnson, K. P., B. L. Williams, D. M. Drown, R. J. Adams, and D. H. Clayton. 2002. The population genetics of host specificity: Genetic differentiation in dove lice (Insecta: Phthiraptera). *Molecular Ecology* 11:25–38.

Keirans, J. E. 1975. A review of the phoretic relationship between Mallophaga (Phthiraptera: Insecta) and Hippoboscidae (Diptera: Insecta). *Journal of Medical Entomology* 12:71–76.

Kethley, J. B., and D. E. Johnston. 1975. Resource tracking patterns in bird and mammal ectoparasites. *Miscellaneous Publications of the Entomological Society of America* 9:231–36.

Lee, P. L. M., and D. H. Clayton. 1995. Population biology of swift *(Apus apus)* ectoparasites in relation to host reproductive success. *Ecological Entomology* 20:43–50.

Lincoln, R. J., G. A. Boxshall, and P. F. Clark. 1982. *A dictionary of ecology, evolution and systematics.* Cambridge: Cambridge University Press.

Lindholm, A. K., G. Venter, and J. and E. A. Ueckermann. 1998. Persistence of passerine ectoparasites on the Diederik Cuckoo *Chrysococcyx caprius*. *Journal of Zoology* 244:145–53.

Merilä, J., and D. Wiggins. 1995. Interspecific competition for nest holes causes adult mortality in the Collared Flycatcher. *The Condor* 97:445–50.

Morand, S., M. S. Hafner, R. D. M. Page, and D. L. Reed. 2000. Comparative body size relationships in pocket gophers and their chewing lice. *Biological Journal of the Linnean Society* 70:239–49.

Moyer, B. R., D. M. Drown, and D. H. Clayton. 2002. Low humidity reduces ectoparasite pressure: Implications for life history evolution. *Oikos* 96.

Nelson, B. C., and M. D. Murray. 1971. The distribution of Mallophaga on the domestic pigeon *(Columba livia)*. *International Journal for Parasitology* 1:21–29.

Nelson, J. S. 1994. *Fishes of the world.* New York: John Wiley and Sons.

Nowak, R. M. 1999. *Walker's mammals of the world.* 6th ed. Baltimore: Johns Hopkins University Press.

Page, R. D. M. 1990. Component analysis: A valiant failure? *Cladistics* 6:119–36.

———. 1994. Parallel phylogenies: Reconstructing the history of host-parasite assemblages. *Cladistics* 10:155–73.

Page, R. D. M., D. H. Clayton, and A. M. Paterson. 1996. Lice and cospeciation: A response to Barker. *International Journal for Parasitology* 26:213–18.

Page, R. D. M., and M. S. Hafner. 1996. Molecular phylogenies and host-parasite cospeciation: Gophers and lice as a model system. In *New uses for new phylogenies,* edited by P. H. Harvey, A. J. Leigh Brown, J. Maynard Smith, and S. Nee, 255–70. Oxford: Oxford University Press.

Page, R. D. M., P. L. M. Lee, S. A. Becher, R. Griffiths, and D. H. Clayton. 1998. A different tempo of mitochondrial DNA evolution in birds and their parasitic lice. *Molecular Phylogenetics and Evolution* 9:276–93.

Paterson, A. M., and R. D. Gray. 1997. Host-parasite cospeciation, host switching, and missing the boat. In *Host-parasite evolution: General principles and avian models,* edited by D. H. Clayton and J. Moore, 236–50. Oxford: Oxford University Press.

Paterson, A. M., R. L. Palma, and R. D. Gray. 1999. How frequently do avian lice miss the boat? *Systematic Biology* 48:214–23.

Paterson, A. M., G. P. Wallis, and R. D. Gray. 2000. Seabird and louse coevolution: Complex histories revealed by sequence data and reconciliation analyses. *Systematic Biology* 49:383–99.

Poulin, R. 1998. *Evolutionary ecology of parasites: From individuals to communities.* London: Chapman and Hall.

Price, R. D., R. J. Adams, and D. H. Clayton. 2000. Pigeon lice down under: Taxonomy of the Australian *Campanulotes* (Phthiraptera: Philopteridae), with a description of *C. durdeni* n. sp. *Journal of Parasitology* 86:948–50.

Reed, D. L., S. K. Allen, and M. S. Hafner. 2000. Mammal hair diameter as a possible mechanism for host specialization in chewing lice. *Journal of Mammalogy* 81:999–1007.

Reed, D. L., and M. S. Hafner. 1997. Host specificity of chewing lice on pocket gophers: A potential mechanism for cospeciation. *Journal of Mammalogy* 78:655–60.

Rózsa, L. 1993. Speciation patterns of ectoparasites and "straggling" lice. *International Journal for Parasitology* 23:859–64.

Secord, D., and P. Kareiva. 1996. Perils and pitfalls in the host specificity paradigm. *Bioscience* 46:448–53.

Sibley, C. G., and J. E. Ahlquist. 1990. *Phylogeny and classification of birds: A study in molecular evolution.* New Haven, Conn.: Yale University Press.

Sibley, C. G., and B. L. Monroe. 1990. *Distribution and taxonomy of birds of the world.* New Haven, Conn.: Yale University Press.

Simberloff, D. 1990. Free-living communities and alimentary tract helminths: Hypotheses and pattern analyses. In *Parasite communities: Patterns and processes,* edited by G. W. Esch, A. O. Bush, and J. M. Aho, 21–40. London: Chapman and Hall.

Stock, T. M., and J. C. Holmes. 1988. Functional relationships and microhabitat distributions of enteric helminths of grebes (Podicipedidae): The evidence for interactive communities. *Journal of Parasitology* 74:214–27.

Stork, N. E., and C. H. C. Lyal. 1993. Extinction or "co-extinction" rates? *Nature* 366:307.

Timm, R. M. 1983. Fahrenholz's rule and resource tracking: A study of host-parasite coevolution. In *Coevolution,* edited by M. H. Nitecki, 225–66. Chicago: University of Chicago Press.

Tompkins, D. M., and D. H. Clayton. 1999. Host resources govern the specificity of swiftlet lice: Size matters. *Journal of Animal Ecology* 68:489–500.

INDEX